PROCEEDINGS OF THE SCHOOL OF UNDERGROUND MINING, DNIPROPETROVS'K/YALTA, UKRAINE, 2-8 OCTOBER 2011

Technical and Geoinformational Systems in Mining

Editors

Genadiy Pivnyak
Rector of National Mining University, Ukraine

Volodymyr Bondarenko
Department of Underground Mining, National Mining University, Ukraine

Iryna Kovalevs'ka
Department of Underground Mining, National Mining University, Ukraine

CRC Press
Taylor & Francis Group
Boca Raton   London   New York   Leiden

CRC Press is an imprint of the
Taylor & Francis Group, an Informa business

A BALKEMA BOOK

First issued in paperback 2017

CRC Press/Balkema is an imprint of the Taylor & Francis Group, an informa business

© 2011 Taylor & Francis Group, London, UK

Typeset by Olga Malova & Kostiantyn Ganushevych, Department of Underground Mining, National Mining University, Dnipropetrovs'k, Ukraine

Published by: CRC Press/Balkema
P.O. Box 447, 2300 AK Leiden, The Netherlands
e-mail: Pub.NL@taylorandfrancis.com
www.crcpress.com – www.taylorandfrancis.co.uk – www.balkema.nl

ISBN 13: 978-1-138-11244-5 (pbk)
ISBN 13: 978-0-415-68877-2 (hbk)

Technical and Geoinformational Systems in Mining – Pivnyak, Bondarenko & Kovalevs'ka (eds)
© 2011 Taylor & Francis Group, London, ISBN 978-0-415-68877-2

Table of contents

Technical and Geoinformational Systems in Mining – Pivnyak, Bondarenko & Kovalevs'ka (eds)
© 2011 Taylor & Francis Group, London, ISBN 978-0-415-68877-2

Preface

The second collection of articles dedicated to the "School of Underground Mining" conference embraces many important scientific trends such as implementation of new mining methods to extract mineral deposits with high methane content together with new methods of roof management during high rates of the longwall advance. Specific attention is given to mathematical simulation of the support functioning in the development mine workings, creation of 3-D modeling to study stress-strain state of the rock massif and development of new bolt support designs.

Much of work is done in order to simulate and assess economic and ecological risks during undermining land surface together with forecast of dynamic phenomena in regional zones of the Donbass mines. Geoinformational systems in mining, electro-stimulation of chemical reactions in coal and new methods of mine wastes utilization are scrutinized as well.

Consideration is given to rational parameters of ventilation and degassing at production units of deep mines with use of air cooling complex systems.

Taking into account worsening of the mining-geological conditions for conventional extraction of coal, there is much of attention dedicated to borehole underground coal gasification technology at Ukraine's coal deposits. Very intriguing topic is connected with one of the most perspective and abundant sources of energy on the planet – gas hydrates. The question of their prospecting, properties and ways of extraction is also covered in this book.

Examination is given to financial conditions of work and financial strategy of mining industry in Poland Ukraine and other countries.

Genadiy Pivnyak
Volodymyr Bondarenko
Iryna Kovalevs'ka
Dnipropetrovs'k
October 2011

Induction heating in electrotechnology of machine parts dismantling

G. Pivnyak & N. Dreshpak
National Mining University, Dnipropetrovs'k, Ukraine

ABSTRACT: General requirements for induction heating of machine parts connections with a purpose of their dismantling are substantiated. The methodology for the specific surface power and other mode parameters determining that meet formulated requirements is developed. The influence of electromagnetic field parameters on the character of thermal process development is shown.

1 FORMULATING THE PROBLEM

Cylindrical steel connections of machine parts (bushings to a shaft) made with an interference fit are widely used in machine-building and mining. Bushings fittings on the shaft are often made as piles and bandages that fix the position of other parts on the shaft, and prevent their axial movement. Such type connections also include internal bearing rings fittings on the shaft.

While repairing and testing machines it is necessary to perform dismantling. Connections dismantling realized by means of the axial loads using removers accompanied with damage surfaces as emerging surface scratches. After several repairs the shaft becomes unusable. Large numbers of dismantling result in substantial material losses.

Heating details connections permit avoiding undesirable consequences. While heating bushing enlargers and it can be easily removed from the shaft without any surfaces damages. One of the most effective methods of dismantling is induction heating of details connections. Dismantling is realized with minimum time and energy costs to be typical for direct heating systems.

At the same time connections heating modes that lead to interference liquidation and conditions necessary to dismantle the site are not studied enough. Induction heating of cylindrical part connections is characterized by the fact that along with electromagnetic and thermal processes there are temperature deformation processes in the system which results in the bushing extension. This significantly affects the conditions of heat transfer between the bushing and shaft, and determines the level of temperature. Lack of theoretically substantiated mode parameters results in practical use of induction heating systems with unsuccessful constructive decisions and relatively low technical and economic factors.

Thus there is a need for the research aimed at defining the parameters of rational heating mode, providing on this basis with efficient connections dismantling and improve technical and economic factors of induction units (reducing mass, size, and cost).

There is determined basic structure of an investigated object (Figure 1).

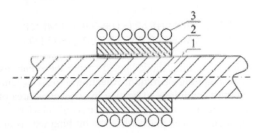

Figure 1. System "inductor-collection": 1 – shaft; 2 – bushing; 3 – coil winding.

A coil winding 1 of an induction installation is located on the surface of the thin-walled equal thickness bushing 2, which is connected to the shaft with interference. Heating modes should be connected to the temperature deformation processes that result in interference liquidation and create conditions for the connection dismantling. In the presence of interference fit a direct contact between the bushing and shaft is possible owing to the protruding hard surfaces. During the heating process of the bushing and its extension the interference is gradually liquidated. In this context the area of direct contact of solids decreases following by significant changes in thermal transfer conditions. The feature of details connection dismantling associated with the contact thermal conductivity changing between the bushing and shaft should be taken into account when analyzing the process of heating.

Specific literature analysis shows that interconnected electromagnetic, thermal and mechanical processes in such system were not previously considered. The criteria of dismantling operation efficiency are not determined. Therefore it is necessary to specify the requirements for the heating process and to determine the conditions for their implementation. It is required to identify available dependences between the parameters of electromagnetic fields and temperature of heating as well as the influence of individual parameters field on the nature of the heating process. This will allow establishing rational heating modes, will provide them with efficient connection dismantling, and will improve technical and economic factors of an induction setup.

Induction heating process control can be done by changing the parameters of inductor current. Variants of sinusoidal and pulsed current are considered. Under such conditions it is expedient to focus on variant implying changing the frequency of current repetition. To consider the variant it is necessary to guarantee a possibility of thermal process modeling under heat sources impulsive effect on the bushing surface.

2 INDUCTION HEATING REQUIREMENT FOR MACHINE PARTS DISMANTLING

Selected heating mode should ensure the reliability of dismantling operation. Attention is paid to temperature conditions of heating in connection area of the shaft and bushing as the process of interference liquidation takes place while the bushing extension. It is important to consider not only individual local areas of the connection but also the whole surface as dismantling implies interference liquidation within the area of the contact surface. Unsuccessfully selected mode may lead to the fact that in some local areas or the within the whole connection surface the interference will not be liquidated and dismantling will not be performed. Thermal heating mode is determined by configuration and parameters of the current magnetic field. Therefore the problem should be solved by means of substantiating the required characteristics of the magnetic field and determination of the heating mode conditions.

Modern tools of dismantling connections have to guarantee high technology of operation primarily determined by time spent for its implementation. This characteristic is particularly important if there repair time is limited. The time of parts heating reduction results in energy saving by means of increasing heating power i.e. the use of high power rated source. This way of the problem solving leads to increased cost of power and lowering its technical and economic factors. This is especially important when using devices of power transforming equipment (frequency converters). Therefore selecting a mode it is necessary to find a compromise that will ensure an acceptable rate of nominal power of energy source with negligible heating time.

Ensuring reliability, feasibility and energy efficiency of the dismantling process, reducing the rate of nominal power of the supply are the requirements that should be met to improve technical and economic factors of induction heating; besides they determine its high performance characteristics. The general requirements for induction heating imply the determining the conditions and ways of their implementation. It can be done on the basis of analyzing the temperature modes of interference liquidation, and studying the picture of magnetic field.

Taking into consideration the variety of fitting types as well as properties of steel products it is determined the required temperature difference ΔT_R between the internal bushing the shaft surfaces.

Depending upon the axial symmetry of the system (symmetry axis of the shaft) it is expedient to obtain required value of the temperature difference simultaneously in all points of mating surfaces. It prevents overheating of certain areas of a connection, and reduces time for interference liquidation across its surface. Hence a need for uniform distribution of heat sources on the surface of a bushing is obvious. Magnetic field with the same intensity at all points of the surface of the shaft is the solution. In the system of induction heating of cylindrical parts connections the size of the bushing heated is limited. Therefore there are end effects that distort magnetic field. Getting the necessary field picture is possible by changing coil winding step. The task is to define the necessary change of this parameter. Considering the variety of design solutions offers conducting physical modeling of electromagnetic processes directly in real objects. It helps to obtain required magnetic field configuration based on the specific measurements of field tension on the heated surface. Measurements are carried out by sectional winding located on the surface of bushings. The values of EMF (electromotive force) induced in each section of a measuring winding allow to estimate the field tension of a bushing area where a curtain section is located. Under equality of EMF each section receives a homogeneous magnetic field. It is used 50 Hz current in an experiment which flows through the inductor and does not result in significant bushing heating interference liquidation. This heating mode applies to use relatively cheap simple low-power equipment. The method of magnetic field forming is developed to reveal the idea and sequence of actions carried out during the experiment. This field application in the active induction setup for connection dismantling ensures constant for the whole surface of the bushing surface value of specific power, and allows using one-dimensional models that simplifies the modeling of either electromagnetic or thermal processes.

3 THE METHODOLOGY FOR HEATING MODE PARAMETERS DETERMINING

It is shown that choice of P_0 level significantly influences the thermal process character (Vypanasenko 2008). It is proposed to set the value of P_0 to provide the required level of ΔT_R in transient heating mode. Herewith, required temperature conditions for connections dismantle are stored in a stationary mode. This requirement for P_0 determination is important for the heating time reduction, increased reliability and efficiency of technological operations.

The modeling of electromagnetic processes in the heating system is performed (Dreshpak 2009). The modeling of method is substantiated, the algorithm of the specific surface power, current frequency and other settings definition that provide the conditions necessary for the connection dismantling are developed.

It is shown that under conditions typical for the induction heating of parts connections with the purpose of their dismantling it is necessary to consider a one-dimensional longitudinal magnetic field that operates in a continuous cylinder (Vypanasenko 2008). It is proved the expediency of analytical method of electromagnetic processes calculation in the system. The method takes into account the features of connections induction heating technology (the one-dimensional electromagnetic field, the limitation of penetration depth with a bushing thickness Δ_b) and provides highly informative and accurate results. In the process of analytical dependences formation for the specific surface power calculating P_0 the assumptions is introduced: heat losses from the external and lateral surfaces are not available.

The value of P_0 is received from the formula

$$P_0 = \alpha_i \cdot \Delta T_R \cdot \frac{R_2}{R_1}, \qquad (1)$$

where R_1 and R_2 – external and internal bushing radii.

Contact thermal conductivity between the bushing and the shaft α_i is defined experimentally identifying its value directly on the object to be dismantled (Patent 43365). During low-temperature (without interference liquidation) stationary heating measurements are made on the lateral surface of internal T_i and external T_e temperatures of its surfaces. Also surface temperature of the shaft T_s is taken. Figure 2 illustrates the points of temperature T measurement.

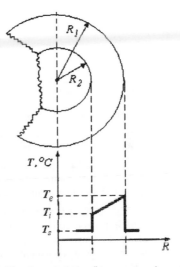

Figure 2. The characteristic of temperature in a stationary connection heating mode.

The value of α_i is received from the formula

$$\alpha_i = \frac{(T_e - T_i)\lambda_{st}}{(T_i - T_s)(R_1 - R_2)}, \qquad (2)$$

where λ_{st} – factor of thermal conductivity of steel.

When choosing the current frequency of induction installation it is offered to consider high energy efficiency of the heating process, limiting the penetration of electromagnetic field trough the shaft and taking into account the default output values of power supplies frequency. Complete attenuation of electromagnetic waves accounting the dependence of relative magnetic permeability μ of the field strength H is at a distance from the surface of a bushing $X_n = 1.68 \cdot \Delta_e$. Then the formulated condition corresponds inequalities

$$X_n \leq \Delta_b; \; \Delta_e \leq \Delta_b / 1.68, \qquad (3)$$

where Δ_b – bushing thickness; Δ_e – depth of penetration of electromagnetic waves, calculated on the basis of the value of μ – bushing surface (μ_e). To limit mode $\Delta_e = \Delta_b / 1.68$ "deep" bushing heating is typical. This mode corresponds to the lower recommended value of inductor current frequency. Figure 3 shows the dependence $f_L(P_0)$.

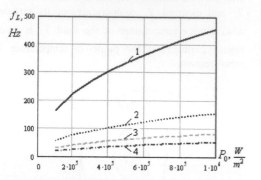

Figure 3. $f_L(P_0)$ dependencies: 1 – Δ_b = 0.005 m; 2 – Δ_b = 0.01 m; 3 – Δ_b = 0.015 m; 4 – Δ_b = 0.02 m.

The initial frequency of the power source is chosen in view of its common values and the condition $f_g \geq f_L$. Figure 3 shows that at the bushing thickness being less than 2 cm it is possible to carry out heating with the industrial current frequency of 50 Hz. Under lower thickness values and also under size restrictions in design it is reasonable to use high frequencies (kHz).

Developed mathematical model is focused on determining the mode parameters P_0, f_g, H_e (H_e is the value of magnetic field tension on the bushing surface). The parameters guarantee temperature conditions for connection dismantling and are used for inductor calculation, nominal power of energy source selection applying known methods.

4 THE INFLUENCE OF ELECTROMAGNETIC FIELD PARAMETERS ON THE CHARACTER OF THERMAL PROCESS DEVELOPMENT

The parts connections dismantling process is directly connected with thermal heating modes. Therefore there was a need to evaluate the effects of electromagnetic parameters on the thermal process, and to confirm their eligibility for the technological operation. This led to a mathematical model creation to study transient and stationary thermal processes that take place in a cross-section of the bushing. The implementation of the one-dimensional longitudinal magnetic field acting in the bushing allow describing the non-stationary process of connection induction heating with the one-dimensional heat-conduction equation in second order partial derivatives. The impact of the bushing extension on the value of the contact thermal conductivity α_i of the parts connection is taken into account and also the possibility of the process calculating in the presence of pulsed current in an inductor is realized.

The modeling pulse current of the inductor is done by periodic input and output of heating sources concentrated in the active layer of the bushing that allows analyzing the possibility of pulse-frequency control for heating.

There are a number of factors that affect the nature of thermal process in the bushing: the specific surface power P_0, the current frequency of inductor f_g, the value of the contact thermal conductivity α_i. A heating modeling indicates that on the increased current frequency of inductor the necessary value of ΔT_R for dismantling realization is achieved at the higher level of temperature on the external surface of the bushing to be explained by more vivid surface effects. This dependence should be considered determining the rational value of current frequency. It is shown that the current frequency influence on the character of the thermal process it is manifested in different values of ΔT delays relative to the start of heating because of the character of heat sources location in a cross-section of the bushing. With $\alpha_i = const$ the increase of a specific surface capacity reduces heating cycle time t_c of details connection. The non-linear characteristic of this dependency with a significant time increase with low capacity values (Figure 4) is distinctive. Here $n = P_d / P_0$ (P_d —estimated power). The reduction during the heating process of the value of α_i is equivalent to the high surface capacity action which ensures heating cycle time reduction.

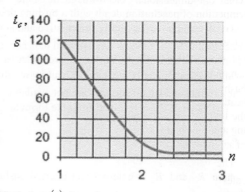

Figure 4. $t_c(n)$ dependence.

The thermal contact conductivity α_i reduction significantly modifies the character of temperature distribution in a cross-section of the bushing. There is a leveling of temperature in the area close to its internal surface (Figure 5). The value of $\Delta T = \Delta T_R$ necessary for dismantling realization is achieved under lower values of surface temperature of the bushing. The increase of heating time is followed by the increase of the speed growth rate of ΔT owing to thermal deformation of the bushing (Figure 6).

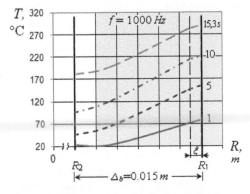

Figure 5. The temperature distribution taking into account $\alpha_b(\Delta T)$ dependences.

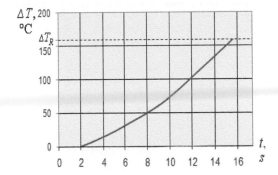

Figure 6. $\Delta T(t)$ Dependence during α_i changing in the heating process.

The slowing-down of ΔT growth at the initial stage of heating depends on significant α_i that corresponds to close contact between the bushing and shaft. The dependence in Figure 6 is obtained by the estimated value of the specific power P_0 that confirms the fact of a thermal mode necessary for dismantling occurrence in a transitional mode of heating. It is shown that decreasing the thickness of the bushing Δ_b the heating cycle time t_c decreases. The increase of P_0 results in more sizeable t_c reduction. The character of these dependences is the result of a contact thermal conductivity action in steel. As a result of inductor current pulse modeling it is shown that the increase of sinusoidal pulses frequency results in the increase of the temperature difference ΔT indicating the possibility of heating temperature conditions operation.

To identify the value of the contact thermal conductivity of the connection area between the bushing and shaft α_i by means of the experimental setup the low temperature (50 °C) heating of the bushing surface was performed. In the stationary heating mode by means of a non-contact pyrometer the temperatures on the lateral surface of the bushing and the shaft surface were measured. The values of α_i were calculated using dependence (2).

The value of α_i obtained experimentally indicates that when there is a tight interference fitting there are favorable conditions for the transfer of heat from the bushing to the shaft. The value of α_i significantly (in two orders) exceeds the value of a heating transfer index from the external surface of the bushing.

If it is necessary to reduce the heating cycle time t_c significantly the use of forced heating mode of details connections is required (Patent 43339). To realize the forced heating mode the frequency of current pulse repetition keeps stable in the period of time that corresponds to the surface temperature increase, and its value is changed up to complete dismantling the connection. The control method allows reducing the heating cycle time by 30 per cent.

5 CONCLUSIONS

The development of methodology for the specific surface power P_0 and other mode parameters on the surface of the bushing determining allows making a reasonable choice of inductor and power source parameters, and abandon the existing in induction heating practice of over-rated power source usage. If the method is available the savings for one installation purchasing is several thousand dollars. The weight and dimensions of power source reduce significantly (by tens of per cent) indicating improved overall parameters of an induction setup.

REFERENCES

Vypanasenko, N.S. 2008. *The specific surface power of induction heating of parts connections made by interference fit determining.* Bulletin of Azov State Technical University. Mariupol: Azov State Technical University. Edition 18. Part 2: 131-136.

Dreshpak, N.S. 2009. Induction *heating modes of cylindrical details connected with interference fit.* Technical Electrodynamics. Kyiv: National Academy of Sciences of Ukraine.Edition, 6: 61-66.

Patent 43365 Ukraine. 2009. IPC B23P19/02 Pivnyak, G.G. & Dreshpak, N.S. *Device for induction dismantling of details*, 15: 4.

Patent 43339 Ukraine. 2009. IPC B23P19/02 Pivnyak, G.G. & Dreshpak, N.S. *The way of induction details dismantling control*, 15: 4.

Mechanism of force interaction of "rock bolt-rocks" system

V. Bondarenko, G. Simanovich & A. Laguta
National Mining University, Dnipropetrovs'k, Ukraine

Y. Cherednychenko
Ministry of Coal Industry of Ukraine

ABSTRACT: The interaction of rock bolts with the blast-hole walls is given. The mechanism of interaction during rock bolt installation and the process of further work based on the load resistance from a rock mass side are described. Analytical-experimental description of four modes of rock bolt (without any expansion shell) interaction with the blast-hole walls, and fully characterized process of rock bolt deformation under influence of an axle load are given. The main criterion of optimal power parameter selection during rock bolt installation is developed.

1 INTRODUCTION

Despite variety of lockless rock bolt's designs, they are characterized (except screw-threaded bolts) by a number of the general principles of force interaction with blast-hole walls:

– bolt fastening is basically carried out by means of friction and cohesion (adhesion) forces. So, the main role of tubular rock bolt's fastening belongs to friction forces; ferro-concrete, polymeric and ferro-polymeric bolts have cohesion forces; and a tubular bolt with a fastening layer has friction and cohesion forces, when extending mixture is applied;

– bearing element (a rod, a cylinder, etc.) of a rock bolt perceives the basic tensile load, directly or through the fastening layer, contacting with blast-hole walls. On these contacts the tangential stresses are operated due to friction and cohesion forces, or their combination;

– lockless rock bolt's interaction with a rock is characterized not only by a direct, but also by a reverse connection of normal tensile stresses in the bearing element with tangential stresses, operating along contact surface of the bearing element with blast-hole walls or the fastening layer, and also fastening layer with blast-hole walls. It is explained that during the deformation process of mine working's bolted mass the bearing element of a bolt is loaded by axial tensile stresses through the rock contact. It leads to changing of the existing stress field of the "bolt-rock" system.

The similarity of main features of bolt force interaction with blast-hole walls is in the very idea of fastening rock mass by lockless rock bolts. It substantiates the single deformation mechanism of bolts with blast-hole walls under the axial loading influence.

2 THE MAIN IDEA

Based on a number of experimental researches (Melnikov 1980; Skobtsov 1973; Shirokov 1971; Roginsky 1968 & Yemelyanov 1978) of rock bolt deformation under the axial loading influence, it is possible to highlight four consecutive modes of bolt interaction with blast-hole walls. The first mode, or the mode of elastic interaction, is characterized by the full contact of a bearing element with rocks (a bearing element with fastening layer, or fastening layer with rocks) along a bolt. In this mode, the displacement of the rock bolt's loaded end is negligible, and maximum of tangential τ_{rz} stresses is on the bolt's loaded end. When limiting state occurs, breaking of a less strong contact, connecting the bearing element with blast-hole walls, happens.

When contact is broken, the second mode takes place – the mode of increasing resistance to axial loading. The maximum of tangential τ_{rz} stresses moves to the sunken bolt end. The tangential stresses operate in the place of lost contact, provided by rough surface of broken contact. There are also stresses of friction in the presence of radial pressure upon the contact, therefore total tangential stress action is raised and increases resistance of the bearing element to axial load. The U displacement of loaded bolt end is grown due to increase of working area of normal tensile σ_z stresses of the bearing element.

As the maximum τ_{rz} moves towards the sunken bolt's end, gradual growth of its resistance to axial loadings occurs, till the moment, when the loss of total tangential stresses will not exceed the total tangential stresses increase on the broken contact area. With equality of these forces, maximum bolt resistance to axial loads will take place.

During further displacement of the maximum τ_{rz} towards the sunken bolt's end, the intensive decrease of bolt resistance to axial loads occurs, because total outgoing tangential stresses on the elastic deformation area considerably exceed the growth of total tangential stresses on a failure contact site. This process characterizes the third mode – the mode of the intensive reduction of bolt resistance to axial load, which finishes when one of the contacts along the full bolt length is broken.

The fourth mode – the sliding mode – causes considerable displacement of a rock bolt, which resists to axial loads due to friction forces along the surface of the broken contacts. The resistance reduction to axial loads in the sliding mode occurs slowly enough since contact sections of bolt's loaded end are gone out of operation in the first place, where the tangential stress value is insignificant.

The revealed modes of the "bolt-rock" system interaction have their peculiarities, and this has to find a reflection in corresponding analytical models for calculation of parameters of rock bolts without expansion shell.

3 THE CALCULATION OF ROCK BOLTS WITHOUT EXTENSION SHELL

For lockless bolt calculation it is necessary to have original data parameters of their installation. There are geometrical, strength and strain parameters of rock bolt elements, and also constructive-technological features of their installation. If the first group of parameters is known in advance and is defined by the material properties, used in different lockless rock bolt constructions, then the second group of parameters has to be defined by the process of their installation in a blast-hole. Thus, it means to install the rock bolts with initial radial P pressure along the contacts (bearing element – fastening layer – blast-hole walls). Ferro-concrete and ferro-polymeric rock bolts, with expanding (during solidification) hardening layer, and also tubular bolts are related to such constructions.

Installation of bolts without extension shell and with initial radial stresses $\sigma_z = P$ along the contact, with the blast-hole walls, in some constructions promotes increase of the bolt fastening in a blast-hole (ferro-concrete and ferro-polymeric rock bolts), and in others it is the main force connection providing bearing capacity (a tubular bolt). Therefore the

problem is to make correct choice of initial radial pressures during the bolting process and, correspondingly, their constructive realization.

As the basis of criterion of calculation of bolt design parameters that provide required initial radial stresses, the condition of achievement of the maximum fastening strength of a bolt in a blast-hole is supposed, which directly depend on radial stresses along the bolt contact with blast-hole walls. The greater radial σ_r stresses along the contact, the greater friction forces that resist to axial load on the bolt. However, it is impossible to infinitely increase radial stresses, as it will lead to blast-hole walls macrodestruction and to abrupt falling of bolt's bearing capacity. Therefore, during the process of radial strain of a bolt, the maximum of radial stresses $\sigma_r = P_{max}$, reaching during bolting, must not exceed the value which creates blast-hole walls macrodestruction. On the other hand, radial stresses along the contact are inevitably decreasing to established P_∞ pressure under influence of rheological factor and due to peculiarities of rock bolt design. That is why, in order to choose rational value of initial radial σ_r stresses on the contact, it is necessary to consider the mechanism of "bolt-rock" system deformation in radial direction.

4 RADIAL DEFORMATION PROCESS OF THE "BOLT-ROCK" SYSTEM

Radial process of deformation of the "bolt-rock" system is visually illustrated in the diagram of stresses (in coordinates of σ stresses – radial strain ε_r) of a bolt material and blast-hole walls (Figure 1). During influence of radial pressures on contact of the $\sigma_r = P$ system, in blast-hole walls triaxial non-uniform stress state occurs, characterized by radial σ_r, tangential σ_θ and axial σ_z components of stresses. As axial component σ_z is average and according to the Mohr's theory of strength does not essentially effect limiting state of rock, then the process of components σ_r and σ_θ changing in blast-hole walls is considered. At a stage of elastic and elastic-plastic deformation, component σ_r is a compressing type, and tangential σ_θ is a stretching type (Baklashov & Kartoziya 1975). As rocks resist to stretching loads poorly, when component σ_θ reaches ultimate rock strength on σ_p tensile (see Figure 1, point 2), discontinuity and radial microcracks formation occur in blast-hole walls.

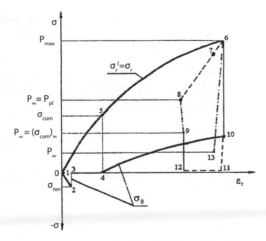

Figure 1. The diagram of lockless bolt interaction with blast-hole walls:——— loading; — — — unloading at $(\sigma_{com})_\infty \geq P_{pl}$ and — · — · — at $(\sigma_{com})_\infty < P_{pl}$.

Tangential stresses disappear (area 2-3) and during the further loading process of blast-hole walls a tangential component equals to zero (area 3-4). Radial stresses simultaneously increase. Therefore, ultimate rock strength on uniaxial compression σ_{com} and blast-hole walls transfer into limiting state (point 5). Thus dilatation process of blast-hole walls is intensively developed, areas 5-6 and 4-10, which are characterized by the occurrence of compression tangential stresses. The essence of this process consists in the following. During compressing of rocks in radial direction, limited by radial microcracks, its (rocks) expansion occurs (according to Poisson law) in tangential direction and that characterizes tendency of closing microcracks. The tangential strain's value of closing radial microcrack $(\varepsilon_\theta)_3$ equals to

$$(\varepsilon_\theta)_3 = 2\mu\varepsilon_r,$$

where μ – coefficient of rock transversal strain; ε_r – radial strain of blast-hole walls.

It is also important to know tangential strain of opening microcrack $(\varepsilon_\theta)_p$. It equals to the plastic strain of the rocks around the cylinder hole (Baklashov & Kartosiya 1975) of radial strain ε_r, that is

$$(\varepsilon_\theta)_p = -\varepsilon_r.$$

Total tangential strain is

$$\varepsilon_\theta = (2\mu - 1)\varepsilon_r. \tag{1}$$

From the formula 1 it follows that when $\mu = 0.5$ (plastic state of blast-hole walls), tangential strain equals to zero and opening microcrack remains constant. However, in limiting rock state (points 4 and 5) the dilatation effect is observed (Baklashov & Kartosiya 1975). It is characterized by coefficient μ of intersection deformation can significantly exceed 0.5, as the complex of experimental investigations (Baklashov 1975 & Bridgeman 1955) of loading of different rock samples has shown. In this case, according to formula (1), positive tangential strain appears, providing closure of earlier formed radial microcracks and leading to occurrence of compressing tangential stresses. And, therefore, it promotes the increase of abrupt resistance to radial loads (points 5-6 and 4-10) without macrodestruction of blast-hole walls.

At a definite stage the process of the bolt's radial strain finishes and the balance state of "bolt-rock" system occurs, during which radial stresses along the contact reach their maximum $\sigma_r = P_{max}$ (point 6). Further, the unloading process of the "bolt-rock" system begins. For ferro-concrete and ferro-polymeric rock bolts this process is identified by unloading blast-hole walls. That is why the unloading process of a tubular bolt is considered, as the most common process for "bolt-rock" system.

At the moment of reaching maximum P_{max} by radial stresses, a rock bolt is loaded by external compressive radial stresses. And its material is under tensile tangential stresses, appearing when a bolt is extended due to explosion power, or pressure inside of a rock bolt. Such state is unstable and in the next moment anchor elastic compression to its own axis begins. The process of which can be divided into two serial stages. At the first stage tangential tensile stresses σ_θ^I of a rock bolt are reduced to zero, blast-hole walls are simultaneously unloaded and radial stresses are reducing on the contact of a system. At the second stage, elastic compression of a bolt by tangential compressive stresses occurs under influence of the remained tangential stresses. So, during this process, three cases of "bolt-rock" system's unloading take place.

In the first and second cases, value P_{max} of maximum radial stresses on the contact is such as even when the system is elastically unloaded (site 6-7), remaining radial compressive stresses exceed value P_{pl}, with which a rock bolt transfers into plastic state due to external pressure. (Pisarenko 1979)

$$P_{pl} = \frac{r_2^2 - r_1^2}{2r_2^2}\sigma_T^I, \tag{2}$$

where σ_T^I – ultimate flow material of the bearing element; r_1 and r_2 – inner and outer radiuses of a tubular bolt.

Plastic deformation of a bolt's bearing element towards its axis begins and unloading of blast-hole walls takes place also (area 7-8). When radial stresses on the contact are reduced to value P_{pl}, then the process of bolt plastic compression ends.

The mechanism of blast-hole walls' unloading process is considered as well. In the works (Bridgeman 1955 & Stavrogin 1979) the dependence of a rock sample volume's change, during uniaxial loading, has been experimentally received. One of the features of such diagram is when a sample is unloaded, that was in a limiting stress state, its volume does not increase at a primary stage but decreases. Therefore, during unloading the increase of sample volume in direction to main stress action is lower, than the decrease of sample volume in lateral direction. This phenomenon can be explained by the fact that during loading-up to limiting state of a sample, a row of microcracks (dilatation) appears in the sample. At the primary stage of unloading, a rock of sample, striving to take initial position, fills part of a volume of these microcracks, and after this the sample is unloaded as a solid body. That is, both during the loading process and unloading process, a rock sample goes through three sequential states: elastic-plastic (decrease of volume during loading and increase – during unloading), plastic (volume is constant) and limiting (increase of volume during loading and decrease – during unloading).

Using mentioned regularities it is possible to make a conclusion that in primary stage of unloading of blast-hole walls from radial σ_r stresses, tangential strain exceeds radial. The opening of rupture radial microcracks occurs and tangential σ_θ stresses disappear (see Figure 1, area 10-11). Stress state of blast-hole walls turns to biaxial again. Their resistance to pressure from rock bolt side is dramatically fallen and causes unstable state appearance. Rock bolt material has the same condition, if radial stresses exceed value P_{pl}, as a result unloading process of the system continues to develop. If ultimate axial σ_{com} compression of rock strength is greater or equals to value P_{pl}, then when $\sigma_r = P_{pl}$, the balance state at the system's contact establishes. Further this state is disturbed by the influence of rheological factors, in particular, by the factor of reduction of ultimate rock compression strength with time. In connection with this, the first case of unloading of the blast-hole walls

rocks is characterized by the fact that ultimate long-time rock strength of uniaxial $(\sigma_{com})_\infty$ compression is greater or equals to P_{pl}. Because, when $\sigma_r = P_{pl}$, longtime balance state of "bolt-rock" system occurs. Thus, at $(\sigma_{com})_\infty \geq P_{pl}$ established P_∞ pressure on the system contact equals to

$$P_\infty = \frac{r_2^2 - r_1^2}{2r_2^2} \sigma_T^I. \qquad (3)$$

The second case is characterized by the fact that longtime ultimate compression strength $(\sigma_{com})_\infty$ is lower than P_∞. In this case during the decrease of radial stresses to value P_∞ "bolt-rock" system does not turn into balance state. Blast-hole walls are plastically deformed, being in limiting state, the surface of a contact moves towards their side, therefore a rock bolt is elastically unloaded and radial stresses on the contact are reduced to P_∞ (area 8-9), and longtime balance state of the system happens. So, at $(\sigma_{com})_\infty < P_{pl}$ established P_∞ pressure on the system's contact equals to

$$P_\infty = (\sigma_{com})_\infty. \qquad (4)$$

The third unloading case of "bolt-rock" system is characterized by the fact that the strain of elastic unloading of the bolt (area 6-7) is enough for unloading of blast-hole walls up to the value of radial stress, which is lower then $(\sigma_{com})_\infty$ and P_{pl} (area 6-13), and system turns into longtime equilibrium state immediately. Settled pressure P_∞ on the contact, for the third unloading case, is determined from formula of radial strain compatibility on the contact of "bolt-rock" system.

For ferro-concrete and ferro-polymeric rock bolts, the settled pressure, as it was mentioned earlier, is determined by unloading process of only blast-hole walls and equals to $P_\infty = (\sigma_{com})_\infty$

Thereby, it is established that radial stresses on the contacts of the "bolt-rock" system's elements, during bolting process, achieve maximum. As a result of rock dilatation this maximum can considerably exceed its uniaxial ultimate compression strength not causing microcracks. Further unloading takes place, during process of which radial stresses are reduced and consequently longtime balance state is settled. It is initial state during studying of static interaction of lockless bolt with a rock mass.

5 RESULTS

Research of the mechanism of "bolt-rock" system's interaction during bolting with initial radial stresses is a basic for the criterion development to choose installation parameters of rock bolts (blasting charge, inner pressure for tubular bolt installation; volume expansion of fastening layer – for ferro-concrete and ferro-polymeric bolts and etc.). A selection of parameters is defined by the method of determination of changing rational interval of required radial pressure on the rock bolt contact with blast-hole walls. Upper boundary of interval is determined by acceptable radial P_{max} pressure, which does not cause microcracks of blast-hole walls taking into account its dilatation. Lower boundary P_{min} equals to settled P_∞ pressure plus the loss of radial stress under elastic unloading of a bearing element (see Figure 1, site 6-7). Possibility of radial stress changing in certain interval favorably affects on regulation efficiency of bolt installation parameters.

Let us consider the definition of rational interval of changing radial σ_r stresses on top P_{max} and bottom P_{min} pressures, according to above stated requirements. For the calculation of P_{max} the exponential equation of rocks' ultimate state, which was experimentally gotten in (Stavrogin & Protosenya 1979), was used

$$\sigma_r - \sigma_\theta = \sigma_{com} exp\left(T\frac{\sigma_\theta}{\sigma_r}\right), \qquad (5)$$

where T – constant coefficient for given rock type (Stavrogin & Protosenya 1979).

Unknown value of tangential stresses in formula (5) is determined by the system of

$$\left.\begin{array}{l} \varepsilon_\theta = (2\mu-1)\varepsilon_r;\ \varepsilon_\theta = \dfrac{\sigma_\theta}{E(\varepsilon_i)}; \\[3mm] \varepsilon_r = \dfrac{1}{E(\varepsilon_i)}\left[\sigma_r - \mu(\sigma_\theta + \sigma_z)\right]; \\[3mm] \sigma_z = \mu(\sigma_z + \sigma_\theta);\ \mu = \mu_0\, exp\left(-\Gamma\dfrac{\sigma_\theta}{\sigma_z}\right), \end{array}\right\} \qquad (6)$$

where μ – coefficient of transversal rock strain; $E(\varepsilon_i)$ – function of dependence of rock's elasticity modulus on strain intensity ε_i; μ_0 and Γ – constant coefficients for given rock type (Stavrogin & Protosenya 1979).

The first equation links radial and tangential strains during the dilatation of blast-hole walls. The second and the third equations are physical, and the fourth describes the condition of flat blast-hole walls' strain in axial direction. The fifth equation defines the dependence of coefficient μ, transversal rock strain, on a type of stress state of a rock sample (Stavrogin & Protosenya 1979). During solving of system (6) it is taken into account that

$\dfrac{\sigma_z}{\sigma_r}$ ratio, as a rule, is bigger, than 0.5, when in line

with data work (Kuznetsov 1973) coefficient μ strives to 0.5. As a result formula for the calculation of tangential with radial components $C = \dfrac{\sigma_z}{\sigma_r}$ is given

$$C = \frac{1.5\mu_0 e^{-2c} - 0.75}{0.75 - 0.5\mu_0 e^{-2c} + 2\mu_0^2 e^{-2ac}}. \qquad (7)$$

Then maximum acceptable radial pressure on the rock bolt's contact with blast-hole walls is determined by formula

$$P_{max} = \frac{\sigma_{com}}{1-C} exp\, TC. \qquad (8)$$

For the definition of interval's lower boundary of rational changing maximum radial σ_r stresses, the equation of compatibility of radial displacement on the bolt contact with blast-hole walls in the system's unloading process is done

$$U_1 = U_2 + U_3, \qquad (9)$$

where U_1, U_2 and U_3 – radial dislocations during unloading of blast-hole walls of a bearing element and elastic loading of a bearing element accordingly.

Taking into account, during the process of unloading blast-hole walls, as a material of a bearing element, elastically deform. For the calculation of dislocations U_1, U_2 and U_3 known formulae were used

$$\left.\begin{array}{l} U_1 = \dfrac{1+\mu}{E} r_3 \left[P_{min} - P_\infty\right]; \\[3mm] U_2 = \dfrac{1-\left(\mu^I\right)^2}{E^I} r_2\left[\sigma_T^I + \dfrac{2r_2^2}{r_2^2 - r_1^2} P_{min}\right]; \\[3mm] U_3 = \dfrac{1-\left(\mu^I\right)^2}{E^I} r_3\left[2\left(1-\mu^I\right)\dfrac{r_2^2}{r_2^2 - r_1^2}\right]P_\infty, \end{array}\right\} \qquad (10)$$

where E^I and μ^I – module of elasticity and Poisson coefficient of a bearing element's material.

Jointly solving the equation (9) with the system (10), formula for the calculation of P_{min} is received

$$P_{min} = \left(\frac{1-\left(\mu^I\right)^2}{E^I} r^2 \sigma_T^I + P_\infty \left\{ \frac{1+\mu}{E} r_3 + \frac{1+\mu^I}{E^I} r_2 \times \right. \right.$$

$$\left. \left. \times \left[2\left(1-\mu^I\right) \frac{r_2^2}{r_2^2-r_1^2} - 1 \right] \right\} \right) \times$$

$$\times \left[\frac{1+\mu}{E} r_3 - 2 \frac{1-\left(\mu^I\right)^2}{E^I} \cdot \frac{r_2^3}{r_2^2-r_1^2} \right]^{-1} \quad (11)$$

Formula (11) is just for the determination of P_{min} parameter in the first and second cases of "bolt-rock" system's unloading, i.e. when settled P_∞ pressure on the contact is defined by formula (3) or formula (4). If mechanical parameters of "bolt-rock" system are such that settled pressure P_∞ is less than value of longtime rock compression strength $\left(\sigma_{com}\right)_\infty$ and the beginning of plastic bolt's phase P_{pl}, then P_{min} should be defined by formula

$$P_{min} = \left\{ P_{max} \left[\frac{1+\mu}{E} r_3 - 2 \frac{1-\left(\mu^I\right)^2}{E^I} \cdot \frac{r_2^3}{r_2^2-r_1^2} \right] - \right.$$

$$\left. - r_2 \frac{1-\left(\mu^I\right)^2}{E^I} \sigma_T^I \right\} \left[\frac{1+\mu}{E} r_3 + \frac{1+\mu^I}{E^I} r_2 \times \right.$$

$$\times \left(2r_2^2 \frac{1-\mu^I}{r_2^2-r_1^2} - 1 \right) \right]^{-1} . \quad (12)$$

Thus, rational interval of changing $P_{max} - P_{min}$ radial pressure has been defined. It is the main criterion of optimal power parameters in the bolting process.

REFERENCES

Melnikov, N.I. 1980. *A rock bolt*. Moscow: Nedra: 252.
Scobtsov, B.S. & Afanasiev, U.S. 1973. *Research of stresses in ferro-concrete rock bolts*. Physicotechnical mining problems of minerals, 3: 29-34.
Shirokov, A.P. 1971. *Practice and theory of rock bolt application*. Mine working's fastening on the Far East' mines. Prokopyevsk: city topography: 6-51.
Roginskiy, V.M. 1968. *Ferro-concrete rod's testing with strain sensors of resistance*.Mining journal, 12: 69-70.
Yemelyanov, B.I. 1978. *Tensometric investigations of steel-polymeric rock bolt operation*. Stability and fastening of mine workings. Leningrad's mining university, 5: 52-54.
Baklashov, I.V. & Kartosiya, B.A. 1975. *Rock mechanic*. Moscow: Nedra: 271.
Bridgeman, P. 1955. *Research of big plastic deformations and rupture*. Moscow: 350.
Pisarenko, G.S. *Resistance of materials*. 1979. Kiev: High school: 696.
Stavrogin, V.N. & Protosenya, A.G. 1979. *Rock's plasticity*. Moscow: Nedra: 300.
Kuznetsov, V.I. 1973. *About breaking of metallic rings in plastic state*, 4: 567-571.

Technical and Geoinformational Systems in Mining – Pivnyak, Bondarenko & Kovalevs'ka (eds)
© 2011 Taylor & Francis Group, London, ISBN 978-0-415-68877-2

Results of realized new concept of complex coal-gas deposit development

A. Bulat, V. Lukinov & V. Perepelitsa
M.S. Polyakov's Institute of Geotechnical Mechanics, Dnipropetrovs'k, Ukraine

ABSTRACT: The authors have created a new concept of the coal-gas deposit development according to which a combined degassing of both natural and man-caused methane accumulations increases the productivity of the coal mining thanks to the captured coal-mine methane used to produce electric and thermal energy, and improves effectiveness of nature-conservation measures aimed to make ecological situation in the coal-mining region better.

Difficultness of mining-and-geological conditions for the coal seam development in Donetsk coal basin (Donbas) is explained by small thickness of the coal strata, developed small and large erosion, and tectonic deformations. Mining of the deeper horizons in the key industrial geological regions of Donbas folding is complicated by high rock pressure, high gas content, sudden outbursts of coal, rocks and gas, methane blowers and complicated geothermal conditions (Lukinov & Pimonenko 2008). More than 25% of active mines are referred to the highest risk categories by methane content (more than 15 m^3 per 1 ton of daily coal production). Only 12% of active mines are referred to the gas-free mines, and they produce mainly nongaseous anthracites (Dolzhansk-Rovenetsk and Shahtinsk-Nesvetaevsk regions).

Today, gas composition in the Donbas coal deposits accounts 16 components and includes methane, ethane, nitrogen and carbon dioxide content of which is more than 1%. Trace contaminants in the gas mixture with content less than 1% are presented by propane, butane, pentane, hydrogen, hydrogen sulphide, argon, krypton, xenon, helium, neon, oxygen, carbon dioxide.

Most of researchers of the coal-bed gases recognize obviousness of the fact that methane was mainly formed in the process of metamorphism of organic matters in the coal strata and rocks. However, no general consensus has been obtained between the scientists concerning volume of methane formation at the moment when coals transit from one grade of metamorphism to another. According to G.D. Lidin (Lidin 1944), when 1 t of anthracite is created from coal grade D about 200 m^3 of methane is emitted, and according to V.A. Uspensky (Uspenskiy 1954) – only 150 m^3. Kozlov V.P. and Tokarev G.D. (Kozlov & Tokarev 1961) said that volume of emitted methane was 251 m^3 per 1 t of anthracite.

Article (Kravtsov 1968) shows clear dependence of changes of key gas components – methane, nitrogen and carbon dioxide – on the depth of the coal bedding. It should be noted that such clear definability of gas zoning in the coal stratum is explained by counter flows of metamorphogene gases and atmospheric gases and also, as we believe, by physical and chemical peculiarities of the coal structure (coals is a natural sorbent of methane) and its low filtering properties. This our opinion is obviously confirmed by less clearly definable gas zoning in the rocks and different locations of the gas zones in coal strata and adjoining deposits.

Gas composition in the coal seams and rocks is identical though distribution of the gases essentially differs and depends on geological conditions and factors.

By impact on gas content in the coal-bearing thicknesses, main geological factors include the following: 1) gas formation and burial; 2) gas accumulation and reservation; and 3) degassing of the coal-bearing deposits. At the same time, it is obvious that all these factors were under direct or indirect impact of tectonic processes that had happened in the region at various stage of its development.

Long-term researches of gas content in Donbas coal-bearing deposits help us to reveal and study regional and, partially, local changes of gas characteristics in the coal seams and rocks; these changes could have far-reaching impact on organization of degassing processes and methane recovery. It was found (Brizhanev 1979; Dmitriev & Kulikova 1982; Zabigailo & Shirokov 1972; Skochinskiy & Lidin 1948) that gas in the coal seams and rocks could be in free, aqueous and retained states; content of free gas depends on porosity and fracturing of the rocks and coals and thermodynamic conditions of the coal and rock bedding.

Regional factors include depth of the coal-seam bedding and rate of the seam metamorphism, while local factors include rate and type of tectonic deformations and lithologic and facial characteristics of the rocks.

Articles (Kravtsov 1968; Zabigailo & Shirokov 1972; Kravtsov 1980) describe impact of paloestresses on gas content in the coal-bearing thickness: deformations caused by stretching (faults) are permeable and make degassing of the coal-bearing deposits easier, while in compression zone overthrusts are formed: they are gas-impermeable and force gas to be accumulated and reserved in adjoining zones. According to A.M. Brizhanov (Brizhanev 1979), gas content depends on time when deformations were created and proceeds from the following properties of the deformations: post-sedimentation faults are permeable, and con-sedimentation faults are impermeable; overthrusts are characterized by changeable permeability that depends on lithology of adjoining rocks.

Thus, it is generally recognized that, in regional terms, natural gas content in a massif depends on rate of catagenetic transformations and forms pattern of gas distribution; and, in local terms, gas content depends on 1) lithologic and facial conditions; 2) types, parameters, time and conditions of breaking deformations formation; and 3) local folding.

Recovery of mine methane as a hydrocarbon input plus supported by safety conditions for mining operations, reduced harmful emissions of mine methane into atmosphere and improved ecological situation – all these are the strongest business case for practically each coal-mining regions of the world. To solve this problem, it is necessary to understand impact of natural and man-caused factors on methane resource distribution in the coal-bearing thickness and principles of accumulation of this valuable energy resource in the coal thickness.

Methane resources in the coal-bearing thickness primary depend on geological conditions and factors which form regional patterns of methane-content changes in the rocks and coals. Key methane accumulators in the coal-bearing thickness are coals and sandstones. Character of the gas-content changes in coals and sandstones depends on level of their catagenesis. In coal-bearing thicknesses with weakly-metamorphized coals methane is mainly located in sandstones, and in areas with OS, T and PA coal grades methane is mainly accumulated in the coals.

As opposite to the natural gas deposits, main portion of methane in the coal deposits is not locally concentrated but is widely dispersed throughout the

whole coal-rock massif. In the coal strata, coal interlayers, coalified organic sediments and dispersed organic substances, methane is mainly available in retained (but not in sandstones), free or dissolved-in-water state.

By conditions of formation of methane accumulations suitable for commercial recovery, the gas can be divided into the following groups:
– methane in the coal strata and rocks outside the active mine fields;
– methane inside the coal-rock massif of the active mines;
– methane in coal-and-rock massif and tunnels of the closed mines.

Methane from the coal strata outside the mine field can be extracted by various methods: for example, by hydraulic fracturing of the stratum or by directed multilateral drilling of wells with further hydraulic fracturing and dewatering. In international terminology, methane extracted by these methods is called CBM, or coal bed methane. Technology of the CBM extraction is very complicated and required essential funds. In order to provide good effectiveness of this technology it is necessary to drill hundreds of wells that should operate simultaneously. One more problem arises: big volumes of salt water that should be treated and utilized.

Experience of other coal-producing developed countries with commercial recovery of methane shows that for proper extraction of methane it is necessary to increase permeability of the coal-and-rock massif and to disturb sorption balance in order to stimulate desorption of methane from the coal and force the gas to move.

Today, this technology is successfully applied in the USA and Canada though operators in Germany, Russia and Ukraine have failed to obtain expected results.

However, there are some fields in Donbas within western, southern and northern boundaries where weakly-metamorphized coals and, accordingly, sandstones with favourable collecting properties are developed and where methane deposits are available for commercial production with no additional efforts. Actually, the fields can be considered as small deposits of natural gas. They are confined to positive tectonic structures: Leventsovskaya structure in West Donbas, Lavrentievskaya structure in South Donbas geological industrial region, Matrosskaya structure in Lisichansk region and some others, mainly, in northern and southern zones of the small folding. Unfortunately, number of these zones is not great.

Sandstones bedded in the regions with developed medium-metamorphized coals feature low permeability (less than tenth and hundredth fractures of millidarcy), and such permeability impedes formation of methane accumulations in the sandstones. Here methane is dispersed in the pores, and the only source to extract the gas from without additional measures is tectonically created collecting fracture reservoirs where free methane accumulations could be formed.

In 2005, our Institute IGTM of NASU worked out and approved a as normative document a technique to predict tectonically deformed zones favourable for formation of free-methane accumulations in coal-and-rock massifs not disturbed by mining operations. The technique helps to detect promising areas for methane extraction in boundaries of so-called opened anticline structures with parameters specified by local structure maps made in the form of projections of the stratum hypsometry on approximating surface. We determined some structures promising for preliminary degassing in the Butovskaya Mine field and in one district of Kalmiuskiy Colliery that is a reserve unit of A.F. Zasyadko's Mine. Here, basing on our recommendations, the operators set location of and installed equipment for drilling a ЗГ – 2 well.

Ukrainian mines and Donbas coal mines in particular have obtained good experience in methane recovery with simultaneous coal mining through investigating man-caused processes that influence on changes of physical properties of rock-and-coal massif and, in particular, the massif fracturing, permeability and gas-bearing.

When coal is extracted the rock pressure in the coal strata and adjoining rocks is dropped, the coal strata and adjoining rocks become fractured and soft, and, accordingly, permeability increases. These processes are followed by formation of man-caused collecting reservoirs and, consequently, concentrations of free and liberated methane in the reservoirs. This gas is called CMM, or coal mine methane, and it can be extracted by current degassing and drilling of underground and surface wells in the active mine fields.

This is exactly the line that is directly associated with the needs of Ukrainian coal industry and to which our Institute of Geotechnical Mechanics (IGTM) orients our researches. In this filed we closely cooperate with other trade and academic institutes, Safety Department of Ministry of Energy and Coal Industry of Ukraine and some other organizations among which the A.F. Zasyadko's Mine should be specially noted. Only for the last 5 years, we worked out and approved as normative documents: schemes and methods of gas-emission control in the working areas of coals mines (2006); technique to predict man-caused methane accumulations in disturbed coal-and-rock massif (2007); procedure for applying advance degassing method in the coal strata (2010).

Practice of advance coal strata degassing is our absolutely new technology based on the laws of methane flow distribution in the border between mined area of and not-disturbed coal-and-rock massif and makes it possible to capture in advance additional volumes of methane. For example, this practice helped the A.F. Zasyadko's Mine to produce additionally more than 4 mln. m^3 of methane.

Our new degassing technologies increase methane content in composition of gas recovered with the help of underground degassing wells, extend output of methane production, and improve safety of the coal mining.

We also worked out a method to estimate density of methane resources in the roof rocks of the mining coal stratum. The method takes into account geological, mining and technical factors that influence on formation of methane accumulations in the undermined coal-and-rock massif. For example, methane resource density in undermined roof rocks in mines of Makeevugil Company varies between 40 m^3 / m^2 and 140 m^3 / m^2 depending on metamorphism rate of produced coal, extracted coal thickness, depth of mining operations, method of the roof control and some other parameters. Volume of methane resources in these mines is between 9 mln. m^3 and 35 mln. m^3 at mining column with length 1000 m and longwall with length 230 m. Extractable methane reserves are approximately half of estimated resources and vary between 6 mln. m^3 and 17 mln. m^3.

Practice of mining high-productive longwalls with methane content more than 15 m^3 / min. shows that up to 80% and even more methane is emitted into the mining district from the roof undermined coal-and-rock massif. Due to divergent parameters of deformation, behaviour of rocks with different lithologic characteristics is not the same at the similar undermining conditions: undermined rocks with greater ultimate tension deformation are less soft and with lower permeability, and vice versa. Besides, in the process of undermining each geological object in the coal-and-rock massif should be considered from the point of its gas output (in case of gas-bearing rocks) and from the point of gas filtered though it (in case of both gas-bearing and non-gaseous rocks). To estimate possible resources of methane in the roof coal-and-rock massif we worked out a method that 1) specifies interval space for the rock unloading towards the roof when the seam has been extracted; 2) calculates distance to the seams with screening properties and collecting seams; and 3) specifies parameters of degassing zone and zone of gas recovery.

The authors have created an absolutely new concept of combined degassing of the coal-and-rock massif. The concept is based on previously designed methods, practices and technologies tested in-situ in mines and assumes that these methods, practices and technologies should be applied with taking in account all natural geological, mining and technical conditions. Processes of the coal mining and combined degassing (that ensures safety operation of the mine) are sequenced in time and space. Usage of mine methane to produce electricity and thermal energy essentially reduces methane emissions into atmosphere and serves as nature-conservative measure that improves ecological situation in the coal-producing region.

Concept of combined degassing assumes combination of fully and/or partially applied: "gas horizon" technology; technology of advance degassing; methane capture by degassing wells drilled from the surface and mine tunnels; recovery and usage of methane degassed from the mined longwall faces; estimation of methane resource distribution in coal-and-rock thickness; and maximal utilization of captured methane.

Today, the largest scale application of the concept is realized in the A.F. Zasyadko's Mine (Lukinov, Perepelitsa, Bokiy & Efremov 2010). Here, for the time period of the concept operation from 2006 till 01.01.2011, co-generation plants generated 735312 MW-h of power and 232177 Gcal of heat, and about 210 mln. m^3 of mine methane was used. Besides, from 2004 till 01.01.2011 more than 13 mln. m^3 of methane was used as fuel for cars: this gas was captured through surface degassing wells. Totally, about 223 mln. m^3 of methane were used from 2004 till 01.01.2011 providing reduction of harmful gas emissions in amount equivalent to 3576149 tons of CO_2 which, in accordance with Kyoto Protocol, were officially registered, and the Mine received essential compensation.

Gases received from different sources of degassing feature different qualitative composition. Gas captured through the surface wells contains more than 90% of methane and is used as fuel for cars and igniter for gas reciprocators. Gas from underground wells contains 30-50% of methane and together with gas pumped from the mined districts of the mines with methane content less than 30% is used to generate electrical and thermal energy and in boiler houses.

It should be noted that a strong trend to reduce amount of the coal mines is observed in Europe and in the world, and the same happens in Ukraine. There are a great number of closed old mines in Ukraine, especially in Donbas region, thus, problem of methane extraction from the coal-and-rock massifs and mine tunnels in the closed mines is a question of the day from both economic and ecological points of view.

Methane that escapes from disturbed coal-and-rock massifs into goafs of the mined longwalls and fills mine tunnels in the closed mines is extracted when coal production has been finished; to this end, wells are usually drilled into the old tunnels via which methane is pumped out by vacuum pumps. Such methane is called CAM, or coal abandoned methane. Today, CAM is successfully recovered and utilized in the USA, Germany, United Kingdom and in some other countries where methane-air mixture with methane content more than 50% is used to generate power. Ukraine does not have such good experience in CAM recovery and utilization, however, successful application of this practice in other countries is a witness of its livability as well as economic and ecological effectiveness of CAM recovery and utilization.

Density of remaining methane resources is estimated in each concrete case with taking into account mining and geological conditions and all factors of the deposit development. To our estimates, density could be more than 50% of the mining reserves and could reach 60 mln. m^3 per 1 km^2 of the stripped area in abandoned mine.

Methane that fills tunnels in abandoned mines creates threat of explosion for buildings and objects and poisoning for people in areas where methane escapes to the surface. Coal in Ukraine has been mined for more than 200 years, and hundreds of mines were closed during this period of time especially during the last twenty years resulting in depressive development of the regions.

Realization of the projects on recovery and utilization of methane from the closed mines will, from one hand, involve additional energy resources into fuel-and-power system of Ukraine and, from the other hand, ensure safety life and better ecological situation in the old coal-producing regions of Ukraine.

REFERENCES

Lukinov, V.V. & Pimonenko, L.I. 2008. *Tectonics of Methane-Coal Deposits in Donbas.* Kyiv: Naukova dumka: 352.
Lidin, G.D. 1944. *Zonal Distribution of Natural Gases in Donbas.* News of AS of the USSR, 6: 337-345.
Uspenskiy, V.A. 1954. *Material Balance of the Processes in Metamorphism of the Coal Strata.* News of AS of the USSR, 6: 94-101.
Kozlov, V.P. & Tokarev, L.V. 1961. *Scale of Gas Formation in Sediment Thicknesses (Donetzkiy Basin as an Example).* Soviet Geology, 7: 19-33.
Kravtsov, A.I. 1968. *Geological Conditions of Gas-Content in Coal, Ore and Non-Metallic Mineral Deposits.*

Moscow: Nedra: 331.

Brizhanev, A.M. 1979. *Regularity of Gas-Content Changes with Changed Depth of Mining in Donbas*. Gas-Content of the Coal Basins and Deposits in the USSR. Moscow: 98-101.

Dmitriev, A.M. Kulikova, N.N. & Bodnya G.V. 1982. *Problems of Gas-Content in the Coal Deposits*. Moscow: Nedra: 263.

Zabigailo, V.E. & Shirokov A.Z. 1972. *Geological Problems of Gases from the Coal Deposits*. Kyiv: Naukova dumka: 172.

Skochinskiy, A.A. & Lidin, G.D. 1948. *Classification of Emitted Methane from the Coal Mines*. News of AS of the USSR, OTNH, 11: 1741-1751.

Kravtsov, A.I. 1980. *Impact of Geological Factors on Natural Gases Distributed in the Coal Seams and Adjoining Rocks*. Gas-Content of the Coal Basins and Deposits in the USSR. Moscow: Nedra. V.3: 74-101.

Lukinov, V.V., Perepelitsa, V.G., Bokiy, B.V. & Efremov, I.A.. 2010. *Creation of Effective Energetic Complex on Mine Methane Recovery and Utilization*. Geotechnical Mechanics. Dnipropetrovs'k: Issue 88: 3-8.

Substantiation of chamber parameters under combined open-cast and underground mining of graphite ore deposits

V. Buzilo, T. Savelieva, V. Saveliev & T. Morozova
National Mining University, Dnipropetrovs'k, Ukraine

ABSTRACT: The move from open-cast to combined mining method of developing Zavalyevsky deposit of graphite is considered in the paper. The selection of chamber dimension satisfying conditions of long-term stability is one of the questions for scientific substantiation of the move to combined mining method of deposit development. With this purpose physical and mechanical properties of graphite and enclosing rocks are determined. Stress and strain condition of rocks around the chambers using the method of finite elements is studied.

1 INTRODUCTION

The main consumers of graphite are enterprises of metallurgical, electrical engineering, mechanical engineering, aircraft and other branches of industry. One of the most important raw material base supplying graphite is Zavalyevsky deposit of graphite (Ukraine).

Nowadays the development of deposit is carried out by the open-cast method. The Southern-Eastern part of the deposit is worked-out. The depth of the quarry is 100 m and it is very close to its boundaries.

A great part of commercial reserves is within the pillars under the bed of the Yuzhny Bug river, dressing mill and the village. While deepening a quarry it's boundaries should be moved. Besides, it is necessary to move aside dressing mill and the bed of the Yuzhny Bug river. For further open-cast mining operations it is required to construct a dam and offtake bed channel, relocate dressing mill and amortize the land of 400 ha. Total cost of underground mining of graphite is muck higher then it was expected. Therefore, it is required to work-out the deposit by combined mining method, e. i. to carry out further development of the quarry up to the level of 19 m. For further deepening reserves of graphite ores are worked-out in designed contour of Southern-Eastern quarry without wall cutback. After that the transition to underground mining with dressing mill reconstruction is performed. It's capacity is up to 62.0 thousand ton of graphite per year. At the same time the thickest and the richest ore bodies of the Northern part of Zavalyevsky deposit are worked-out.

Taking into account the demands as for Earth surface protection, steep dip of ore body, available information concerning mechanical properties of ore and enclosing rock, attempts to avoid significant losses and delution under the safe condition of stoping, it is recommended to apply level and chamber mining method. In this case ore is broken by deep boreholes and subsequent filling of worked-out area is carried out by consolidating stowing.

At the same time three chambers are worked-out. Chamber stowing is carried out after finishing extraction. There are three stages of filling worked-out chambers by consolidated stowing:

– filling chamber bottom with the mixture of higher content of binding substance with the aim to obtain hardened stowing on the level of the dam;

– filling chamber from the bottom till the top stopping on the level of the dam;

– additional chamber stowing with higher content of binding substance.

Applying chamber method of mining determines the necessity of substantiating chamber and pillar dimension meeting the requirements of long-term strength. For this purpose it is required to carry out the study of physical and mechanical properties of graphite and enclosing rock.

2 DETERMINING PHYSICAL AND MECHANICAL PROPERTIES OF GRAPHITE AND ENCLOSING ROCK

Study concerning determination of breaking stress under one-axle compression and on tension was carried out on regular form samples according to the standard methodic (Baron, Loguntsov & Nozin 1962).

According to this methodic cylindrical form samples with the diameter and height ratio equal one were used during the study. Diameter of the samples was from 50 till 51 mm for graphite and from 57 till 58 mm for enclosing rock. Sample processing was carried out in such a way that fluctuation from the end parallelism was not more than 0.05 mm, from end perpendicularity to the cylinder element was also 0.05 mm. Tests were performed on the press PG-100A.Ball centering device was used for strict load centering. Loading was performed gradually increasing the load up to sample destruction .Value of destructive load was fixed.

Strength under tension is determined by Brazil method on regular form samples: diameter of the cores is 57-58 mm. Height of the cores is equal to the diameter. Determination was carried out by the method of diametrical compression. It means that cylindrical samples were broken by the forces applied along diametrically opposite elements.

Characteristic of the rock hardness under displacement is its shift resistance created by two physical factors: internal friction and cohesion. Internal friction can be comparatively easy calculated as it represents the interacting force under deformation taking place between mineral particles. This deformation is proportional to regular stresses caused by external load.

Cohesion is that part of shift resistance which is not connected with stresses caused by external load. It is determined by only molecular binding forces and cannot be constant value.

So, complete resistance of the rock to the shift is expressed by the sum of internal friction caused by external forces and cohesion. It was determined on the regular form samples (Turchaninov, Medvedev & Panin 1967). Sample test was carried out on the press with the help of inclined cores with angles of $a_1 = 35°$ and $\alpha_2 = 55°$.

It is known that the rate of distributing elastic waves depends on the module of material elasticity. Therefore, nowadays method of determining rate of compression and cross elastic wave distribution – sound dynamic method – is widely used to determine elastic values of rock properties (elasticity module and Puasson coefficient).

At present this method is one of the most perspective as it is much cheaper and simpler then other well-known methods (particular statistic ones).Besides, it enables to determine rock properties within massif. That makes it unique while performing field operations. Elastic values of rock properties obtained under statistic and dynamic loads are different due to various character of rock deformation.

Ultrasound defectoscope UK-10P, selective amplifier EGU-60, audio generator LEG-60, vacuum tube voltmeter, conductors made of zirconate-titanate barium with $\delta = 10000$ were used to determine Yung module and Puasson coefficient using dynamic method. Resonance of compression and cross waves were determined with the help of these devices. The rate of compression and cross wave transmission was determined according to this resonance.

The value of dynamic elasticity module was determined after finding out the rate of compression and cross wave transmission within the samples.

Table 1 shows averages of determining physical and mechanical rock properties.

Estimation of result reliability of experimental study using the method of Monte-Karlo probability theory was carried out (Ventsel 1972).It was stated that the level of confidence is equal to 0.84-0.99. It enables to confirm that the amount of tests taking into account while determining physical and mechanical rock properties is enough to be considered as a true one.

Table 1. Mechanical properties and elastic characteristics of graphite ore and enclosing rock.

No	Rock	Compression		Extension		Dynamic method			Module of shifting, 10^{-10} Pa
		Amount of samples	σ_{com}, MPa	Amount of samples	σ_{ten}, MPa	Amount of samples	Elasticity module, 10^4 MPa	Puasson coefficient	
1	Graphite	9	20.2	6	1.0	8	1.55	0.26	0.61
2	Quartzite	8	60.3	7	8.8	8	5.42	0.21	2.24
3	Calciphyre	8	40.1	7	6.7	8	4.79	0.25	1.81
4	Gneiss	7	31.4	6	6.8	8	4.51	0.32	1.68
5	Calc-silicate hornfels	8	64.1	7	10.3	6	7.10	0.22	2.92

3 SUBSTATIATING PARAMETERS OF HEADING-AND-STALL MINING METHOD

Development with consolidated stowing means that sizable chambers are worked-out beforehand. After that worked-out chambers are filled in consolidated stowing. The areas (pillars) left between chambers are worked out in three months after complete stowing consolidation. Pillar size is usually equal to the chamber size.

The study of rock stress-strain condition using the method of finite elements was carried out to substantiate chamber and pillar sizes (Zienkiewicz & Taylor 2000).

Ore and enclosing rock are considered as elastic body and the problem is described in terms of elasticity theory. Chamber roof and interchamber pillars are loaded by the press force of overlying rock. These forces are perpendicular to chamber direct axis. All cross-sections are in the same conditions and there is no shifting in longitudinal direction. As cross-section conditions are the same, it's enough to consider thin layer between two sections. The distance between them is equal to one. The system is in condition of the plane deformation.

Chamber location is periodic in both sides. Due to the symmetry only the area between symmetry axis is considered.

The problem was solved under the following boundary conditions:

– the absence of horizontal point shifting of the left and the right boundaries due to the symmetry;

– external forces are not applied to the right and the left boundaries as well as to the internal points

(gravity forces are not taken into account), rigid fixing prevents area turning as a whole one;

– the weight of overlying rock is replaced by concentrated forces applied to the points of horizontal line.

Study of stresses within chambers and chamber roof was carried out to the following tasks.

Task 1. Chamber width is 10 m, height is 48 m, pillar width is 10 m. Chamber shape is symmetric. Due to the calculation stresses in the gravity center finite elements were obtained. Stress epures were build according to them (Figure 1).

Maximal tension stresses in the center of chamber width is $7 \cdot 10^5$ Pa. Strength margin in the center of the roof is 1.4.

Maximal compression stresses within the chamber is $- 44.1 \cdot 10^5$ Pa. So, strength margin is 4.5.

Task 2. Chamber width is 10 m and the height is 48 m, pillar width is 10 m. Chamber shape is not symmetric. Stress epures are built taking into account the results of calculation (Figure 2).

Maximal tension stress in the center of the chamber roof is $7.5 \cdot 10^5$ Pa. Strength margin is 1.3.

Maximal compression stresses is $- 43.2 \cdot 10^5$ Pa. Strength margin is $- 4.6$.

Task 3. This strength margin within the roof is insufficient for durable construction stability. The change which the roof in the shape of basket described by two radius brings in to the stress field is considered. Pillar width and chamber span are the same – 10 m. Stress epures are built according to the calculation data (Figure 3). In this case strength margin is sufficiently increased and it is 5 in the center of the roof as well as within the pillars.

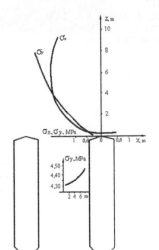

Figure 1. Stress epures around the chambers. Variant 1.

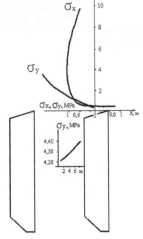

Figure 2. Stress epures around chambers. Variant 2.

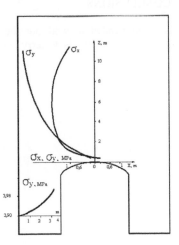

Figure 3. Stress epures around chambers. Variant 3.

Task 4. The width of chambers and pillars should be chosen to ensure safe support of the roof with the help of enterchamber pillars.It is required to find out optimal size and shape of the roof meeting the demand of stability as well as increased graphite extraction.

Stress distribution around chambers with the width of 14 m and pillars with the width of 6 m were considered to choose optimal pillar and chamber sizes Stress epures were build taking into account calculation (Figure 4). In this case tension stresses in the center of the roof is $2.99 \cdot 10^5$ Pa, that corresponds to the strength margin equal to 3.3. Compression stresses within the pillars is $52.47 \cdot 10^5$ Pa, strength margin is 3.7.

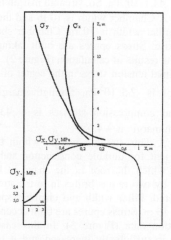

Figure 4. Stress epures around chambers. Variant 4.

4 CONCLUSIONS

It is recommended to work-out the deposit using combined method, that is, to carry out further quarry development up to the point of 19 m and transition to level-chamber mining method filling the goaf with consolidated stowing.

The choice of system development parameters was based on the study of stress distribution character according to four schemes in Figure 1, 2, 3, 4 in the form of epures. Two first calculation schemes showed that strength margin within the roof was 1.5 that is not enough for long-term construction reliability. That is why these schemes cannot be recommended to apply. According to the third scheme where distance between chambers is 10 m and arch is outlined by the basket curve, strength margin within roof and pillars is sufficiently increased and is 5. According to the fourth scheme where distance between chambers is 14 m and pillar width is 6 m, strength margin correspondingly is 3.3 and 3.7.

Taking into account that heading-and-stall method with stowing worked-out area will be used in the quarry and to increase graphite extraction it can be recommended to apply the following parameters: chamber width is 14 m, pillar width is 6 m, chamber height is 48 m, the shape of the roof is basket.

REFERENCES

Baron, L.E., Logutsov, V.M. & Nozin, E.Z. 1962. *Determining rock properties.* Moscow: Gosgornehizdat: 332.

Turchaninov, E.A., Medvedev, R.V. & Panin, V.E. 1967. *Up-to-date methods of complex determining physical rock properties.* Moscow: Nedra: 199.

Ventsel, E.S. 1972. *Operation study.* Publish house. Sovetskoe Radio: 306.

Zienkiewicz, O.C. & Taylor, R.L. 2000. *Finite Element Method.* Volume 1 – The Basis. London: Butterworth Heinemann: 712.

Zienkiewicz, O.C. & Taylor, R.L. 2000. *Finite Element Method.* Volume 2 – Solid Mechanics. London: Butterworth Heinemann: 480.

The problem with increasing metal-content of a development working's combined support

I. Kovalevska & V. Fomichev
National Mining University, Dnipropetrovs'k, Ukraine

V. Chervatuk
OJSK "Pavlogradugol", Pavlograd, Ukraine

ABSTRACT: The results of researches of prolonged flexible hip-roof support (PFHS) combining with roof bolts in the in-seam working are given. The character of distribution of vertical and horizontal components of stresses in the elements of "rock mass-support" system, involving two specific profiles of PFHS (SVP-27 and SVP-19), was examined by computer modeling, using finite element method. Inessential difference of components' values of stresses in load-bearing elements of the system was detected by analyze of stress diagrams. That caused to substitute specific profile of a frame to SVP-19 and to successfully carrying out mine tests with decreasing up to 30% metal-content of a support.

1 INTRODUCTION

The problem of decreasing combined support's metal-content of mine workings is considerably linked with the fullest accounting of laws of changing strain-stress state (SSS) both for rock-contained mine opening and directly for support elements, depending on constructive, technical and force parameters of its interaction. Therefore, using finite element method in geomechanical models the investigations of influence of PFHS specific profile's nominal size in combination with roof bolts, strengthening the roof of a working, were carried out (Bondarenko 2006 & Bondarenko 2007). Comparison of two frame-support specific profiles, SVP-27 and SVP-19, has been done, based on SSS analyze of "layered massive-support of mine working" system. It is done due to vertical σ_y and horizontal σ_x stresses with analyze of stress field in pairs with SVP-27 frame and SVP-19 frame, to determine significant differences and to explain reasons of their appearance. At first, comparison of σ_y field was carried out in near-the-contour rock mass, and then in the frame support and bolts. So far, the level of stresses in mentioned elements of the system differ by one order, so two colored σ_y scales were separately used for near-the-contour rocks and for a support, with illustration of them by diagrams for SVP-19 profile.

2 RESULTS

Directly in the roof of mine working the zone of unloading occurs, similar to the shape of Protod'yakonov's arch of natural balance (Figure 1a). The zone size of rock ultimate state is usually evaluated by the condition of appearing tensile stresses ($\sigma_y \approx 0$). Every mine rocks can badly resist to this stresses, especially layered and weak rocks. Following mentioned condition the sizes of ultimate state area of roof rocks are practically even: 2.02 m width and 0.43 m height for SVP-27; 2.09 m width and 0.44 m height for SVP-19. So, if to change from SVP-27 frame to SVP-19 frame zone's width will be increased only in 3.5%, and height – 2.3%. In both versions quite limited areas are observed (till 0.2…0.3 m), actions of tensile σ_y reach up to 1 MPa in places between neighboring bolts. Enumerated facts can prove almost similar diagram of vertical σ_y stress distribution in the roof, both as qualitative and quantitative meanings.

In the walls of mine working the zone of abutment pressure with increased compressing σ_y stresses is formed. These stresses exceed resistance to compression of mudstone (13 MPa) and siltstone (13.5 MPa) directly in the roof and in the foot of C_6 coal-seam accordingly. Sizes of this zone are enough wide and reach the whole layer of siltstone in depth, directly in the foot (both for SVP-27, and as for SVP-19). Height of zone (from opening's foot) is 2.48 m for SVP-27 and 2.54 m for SVP-19 (increasing till 2.4%), zone's width of distribution is about 2.04 m for SVP-27 and 2.08 m for SVP-19 (increasing till 2.0%).

(a)
(b)

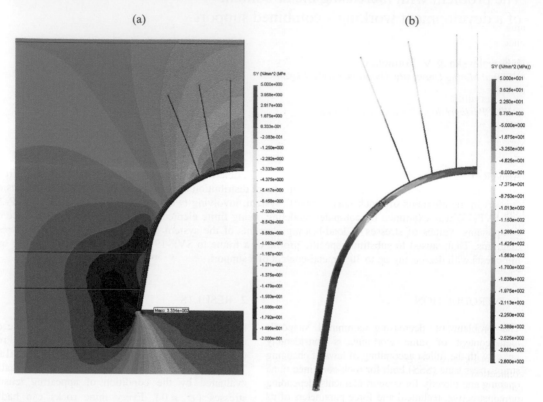

Figure 1. Diagram of vertical σ_y stresses in the "layered massive-working's support" system (a) and in the frame support with bolts (b), especially for SVP-19 using.

Maximum σ_y, about 20…25 MPa, is created in the area of a frame-leg and has local character. Thereby, in the lateral part of a working the field of compressive σ_y stresses is almost identical for both nominal sizes of SVP profile.

The zone of unloading is located in the foot of a working, achieving not only siltstone layer in depth, but and lower layer of sandstone. Tensile σ_y stresses, up to 3 MPa, forms the zone of unstable rocks as the opposite shape of arch along all immediate foot. Ultimate state zone with $\sigma_y \approx 0$ criterion is expanded in the whole width of a working, and is 2.13 m for SVP-27 and 2.17 m for SVP-19 in depth (increasing till 1.9%). Therefore, any considerable differences of σ_y diagrams between both versions of SVP are not observed in working's foot.

Analyze of σ_y diagram in the frame and rock bolts of SVP-27 and SVP-19 profiles led to follow-ing results. In the roof bar of a frame reduced com-pressing σ_y (till 40…60 MPa) operate with passing in the tensile area (till 10 MPa) in the lock of an arch only. This picture (both in qualitative and in quantitative meanings) is almost identical for both versions of SVP. It points to influence of relatively not great load in the roof, making conditional stable state of a roof bar (Figure 1b).

Frame legs are under influence by higher com-pressing σ_y stresses: in the part of yieldability-lock, of 0.8 m long, σ_y is increased from 115 MPa to 240 MPa; further, in the rectilinear part of a bar σ_y is exceeded ultimate tensile strength of steel St.5 ($\sigma_T = 270$ MPa), reaching 280 MPa; in the lower part of a frame-leg, of 0.4 m high, σ_y is decreased to 150…240 MPa, and in the contact with a step-bearing the maximum occurs again, $\sigma_y = 231.1$ MPa for SVP-27 and 333.4 MPa for

SVP-19. However, an action area of contact stresses is limited in the leg's end and a step-bearing. And sufficient σ_y difference is explained by far lower area of contact for SVP-19. Given sufficient difference is single in σ_y diagrams of a frame support, made of SVP-27 and SVP-19, and whole contour of a frame has almost constant strain state, with factor of vertical σ_y stress action.

The analogous conclusion can be drawn due to analyze of σ_y in a rock bolt support. For both SVP versions the transformation of compressing stresses $\sigma_y = 60...110$ MPa (in the sunken part of a bolt) to tensile $\sigma_y = 0...50$ MPa, in closer-to-mine working part of a bolt, is clearly observed. This fact points to bolt resistance to movements of near-the-contour rocks in direction to working interior. These rocks are in ultimate state and fully coordinated with existent researches of rock strengthening by bolts. σ_y value is far away from ultimate tensile strength of bolt armature ($\sigma_T = 240$ MPa), that highlights to its steady state, because of vertical σ_y stresses.

Thereby, due to results of comparison of σ_y diagrams for SVP-27 and SVP-19 it is enough to affirm about its identity. Increased contact stresses, in the connection of a frame end and a step-bearing, can not considerably influence on bearing capacity of a SVP-19 frame, and single local plastic zone is self-balanced due to well-known fact of load redistribution to elastic areas during plastic material flow because of local contact stress action. Comparative analyze of distribution field of horizontal σ_x stresses of near-the-contour rocks during installing frames, made from SVP-27 and SVP-19 (Figure 2), gave following results. In the roof of mine working an area of increased compressing σ_x stresses (with concentration coefficient 1.3...1.9) is formed lengthwise bolt and on width is quiet exceeded the distance between locks of yieldability. But, σ_x value is far lower than resistance to rock compression of an immediate roof. These rocks are totally in steady state, following to action factor of σ_x. Attract attention local σ_x concentration in the place of yieldability's lock, which can be explained by the process of its actuation.

In the walls of working (mainly in C_6 coal seam) along height up to 2 m and about 1 m wide an area of unloading with changing σ_x range of compressing 1 MPa value and tensile 1 MPa value is situated. Tensile σ_x intensifies an area of ultimate state,

however, due to short depth of this area (till 1 m) sufficient blowing up of foot rocks should not be expected. All mentioned characteristics of σ_x distribution in near-the-contour rocks are almost the same as in qualitative and in quantitative meanings for frames, made of SVP-27 and SVP-19.

Features of σ_x stress field in a frame support lead to following. In the roof bar compressing σ_x act from 90 MPa, in the place of lock of yieldability, to 145...160 MPa in the place of lock of support's arch, made of SVP-27. For SVP-19 specific profile (Figure 2b) increased $\sigma_x = 180...200$ MPa work in the central part of exterior side of arch. It points out more intensive unloading of light-weight specific profile from roof side. However, level of σ_x is far from ultimate tensile strength of St.5 and a roof-bar is in elastic state. The lock of yieldability is most unloaded ($\sigma_x = 20...35$ MPa), as it should be when it goes off. The top (curvilinear) part of a frame is under relatively insignificant compressing ($\sigma_x = 40...75$ MPa) stresses, and its rectilinear part is almost fully unloaded with converting from little compressing σ_x stresses to tensile about 20 MPa.

Given combination of high σ_y and low σ_x (opposite symbol) leads to growth of current σ stress values (or intensity of σ stresses). It favors appearing an area of plastic material state in the rectilinear part of a frame. In the place of supporting a quite limited area of contact tensile $\sigma_x = 30...50$ MPa actions is formed. Also these actions increase current σ stresses and generate appearance of plastic zone. Point contact of frame-leg end with a step-bearing causes appearing maximum of tensile stresses $\sigma_x = 335.3$ MPa (SVP-27) and $\sigma_x = 252.2$ MPa (SVP-19). Steel St.5 is in the area of strengthening and can fully resist on given stresses. However, extremely small volume of action of contact maximum values contributes to load redistribution to elastic areas and considerably not to influence on bearing capacity of a frame. Overall, except to increased stresses in the exterior side of the central roof-bar part (SVP-19), σ_x field has almost the same qualitative and quantitative measures for SVP-27 and SVP-19.

Not great horizontal σ_x stresses effect rock bolts: in sunken part – generally compressing stresses about 20...35 MPa, in the near-the-contour – tensile stresses about 15...20 MPa, because of more intensive moving of near-the-contour massive inside the mine working, and bolts resist to this process.

(a)   (b)

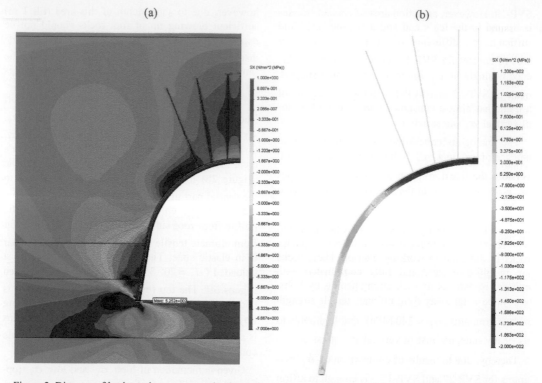

Figure 2. Diagram of horizontal σ_x stresses in "layered massive-support" system (a) and in frame support with bolts (b) for using SVP-19 specific profile.

Figure 3. General view of "Ubileinaya" mine's joining #2, supported by PFHS-15 made of SVP-19 profile in combination with rock bolt support in the roof of mine working.

Therefore, analyze of distribution of horizontal σ_x stresses field in "layered massive-working's support" system does not detect (except of some quite limited areas) sufficient differences for frames, made of SVP-27 and SVP-19 specific profiles.

Mine testing (Figure 3) allows to appreciate in detail rock pressure manifestations in development workings, driven along the C_6 seam of "Ubileinaya" mine. These workings are supported by combination of rock bolts with frame supports, made from SVP-27 (base version), and using the same combination, duy to installing light-weight PFHS-15.0 frame support made of SVP-19.

The fact of increased rock pressure and rock contour displacements of working's walls was experimentally determined by computer modeling and mine researches and geomechanical explanation was given. The most important thing is in over 3.5 times increase of rock movements in the opening's walls. It is more than value of approaching a roof with a foot, that is explained by specific process of weak lateral rock extrusion of Western Donbass. These rocks are located between load-bearing rock plate, strengthened by bolts in the roof and harder (about 3 times) coal bed in the foot.

Comparison of results of modeling and mine measurements is confirmed by validity and reliability of researches of geomechanical processes in "layered massive-working's support" system. So, non-loading of frame roof-bars and absence of any considerable activation of yieldability's locks are confirmed, and disagreement of calculated and measured values of approaching a roof and a foot is 7.4...10.7%. Also increased lateral loading on frame-legs is experimentally confirmed – divergence of mine measurements of horizontal rock contour movements and results of modeling are 13.2% on the level of frame's arch end and 23.4% in the place of lower frame-leg part, that is full satisfactory inaccuracy of geomechanical process calculations in "layered massive-working's support" system.

3 CONCLUSIONS

Results of computer modeling of "layered massive-working's support" system's SSS allow to investigate the mechanism of frame support and bolts operation due to existing scheme of layered working maintenance. The main point of which is formation of reinforced-rock plate in the roof of mine working for increased rock pressure in its walls, leading to appearing areas of plastic state in frame-legs.

Analyze of "layered massive-working's support" system's SSS revealed little differences (as rule, about several percents) of vertical and horizontal stress component values during installation SVP-27 and SVP-19 frames in combination with rock bolt support in the opening's roof. Installation of more light-weight profile allows to reduce till 30% metal-content of frames and to decrease labor-content to installing it.

In the frame-legs two areas (in the place of arch ends and in lower frame-leg part) of plastic material state are created independently on SVP-number. For removing these areas, strengthening lateral working sides by rock bolts is required to do.

REFERENCES

Bondarenko, V.I., Kovalevska, I.A., Simanovich, G.A. & Fomichev V.V. 2006. *Computer modeling of stress-strain state of fine-layered rock mass around an in-seam working.* Preultimate stage of "rock-support" system deformation. Book 1. Dnipropetrovs'k: System technologies: 172.
Bondarenko, V.I., Kovalevska, I.A., Simanovich, G.A. & Fomichev, V.V. 2007. *Computer modeling of stress-strain state of fine-layered rock mass around an in-seam working.* Ultimate and behind-ultimate states of "rock-support" system. Book 2. Dnipropetrovs'k: System technologies: 198.

Energy saving approaches for mine drainage systems

O. Beshta, D. Beshta, O. Balakhontsev & S. Khudoliy
National Mining University, Dnipropetrovs'k, Ukraine

ABSTRACT: The article is devoted to optimization of mine drainage facilities operation for reducing of energy consumption. Specific features of pumps electric drives and their operating modes are described. Basing on drainage pumps requirements the new approach for energy saving operation is developed. Experimental results proving approach adequacy are given.

1 INTRODUCTION

Energy saving is stated to be a key trend in Ukrainian industry. It is especially important for coal and ore mining enterprises which being the basis of GDP in Ukraine meanwhile suffer severe competition on international market. The main reason for low competitiveness of Ukrainian raw materials is high level of specific energy cost in final product.

Privatization of mining enterprises made their owners thoroughly revise management principles. Modern resource planning strategies are being applied instead of obsolete soviet approaches. The need for energy consumption assessment is realized to be the key factor for energy saving. Only "administrative" measures i.e. rational organization of equipment maintenance brought about 10% energy consumption reduction during the first decade since optimization. Now the potential for further energy saving due to only rational planning is believed to be depleted. It is time to improve technology.

The chart (Figure 1) illustrates energy consumption distribution among main facilities of a coal mine. Obviously exact numbers depend on many factors like water content, temperature and humidity. The distribution pattern varies from mine to mine and even within a day for a certain mines.

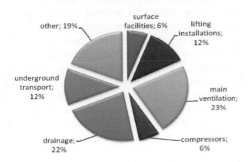

Figure 1. Typical coal mine energy consumption chart.

Amazingly, the share of "useful" energy i.e. connected with coal extraction and underground transportation and lifting it to the surface is only about 30%. The rest of total energy is consumed with "auxiliary" but nevertheless necessary loads – ventilation, drainage, compressors and other facilities.

Mine ventilation and drainage are the most important consumers at any mine. Ventilation must provide necessary oxygen content and methane rarefaction down to safe level. It mainly depends on mine structure and its energy consumption cannot be significantly decreased for safety reasons. And fault of pumping system can cause almost immediate mine flood. So these two facilities must operate stable despite on production level.

Authors had carried out a series of field research at several coal and ore mines of Dnipropetrovs'k and Zaporizhya region. Preliminary results show that mine drainage systems being the key energy consumers are maintained in inefficient modes. So these systems possess the highest potential for energy saving.

2 APPROACH DESCRIPTION

The key concept describing any pumping system is a QH-curve, it shows dependence of pump's flow rate (supply) Q from head H. The dewatering line is characterized with same values – pressure drop as a function of flow rate. The head pressure contains two components: static pressure (simply, the height you need to deliver water on) and dynamic (connected with friction between liquid and tube and other hydraulic phenomena like turbulence) (Pivnyak 2010; report by Xenergy 1998).

Mine dewatering lines are featured with high static pressure and low dynamic component. Another specific feature of mine systems is parallel operation of the pumps. In order to estimate operation points of each pump under parallel operation one must build a resulting pumps' QH-curve and superimpose it to the mains curve.

Figure 2 illustrates scheme and typical QH-curves for mine drainage system in case of parallel pumps' connection.

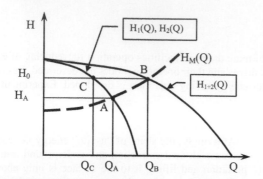

Figure 2. Parallel pumps operation in mine drainage: QH-curves and operation points.

$H_1(Q)$ and $H_2(Q)$ represent individual pumps' characteristics, $H_{1+2}(Q)$ – the resulting curve. $H_M(Q)$ stands for the characteristic of the water mains (the whole dewatering system).

If the pumps have absolutely coinciding characteristics, as it shown at Figure 2, they provide equal flow rates, defined by the resulting head H_0. The value of H_0, in turn, depends on resulting flow rate Q_0 and hydraulic line's characteristic. The "B" point is found as intersection between the resulting pumps' $H_{1+2}(Q)$ and mains $H_M(Q)$ characteristic. Thus, point "A" shows each pump's standalone operation, point "B" represents resulting H and Q under parallel operation. This case each pump operates at its "C" point due to the rise of output head from H_A to H_0 value.

Operating at "C" point means less flow rate for each pump. The maintenance area of each pump must be wide enough to provide stable operation. It also means the increase of hydraulic drop and hence, specific energy consumption. So, the more pumps operates simultaneously the more expensive dewatering is.

The situation becomes even more strained when pumps have different QH-curves (Figure 3).

The "2" curve is peculiar to the worn out pump – it produces less flow under the same backpressure. Simultaneous operation of pumps with different QH-curves results in shifting of operation point of the "weaker" pump down to low flow rate area Q_{B2}. In extreme case it can produce zero supply meanwhile consuming electric power. Therefore, specific energy consumption in dewatering depends not only on pumps' general condition but also on their "matching".

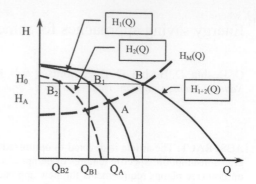

Figure 3. Parallel operation of pumps with different QH-curves.

All the mentioned gives us realization of two major energy saving principles for dewatering systems:
• rational pumps' operation time planning according to inflow rate and time-of-day tariff;
• rational pumps' combination in case of their parallel operation for minimal specific energy consumption.

Implementation of both tasks requires introduction of mine monitoring system for constant survey of hydraulic and mechanic factors. The inflow rate, electric motors' and pumps' and dewatering line condition should be monitored (Europump and Hydraulic Institute 2001).

3 EXPERIMENTAL RESULTS

To estimate energy saving potential in mine dewatering the preliminary experiments were carried out at several coal mines of Dnepropetrovsk region.

The dewatering horizon -225 m contained ten pumps of CSP 300×290 type in three sunk basins (CSP stands for "centrifugal sectional pump", correspondent Russian abbreviation is CNS; 300 represents rated head, m, 290 – rated supply, m^3/ hr).

Special measuring equipment was installed, including autonomous data acquisition systems (ADA) at each pump. ADA has internal power supply and memory storage, providing continuous data collection for 48 hours. Each ADA was synchronized with the rest ones and the basic acquisition module.

Measuring of the flow rate caused certain problems. Ultrasonic flowmeter produced error up to 40% when flow rate reached 1000 m3/hr because of turbulence effect. A special approach was developed basing on water level at the inlet header on the surface. The approach is based on the Bernulli equation

$$Q = \mu \cdot S_0 \cdot \left(2g \cdot k_0 \cdot \left(L - h_0\right)\right)^\alpha ,$$

where μ, α – nonlinear coefficients depending on liquid's viscosity and other parameters; S_0 – area of the inlet; $g = 9.81$ – acceleration of gravity; k_0, h_0 – scale gains; L – level of water in the inlet header.

The actual flow rate was estimated by ratio $\Delta V / \Delta t$, where ΔV is the gain of water volume in the water precipitation pool (measured rather simply) and ΔV is the time period. Thus flowmeter was replaced with much cheaper hydrometric float level meter.

Figure 4 illustrates measuring scheme for each pump and dewatering mains.

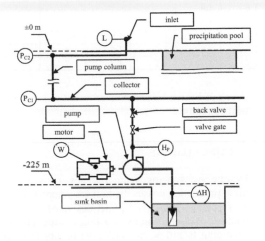

Figure 4. Measuring scheme for a pump and dewatering line.

The following parameters were measured:
- total flow rate Q (measured via water level L);
- each pump's input depression – ΔH;
- each pump's head H_P;
- pressure at the dewatering collector P_{C1} (underground);
- pressure at the pump column P_{C2} (surface);
- electric power consumption W.

During several testing sessions the individual pumps' QH-curves were obtained. Then energy consumption under each pump standalone operation and their various combinations was estimated. As was expected, actual QH-curves and pump's efficiencies differed from each other and from rated values. Figure 5 shows rated and actual QH-curve of one pump under standalone and parallel operation.

The actual QH-curve lies lower than the rated one. It means that for the same backpressure the pump produce less flow rate. For example, under rated backpressure of 290 m the pump produce 210 m³ / hr instead of rated value of 300 m³ / hr.

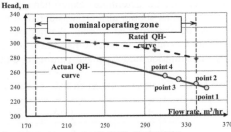

—O— point 1 - standalone operation with Q=350 m3/hr, H=237,08 m
—O— point 2 - operation with 340 m3/hr, H=243 m, simulteneously with pump №8
—O— point 3 - operation with Q=323 m3/hr, H=249 m, simulteneously with pumps №8 and №9
—O— point 4 - operation with Q=310 m3/hr, H=254 m, simulteneously with pumps №8 and №9 and №2

Figure 5. QH-curves and operating points under standalone and parallel operation.

Operating point #1 (standalone maintenance) lies beyond the zone of normal functioning in the area of extra supply, which is can cause cavitation effect.

Point 4 indicates simultaneous operation of four pumps – maximal number of parallel operating pumps. Even at this point the pump has a good reserve to stay in the rated operating zone.

Lower supply under required head means less pump efficiency. Figure 6 illustrates pumps' efficiencies under standalone operation. The value of "total efficiency" includes efficiencies of the pumps themselves and their drive motors.

Obviously, actual efficiency is always lower than the rated value. It depends of how worn the pump is. The pump #2, for example, posses the lowest efficiency of 29%.

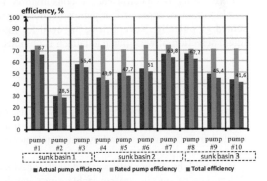

Figure 6. Estimated efficiencies of the pumps.

Meanwhile, most of duty cycle pumps operate in parallel. The resulting efficiency and thus, specific energy consumption, depends not only of these parameters in standalone operation, but also on individual QH-curves.

The energetic performances of dewatering system under all possible combinations of the pumps installed cannot be determined experimentally. Certain restrictions should also be considered – the allowable number of gear starting, the pumps under repair and so on. So, in order to forecast specific energy consumption under pumps' parallel operation, simulation is required.

Table 1 shows specific energy consumption (kWh / m^3) under parallel operation of two of possible ten pumps.

Table 1. Specific energy consumption under operation of pumps in pairs.

pump #	#1	#2	#3	#4	#5	#6	#7	#8	#9	#10
#1	1.0									
#2	2.5	2.2								
#3	1.2	2.6	1.1							
#4	1.5	2.7	1.5	1.4						
#5	1.4	2.5	1.4	1.6	1.3					
#6	1.5	2.7	1.5	1.6	1.6	1.4				
#7	1.2	2.3	1.2	1.6	1.4	1.6	1.0			
#8	1.1	2.4	1.2	1.6	1.4	1.6	1.2	1.0		
#9	1.5	2.6	1.5	1.6	1.6	1.6	1.7	1.5	1.4	
#10	1.6	2.8	1.7	1.7	1.7	1.7	1.8	1.7	1.7	1.5

The diagonal of the table shows specific energy under standalone pumps operation. The cell with coordinates, for instance, 3, 2 indicates specific energy consumed by dewatering system when pumps #2 and #3 are operating simultaneously and so on.

The pump #2, possessing the lowest efficiency in standalone operation, "spoils" the system being launched together with any of the rest of pumps. Generally, the level of specific energy demand varies within 1.1-2.8 kWh / m^3, showing us how important is to chose correct combination of the pumps. Considering average water inflow of about 20000 m^3 / day gives us a huge potential for energy saving.

Forecasting of specific energy consumption for combinations of three or four pumps cannot be presented graphically. Nevertheless, the need for intelligent selection of what pump to launch is obvious.

4 FURTHER RESEARCH

A special system should be developed to assist dewatering dispatcher service. It should later be transformed in automated control system implementing energy saving principles.

It should be taken into account that individual QH-curves of pumps are not stable since wearing goes continuously. Faults occur, pumps and electric motors and shutoff valves are being replaced for certain technological reasons. So, the system to be developed must be adaptive to variations of industrial conditions.

5 CONCLUSIONS

Dewatering systems being the key consumers in mine production posses the highest potential for energy saving. Two basic principles must be implemented: rational pumps' operation time planning according to inflow rate and time-of-day tariff and rational pumps' combination in case of their parallel operation for minimal specific energy consumption.

REFERENCES

Pivnyak, G., Beshta, A. & Balakhontsev, A. 2010. *Efficiency of water supply regulation principles*. New Techniques and Technologies in Mining. Proceedings of the school of underground mining. © CRC Press/Balkema, Taylor & Francis Group, London.

United States Industrial Motor Systems Market Opportunities Assessment, report by Xenergy for Oak Ridge National Laboratory and the U.S. Department of Energy. 1998. http://www.oit.doe.gov/bestpractices/pdfs/mtrmkt.pdf.

Pump Lifecycle Costs: A Guide to LCC Analysis for Pumping Systems, Europump and Hydraulic Institute. 2001. http://www.oit.doe.gov/bestpractices/pdfs/pumplcc_100.pdf.

Technical and Geoinformational Systems in Mining – Pivnyak, Bondarenko & Kovalevs'ka (eds)
© 2011 Taylor & Francis Group, London, ISBN 978-0-415-68877-2

Development of methods for utilization of thermal energy in the underground gasification of coal mining

G. Gayko & V. Zayev

Donbass State Technical University, Alchevsk, Ukraine

ABSTRACT: The methods of heat extracting out of the coal layer combustion zone and surrounding rocks during underground coal gasification are developed. New technology allows obtaining additional electricity another than generating gas by using a liquid coolant.

1 INTRODUCTION

Global trends in energy indicate expansion of underground coal gasification, and actively seeking ways to improve the effectiveness of these technologies. In today's world there are about 20 major stations of the underground gasification of coal, and every year we introduce 1-2 new stations. Leader in this technology is China (in 2007 in the country 6 stations worked and 4 more were built) (Kondyrev 2007), 1-2 stations operate in Australia, Canada, USA, South Africa, Uzbekistan (in Angren lignite mine – since 1963). Interest in introducing technology of underground thermochemical conversion of coal comes from such countries as Belarus, Bulgaria, Britain, India, Kazakhstan, Poland and Ukraine, where the evaluations of coal deposits were made in terms of their prospects for the gasification, the projects of building new plants were justified.

It is noteworthy that the Ukraine in the last century has been a pioneer in the development of technology of underground gasification of coal seams; Lisichansk and Gorlovka for decades successfully operated Podzemgaz stations. In the 80th years of the twentieth century the special geological surveys were conducted, and coal gasification projects were designed foe Dneprovskiy lignite basin, as well as a comprehensive assessment of the suitability of Donbass fields for underground gasification was given (Kolokolov 2000). Significant contribution to the development of new technologies have been made by schools of the Russian Federation (Vladimir Rzhevskii, W. J. Ahrens, J. D. Dyad'kin, A. Ruban, etc.), but using of rich deposits of natural gas has weakened the practical interest to the construction of underground coal gasification plants on Russian territory.

Despite the obvious advantages of underground thermochemical conversion of coal (exploitation of deserted fields, involvement in the development of off-balance sheet inventory, solving a number of environmental problems), the global experience of operating stations "Podzemgaz" identified a number of unresolved issues that impede widespread use of these technologies. Disadvantages include: relatively low heat of combustion of the resulting gas generator, the difficulty in control of the processes of gasification and significant (up to 30-50%) loss of heat in the interior (Gluzberg & Serov 1985). The latter factor causes the lack of effectiveness of the technologies and simultaneously indicates significant opportunities to improve it.

One of the areas of recycling thermal energy at the thermochemical processing of coal seam gas is the use of coolant (water vapor), which fillsburnt-out space after (or during) gasification (development SPGGI). Unstable characteristics of the coolant caused its use only in the heat exchangers, and the complexity of the extraction and separation of steam from the generator gas impede widespread use of this method.

The original method for extracting heat from the combustion zone of the coal face is proposed in KuzGTU (Kemerovo) (Patent RF #2278254). Method is based on building coal-fired units and their burning with the simultaneous selection of the burning heat of bed by steam generators, which should move the slope after the fiery front. Scope of this method involves mining method burning coal seams, however, to ensure the movement of steam generators for biases is extremely difficult due to the uncertainty of the line of fire and slaughter of the complexity of maintenance of slopes directly adjacent to the combustion zone.

2 DISPOSAL OF THE USE OF LIQUID COOLANT

In DonSTU concept of mine-power station, assuming an underground combustion (gasification) coal seams with a liquid coolant circulating in the collector pipe in the ground coal seam is designed (Gayko & Kasyanov 2008). Recent advances in the field of geothermal energy technologies open up entirely new opportunities for electricity generation using a liquid coolant, bringing the efficiency of hydro turbines and high-temperature steam power plant units. Modern modular geothermal power plants are used as the working fluid superheated water ($T = 110 - 250$ °C), have compact dimensions (for "Tuman-2"– $10.5 \times 3 \times 3.5$ m) and consume the amount of coolant from 10 to 40 m^3 per hour. Method of DonSTU first combined the principle of the circulation of hot water in a sealed tube collector, placed in the soil of burnt coal seam using hydro turbines to generate electricity for the modular stations. The developed concept could be rather promising for mine thermochemical conversion of coal seams, or their parts (whole, areas in difficult GSU, etc.).

Since the highest prevalence and cost-effectiveness of downhole are demonstrated by ways of underground gasification of coal seams, the authors posed the problem of adaptation of technical ideas embodied in the concept (Gayko & Kasyanov 2008), for downhole technology. For this purpose, we developed methods of utilization of thermal energy for the horizontal (flat) (Gayko & Zayev 2010) and steep-slope (Patent Ukrainy #54138) coal layers.

On flat layers the method is carried out as follows (Figure 1). In accordance with the method of gasification (or burning) coal layers from the surface there are drilled air-supplying system and gas-escape holes 1, which reach the coal layer 2. Additionally, there drill penetrating wells 3 with pass cavity 4, and the wells are in the soil of the coal layer and come to the surface. In penetrating wells three tight line 5 are brought, which are formed by the connection (for example, by welding) hard segments of the pipeline 6 with flexible intermediate elements 7 made in the form of corrugated hoses. Length of straight rigid pipe section 6 is determined considering the curvature of the penetrating hole 3 and the size of the pass cavity 4 in which the transition of the pipeline in the plane of the soil carbon reservoir is performed.

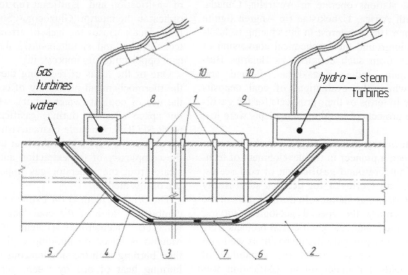

Figure 1. Scheme of utilizing method of heat power.

After installation of the pipeline the process of gasification of coal layer is performed, removing the resulting producer gas for gas turbine 8. Simultaneously, fluid (water) heat transfer is brought to the input of the pipeline 5, adjusting the pump speed of the coolant, whose temperature at the outlet should be 150-200 °C (the optimal parameters for hydro turbine 9). Electricity obtained on the gas and hydro turbines is fed through power lines to 10 users.

For steep-slope layers a method of generating electricity (Figure 2, 3) is developed. On the surface coal layer 3 air-supplying 1, gas-outlet 2, and heat exhaust wells 4 are drilled. The latter are drilled in the soil layer 3 and are equipped with airtight standpipe 5 with a closed end. In the standpipe they place five feeding sleeves 6, forming a system of "pipe in pipe". Using the supply hose 6 liquid heat-carrier is fed to the bottom of the standpipe and going through it the heat-carrier passes combustion zone reservoir 7 in which the coolant is heated to a temperature of 250-300 °C and transported to the wellhead, where it is fed to hydroturbines with electric generators 8.

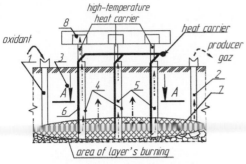

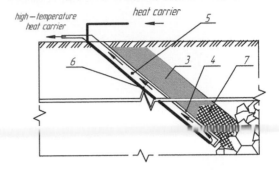

Figure 2. General scheme of providing the method.

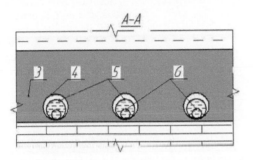

Figure 3. Cross section of a coal layer with a heat-wells, tube set and the supply hose.

Thus, in addition to the produced generator gas from underground coal gasification, the developed methods allow to extract a great part of thermal energy from the combustion zone, which is transported using a liquid coolant along airtight pipes to hydro turbines. This allows us greatly improve the general efficiency of power production and develop new perspectives for the introduction of borehole technology at "Podzemgaz" stations.

3 CONCLUSIONS

1. Global trends in power industry indicate expansion of underground coal gasification, and actively seeking ways to improve the efficiency of these technologies, primarily aimed at reducing heat losses in the subsoil.

2. In DonSTU for utilization of thermal power from underground gas generator it was first proposed to use liquid heat-carrier (overheated water) circulating in airtight pipe or sewer system in high temperature combustion zone of a layer.

3. Advances of previous decades in the field of geothermal energy technologies have opened new possibilities for power generation using liquid coolant, bringing the efficiency of new hydro turbines (water at $T = 110 - 250$ °C) to the traditional high-temperature steam-power units (steam at $T = 700$ °C). This allowed us to combine the methods developed by recycling heat from power generation at hydro turbines.

4. The expected economic effect from introducing the new technology when developing one coal block with the dimensions: on the gradient – 200 meters, on stretching – 300 m, seam thickness – 2.5 m will be not less $ 4 million. These money should be considered as an addition to the cost of generator gas obtained by gasification of coal block.

REFERENCES

Kondyrev, B.I., Belov, A.V. & Mannangolov, D.Sh. 2007. *Developing the technology of underground coal gasification.* Perspectives of coal deposits development in the Far East. Mining informational and analytical bulletin (scientific and technical journal), 1: 297-300.

Kolokolov, O.V. 2000. *Theory and practice of thermochemical technology of producing and conversion of coal.* Dnepropetrovsk: NMU: 281.

Gluzberg, Ye.I. & Serov, V.A. 1985. *Assessment of heat losses from fire area into surrounding massif.* Mining Journal, 11: 59-64.

Patent RF #2278254, MPK E21B 43/295 (2006.01). *Sposob poluczenia elektroenergii pri podziemnem ugleszyganii.* S.A. Prokopienko; Zajawl. 21.12.2004; Opubl. 20.06.2006. Bul. 17.

Gayko, G. & Kasyanov, V. 2008. *The Concept of Mine-*

Power-Station Involving the Underground Burning of Coal Strata. 21ˢᵗ World Mining Congress. New Technologies in Mining. Krakow: AGH: 199-204.

Gayko, G.I. & Zayev, V.V. 2010. *New method for producing power during underground coal seam gasification (burning)*. Donetsk Bulletin of Shevchenko Scientific so-

ciety. Donetsk: Schidnyj wydawnyczyj dim, 29: 64-67.

Patent Ukrainy #54138, MPK E21B 43/00. *Sposib otrymannia elektroenergii pri bezszachtnomu spalenni plastiw pochylogo zaliagannia* / G. Gayko, V. Zayev. – Zajawl. 07.05.2010. Opubl. 25.10.2010. Bul. 20.

Technical and Geoinformational Systems in Mining – Pivnyak, Bondarenko & Kovalevs'ka (eds)
© 2011 Taylor & Francis Group, London, ISBN 978-0-415-68877-2

Engineering support of BUCG process in Solenovsk coal deposits

V. Falshtynskyi & R. Dychkovskyi
National Mining University, Dnipropetrovs'k, Ukraine

M. Illiashov
SC "Donetsksteel", Donetsk, Ukraine

ABSTRACT: In the article characteristics of fitness criteria to BUCG for thin Solenovsk seams are grounded. The calculations of thermal and material balance for coal seam gasification process are executed. The construction, method of underground gas generator preparation, and sequence of coal seam gasification for area №1 are designed.

1 INTRODUCTION

The technology of borehole underground coal gasification (BUCG) is an enterprise for the generation of electric and thermal energy, passing chemical products, fuel and fluid gases in place of coal seams location. Installation of this technology will give a capability to explore uneconomical coal reserves and local deposits of solid fuel in difficult geological conditions.

As compared to traditional mining during BUCG it is possible to except hard work of miners, use uneconomical and unconditional coal reserves. The products of gas combustion are not contained by the oxides of carbon, particulate matters, sulphurous anhydrite.

The methods, technological schemes and constructions of underground gas generators designed in the National mining university allow to manage the process of underground coal gasification, providing remaining the thermo-chemical balance of conversions and physical processes at coal seam gasification. Enterprises using the BUCG technology provide automation of production processes. The final product of such generation becomes not coal, as a feed for further conversion, but kilowatts of thermal, electric energy and chemical row materials. Closed-circuit operation in chemical manufacturing effects a substantial saving in starting materials and utilities.

Industrial enterprises on BUCG technology was first built and exploited in the USSR in 1928. (Lisichanska, Gorlovska, Yuzhno-Abenska, Podmoskovn, Shatska and Angrenska station of "Pidzemgaz"). The stations after a closing were blocked at the end of 1960, Yuzhno-Abenskay – 1996. Angrenskaya works from 1964 until present time.

On the modern stage of underground coal gasification development the leader in this direction is China (8 experimental and experimentally-industrial underground gas generators) Such country as Australia, Poland, South Africa, USA, Spain, Belgium, England, Slovakia also conducted several experiments.

2 CRITERIA OF COAL SEAM FITNESS TO BUCG IN THE AREA #1

Based on the evaluation of practical material of coal seam gasification on the "Pidzemgaz" stations (Lisichanska, Gorlovska and Yuzhno-Abinska), conducted investigations on an experimental mine gas generator and stand unit options, the generalized dependences of criteria are certain for 12 area of Solenovsk coal deposit, for the carrying out industrial experiment the area #1 was chosen (Kolokolov 2000 & Skafa 1960).

Area #1 located on the field of Solenovsk coal deposits – 1, 2, 3, Krasnoarmiysk coal district, Donetsk region, joins to the north-eastern bent of the Ukrainian crystalline rock mass and extends along on a southeast beads of the Donetsk ridge. A department presents a C_6^1, C_6, C_5^1, C_5, C_4^2.

To the down-dip and to the rise the series of strata are limited by Shevchenkivskiy fault #1 and Kirillovskiy fault, along the strike – Shevchenkivskiy fault #3. The size of area to the down-dip $H = 1410$ m, on the rise southward $S = 827$ m. on the north $S = 3000$ m. General productive coal reserves make $Z = 4786.8$ thousand tons. The stratification depth of coal seams $H = 72-221$ m, power $m = 0.5-0.9$ m, angle of inclination $\alpha = 10-19°$.

The criteria of strata formation fitness located in area №1 for underground coal gasification covered on basal factors: to mining-and-geological, hydrogeological and technical. The scopes of area condition the presence of natural screens (disjunctive dislocations). The stratification depth of coal seams enables to provide efficiency and fail-safety working. Coal seam power are within the limits of 0.7-0.9 m, that is conditioned by a lower bound in the criteria of fitness of coal seams to BUCG. Containing rocks (77.4% claystone and siltstone), penetration capabilities within the limits 0.71-1.06 Darcy, providing impermeability and efficiency of process at the penetration capability of coal seams 0.38-0.62 Darcy.

In these terms the expected inflow of water in an underground gas generator will make 1.2-3.4 m^3 / t (on a hydrogeological factor this area requires supplementary explorations).

At existing technological and engineering developments assurance of effective and fail-safety of coal seam gasification process is possible on this area.

The criteria of fitness to in-situ coal seam gasification on the area #1 resulted in Table. 1.

Table 1. Basic suitability criteria of hard coal seams underground gasification within the area #1 (C_6^1, C_6, C_5^1, C_5, C_4^2).

| Coal grade G | Seam thickness, m | Coal seam ash content, A^C, % | Wall rocks (roof, bottom); total | | | Sulphur content in the seam, S, % |
			Thickness of clays or other low permeable rocks in roof; h^1, m $h^1 / m > H$	Thickness of clays or other low-permeable rocks in bottom; h, m $H \geq 2.0$ m	Distance from the seam roof to separate highpermeable layers or undrained water-bearing horizons h_2; $h_2 > h$ h – height of the fractures zone, m	
C_6^1	0.9	6.9-12	14.3 > 8.1	9.6 > 2	24.5 > 10.8	1.9
C_6	0.7	6.2-18	12.5 > 6.3	7.3 > 2	11.2 > 8.4	1.9
C_5^1	0.75	10-21	13.2 > 6.8	5.5 > 2	15.75 > 9.0	1.1
C_5	0.7	5.9-16	10.1 > 6.3	6.2 > 2	11.4 > 8.4	1.9
C_4^2	0.55	9.2-17	18.4 > 5.5	7.8 > 2	22.6 > 67	2.5

Continuation of Table 1

| Coal grade G | Minimal safe mining depth (H, m) and seam dip angle from $\alpha = 0°$ to $45°$ (wings, mould); $H/m \geq 15$, $n = 1$ $H \geq m \cdot n$ | Tectonic abnormalities; $L_n \geq L_g$ | Specific water inflow, m^3 / t into reaction channel of the gas generator considering BUCG process intensity, (not more than 1.6-3.4 m^3 / t) | | Moisture content of BUCG gas, g / m^3 | | Ratio of coal and rock gas-permeability |
			Q_{air}	Q_{oxygen}	Q_{air}	Q_{oxygen}	
C_6^1			4.4 t / h	2.8 t / h	445	238	21-38
C_6	69.8 m > 15; $H = 15$ m	Boundaries of the area. Disjunctive abnormalities	4.15	2.17	429	234	17-29
C_5^1			4.23	2.25	387	231	18-34
C_5			4.04	2.2	375	220	18-36
C_4^2			3.6	1.98	411	235	20-37

3 MATERIALLY-THERMAL BALANCE OF COAL SEAM GASIFICATION PROCESS, AREA #1

For the calculation of materially-thermal balance of BUCG the program MTBalanse SPGU is utillized designed by the employees of underground mining department from the National mining university (Lavrov 1957 & Falshtynskyi 2010). The algorithm of calculation includes thermochemical conversions of solid fuel in a gas and condensed fluid in the conditions of elementary composition of coal seam, external water inflow and thermal balance of underground gas generator is foreseen in it. A program algorithm is presented on a Figure 1.

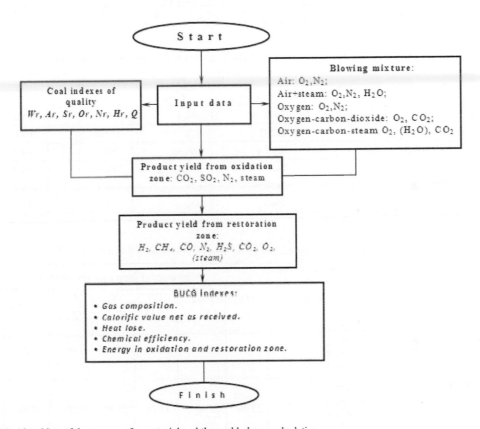

Figure 1. Algorithm of the program for material and thermal balance calculation.

The program of calculation takes into account such terms of process:

1. A change a technogenic situation in rock masss which contain a underground gas generator taking into account mine and geological conditions and technological characteristics of process;

2. Specifics of blowing composition and effect them on the process of coal seam gasification;

3. Change of high-quality and quantitative coefficients of BUCG gas from the grade of coal seam and blowing mixture;

4. Influence of geometrical characteristics of oxidizing and restoration zone of reactionary conduit of gas generator on balance of coefficients of chemical conversions and physical speeds;

5 Influence on thermal balance at coal seam gasification;

6. Effect of ballast gas of gasification process on the high-quality coefficients of underground gas generator;

Technological coefficients of underground gas generator and escape of basic chemical products at BUCG, presented in Tables 2 and 3, characteristics of materially-thermal balance on the area for SPGU #1 presented in Tables 4, 5, 6, 7.

Table 2. Technological indexes of underground gas generator.

Indexes of underground gas generator	COMPOSITION OF BLOWING					
	Air	Oxygen	Oxygen + CO_2 + steam	Oxygen + CO_2	Air + steam	Oxygen + steam
Thermal power	GKal					
	11.97	23.6	21.7	23.72	14.17	22.1
Electrical power	MVat					
	13.9	27.4	25.2	27.5	16.4	25.6
Capacity on gas (CH_4, CO; H_2)	10^6 m^3					
	26.9	68.4	57	80.7	32.2	571

Table 3. Escape of main chemical products during production activities of underground gas generator.

Types of blowing mixture	Escape of chemical products at BUCG (tons)			
	Coal gum	Benzol	Ammonia	Sulphur
O_2N_2	2649	482.4	1050	153.3
H_2O (steam)+ O_2, N_2	26207	476	1132.6	164.6
O_2 (30-62%) N_2	2829.4	609.4	758.3	286.2
O_2 + steam	2624	578.1	782.6	235.3
$CO_2 + O_2$	2858	621	718.4	293.4
$CO_2 + O_2 + H_2O$ (steam)	2665	588	773	277.2

Table 4. Characteristics of materially-thermal balance on the area #1 (coal seam, C_6^1 – 0.9 m).

Type of blowing mixture	Characteristics of blowing m^3 / h	Gas quantity from a gas generator, %						
		H_2	CH_4	CO	N_2	H_2S	CO_2	O_2
Air	6957	4.68	4.46	26.13	60.21	0.3	3.68	0.54
Air + steam O_2 N_2 Steam	7026 1266.2 3541.4 2218.4	15.13	15.07	6.31	52.74	0.49	9.30	0.78
Oxygen + steam O_2 N_2 Steam	6323 2856.8 1429.1 2037.1	23.06	22.45	11.05	21.61	0.69	19.98	1.16
Oxygen + carbon dioxide + steam O_2, CO_2 steam	6118 1963.8 911.6 3242.5	26.58	25.24	13.83	0.27-	0.61	32.34	1.14
Oxygen O_2 N_2	6885 4186.1 2698.9	10.27	9.78	37.31	21.46	0.66	10.26	1.19
Oxygen + carbon dioxide O_2 CO_2 N_2	6 506 4 014.2 969.4 1 522.4	11.68	10.17	52.9	11.73	0.69	11.68	1.15

Continuation of Table 4.

Type of lowing mixture	Speed of coal seam gasification	Efficiency	Lower heat of combustion	Gas discharge from gasgenerator	The humidity of BUCG gas	Quantity of coal gasification
	m / day	%	Mj / m³	m³ / kg of coal	g / m³	t / h
Air	1.94	62.87	4.71	2.84	372	2.5
Air + steam O_2, N_2, steam	2.04	68.21	5.82	2.94	473	2.62
Oxygen + steam O_2, N_2, steam	2.63	78.4	8.5	2.12	369	4.25
Oxygen + carbon dioxide + steam O_2, CO_2, steam	2.3	79.12	9.28	2.19	320	4.2
Oxygen O_2, N_2	2.5	80.5	9.84	1.95	238	5.06
Oxygen + carbon dioxide O_2, CO_2, N_2	2.25	80.03	9.36	2.08	274	4.26

Table 5. Technological characteristics of BUCG process (area #1, coal seam, $C_6^1 - 0.9$ m).

Type of blowing mixture	Expenditure of blowing, thousand				Escape of BUCG gas, thousand			
	m³ / h	m³ / d	m³ / m	m³ / y	m³ / h	m³ / d	m³ / m	m³ / y
Air	6.96	167	5011.2	60134.4	8.83	212	6360	76320
Air + steam	7.03	168.7	5061	60732	9.2	221	6630	79560
Oxygen	6.88	165.1	4953.6	59443	12	288	8640	103680
Oxygen steam	6.32	151.2	4536	54432	10.5	252	7560	90720
Oxygen + carbon dioxide	6.5	156	4680	56160	12.5	300	9000	108000
Oxygen + carbon dioxide + steam	6.12	146.9	4406.4	52876.8	10.1	242.4	7272	87264

Continuation of Table 5.

Type ob blowing mixture	Quantity of coal gasify for set time:				Time of gasification, days	Quantity of gas at expluatation 10^6 m³
	t / hour	t / day	t / year	kg / for the time of gasgenerator expluatation		
Air	2.5	60	21 900	15 330	256.6	53.4
Air + steam	2.62	62.88	22 951.2	16 754.4	240.5	59.6
Oxygen	5.06	1214	44 311	13 293.3	105	31.1
Oxygen +steam	4.25	102	37 230	13 075.3	128.5	33.2
Oxygen + carbon dioxide	4.26	102.2	37 303	13 802	135	40
Oxygen + carbon dioxide + steam	4.2	100.8	66 792	15 452.6	152	36.7

At the oxygen blowing in the range $O_2 = 45-62\%$ (4186.13 m / h), the capacity of underground gas generator is provided gas production $8.4 \cdot 10^6$ m³ / h and electrical power 27.4 MWt, with efficiency equal 80.5% and by the temperature in production borehole $T = 534$ °C.

Blowing carbon dioxide CO_2 – 379.3 – 969.4 m³ / h, provides in combination with oxygen (2092.4 –

4014.2 m³ / h) and steam (1863.8 m³ / h) the receipt of power gas with high-quality coefficients: escape of burning gases of $50 - 80.7 \cdot 10^6$ m³ / y, electrical power 25.2 – 27.5 MWt, with Efficiency equal 79.12 – 80.3% and by the temperature gas in outlet borehole $T = 529$ °C.

Arrangement of blowing mixture O_2 (2856.8 m³ / h) + steam (2037.1 m³ / h), provides gas $57.1 \cdot 10^6$ m³ / y with $N_2 - 23.06\%$, $CH_4 - 22.45$ and $CO - 11.05\%$, such arrangement of burning gases, at oxygen + steam and air + steam blowing (steam = 2218.4 m³ / h, $N_2 - 15.13\%$, $CO - 6.31\%$), enables to provide a technological gas discharge suitable for synthetic gas.

The coefficients of the air blowing provide a power gas discharge with coefficients: escape of burning gases $26.9 \cdot 10^6$ m³ / h, and electrical power 13.9 MWt, with efficiency equal 62.9% and by the temperature gas in outlet borehole $T = 366$ °C, at the heat of combustion of power gas 4.71 MJ / m³.

The pressure in a gas generator at the air and air + steam blowing $P = 0.24 - 0.57$ MPa, at blowing, enriched O_2 ; CO_2 ; H_2O (steam) $P = 0.38 - 1.2$ MPa.

Table 6. Thermal balance of underground coal gasification on the area #1 (coal seam, $C_6^1 - 0.9$ m).

| Indexes | COMPOSITION OF BLOWING | | | | | | | | | | | |
| | Air | | Oxygen | | $O_2 + CO_2 +$ steam | | Oxygen + carbon dioxide | | Air + steam | | Oxygen +steam | |
	Mj/kg	%	Mj/kg	%	Mj/kg	%	Mj/kg	%	Mj/kg	%	Mj/kg	%
Heat of combustion on a working fuel	35.04	97.64	35.04	91.25	35.04	91.25	35.04	91.25	35.04	92.79	35.04	91.25
Entalphy in oxidation zone	0.636	1.772	1.272	3.312	1.272	3.313	1.272	3.312	0.636	1.684	1.272	3.312
Entalphy in blowing	0.208	0.580	2.087	5.434	2.087	5.434	2.087	5.434	2.087	5.526	2.087	5.434
In all:	35.88	100	38.40	100	38.4	100	38.4	100	37.76	100	38.4	100
Heat of gas combustion	13.37	38.32	19.09	49.70	20.12	51.96	19.47	50.83	17.1	44.40	18.02	46.88
Heat lose: Heating of ash and slag, MJ	0.095	0.272	0.095	0.247	0.095	0.245	0.095	0.248	0.095	0.247	0.095	0.247
Warming evaporation of moisture, MJ	0.375	1.074	0.375	0976	0.375	0.965	0.375	0.979	0.375	0.974	0.375	0.974
Heating of containing rocks (roof, ground), MJ	6.310	18.079	5.562	14.482	5.510	14.23	5.146	13.435	5.915	15.365	5.967	15.5
Entalphy of generator gas.	14.74	42.25	13.28	34.59	12.62	32.6	13.21	34.5	15.01	39.0	14.0	36.39
In all:	34.903	100	38407	100	38.724	100	38.304	100	38.495	100	38.497	100
Outlet temperature in gasgenerator °C	522		803		798		767		652		705	
Outlet temperature in production borehole °C	346		441		436		421		360		378	

4 TECHNOLOGICAL SCHEME AND ORDER OF UNDERGROUND GAS GENERATOR PREPARATION IN AREA #1

The gas generator preparation is carried out from a surface. The order of gasification of coal seam descending (Figure 2). For assurance the efficiency of gasification process in area #1 the construction of underground gas generator (Figure 3) is used with long coal walls. The system of gasification by long columns to up-dip $L = 400$ m. The distance between the boreholes, $l = 30$ m.

Preparation of gasgenerator providing by in-seam directional drilling (Figure 3).

Coal seam ignition is porvided through the directional boreholes by binary explosives (Falshtynskyi 2008). Ignitions of coal at application of this method possibly at presence of underwaters and does not require investments on the generation of ignition boreholes.

Control of blowing mixture direction is carried out by flexible pipeline Figure 3. The selective discharge of blowing mixture will provide interference of blowing with fire combustion face. For intensification of process the six arrangements of blowing mixture and heating of blowing are foreseen before a discharge in a underground gas generator to 200 °C, and also the impulsive discharge of main chemical agents (oxygen, steam, carbon dioxide) is provided on combustion face with different time duration. With the purpose of equal combustion face advance the reverse direction of gasification is foreseen.

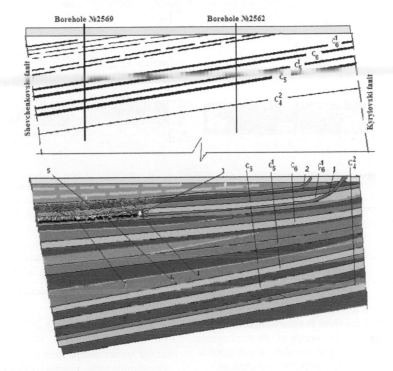

Figure 2. Technological scheme of coal seams on area #1: 1 – inlet borehole, 2 – stowing borehole; 3 – combustion face; 4 – reaction channel of underground gas generator; 5 – a stowing rock mass; 6 – goaf, 7 – ash and slag.

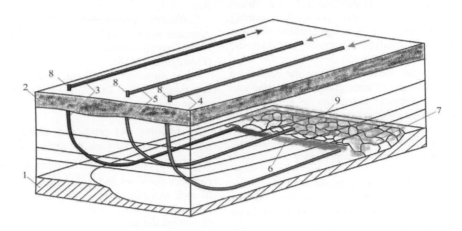

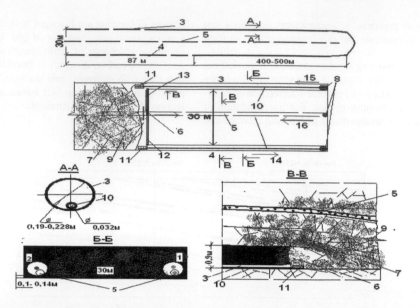

Figure 3. Technological scheme of underground gas generator in the conditions of area #1: 1 – coal seam; 2 – surface; 3 – gas outlet borehole, 4 – inlet borehole; 5 – stowing pipeline; 6 – reaction channel (oxidation zone (blue), restoration (red)); 7 – goaf; 8 – casing head tool; 9 – stowing rock mass; 10 – flexible pipeline; 11 – heat-resistant nose (0.4 m); 12 – restoration zone of reaction channel; 13 – oxidation zone of reaction channel; 14 – direction of BUCG gas; 15 – direction of blowing mixture; 16 – direction of stowing material.

CONCLUSIONS

Efficiency of BUCG is related to the seasonal expenditure of products, so the complex use of BUCG products has to be reached. That foresees the receive of chemical products from the condensed fluid, use the generator gas for the receipt of chemical agents by thermal conversion, and also thermal and generator gas for electric energy on power installations.

The heat recuperation from rocks containing a gas generator and products of BUCG provided by recuperational collections (Falshtynskyi 2010 & Brand 2008) for receipt of electric power. A remaining heat is utillized for engineerings (heating of blowing, process of catalytic conversion) and domestic needs.

Chemical raw material, got from the condensed fluid of power gas of underground gasification can be released chemical enterprises as a feed (coal gum, benzene, ammoniac water, phenols, acetylene, pyridines et.c.) or processed in place in the finished good (grey, surface active agents, solvents, carbon, dyes, polymer cements, naphthalene et.c.) (Falshtynskyi 2009).

Utilization of outcoming smoke from power station is on a principle of the closed cycle. Combustion gases CO_2 act from power-station back in an underground gas generator, where at co-operating with a burning hot carbon passes to burning gas – carbon monoxide (CO) and oxygen of O_2. Adding CO_2 to blowing calorie content of gas, will not yield to conventional gas.

In an underground gas generator direct oxides SO_n and nitrogen NO_n and other toxic components of smoke escape from power station.

Noncombustible mineral particles at coal gasification and carboniferous rocks remain in goaf, because they are not exposed to thermal decomposition.

Sufficient impermeability of underground gas generator is provided by injection stowing of the deformed rocks containing a gas generator and goaf (Falshtynskyi 2010).

As an stowing material can be used offcuts from coal power station. That will provide safety of landscape, fail-safety and efficiency of gasification process.

At building and production activity of underground gas generator on the area #1 on coal seam C_6^1 the investigations are foreseen on the audit of engineer decisions and technological characteristics of rock mass behavior at gasification. Varying the parameters of gasification process with the purpose of receipt the complex-industrial product of gasification from coal seam.

REFERENCES

Kolokolov, O.V. 2000. *Theory and practice of thermo-chemical coal process*. Dnipropetrovs'k: NMU Ukraine: 281.

Skafa, P.V. 1960. *Underground coal gasification*. Moscow: 169.

Lavrov, N.V. 1957. *Physical and chemical combustion of coal*. Moscow: 40.

Falshtynskyi, V.S. 2010. Analytical *determination of parameters of material and thermal balance and physical parameters of coal seam work-out on mine "Barbara", Poland*. Dnipropetrovs'k/Yalta: Proceedings of the school of underground mining: 157-161.

Falshtynsky, V.S., Dychkovskyi, R.O. & Tabachenko, M.M. Patent №35883 UA (2006) E21B 43/25. *Coal seam ignition* #200805265; Pub. 10.10.2008, Bjul. 19.

Liu, SQ, Li, J, Mei, M., & Dong, D. 2007. *Groundwater Pollution from Underground Coal Gasification*. Journal of China University of Mining & Technology, 4. Vol.17.

Falshtynskyi, V.S, Dychkovskyi, R.O. & Tabachenko, M.M. Patent №50867. *Heat recuperation at coal seam gasification* (UA) Pub. 25.06.2010. Bjul. 12.

Brand, J.F. 2008. *UCG pilot study in Secunda, South Africa – the experimental design and operating parameters for the demonstration of the UCG technology and verification of models*. In: Proceedings of the 2008 Pittsburgh coal conference. 30 Sep-2 Oct 2008, Pittsburgh, PA, USA, CD-ROM, Pittsburgh, PA, USA, University of Pittsburgh: 11.

Falshtynskyi, V.S. 2009. *Underground coal gasification technology*. Dnipropetrovsk: NMU: 131.

Falshtynskyi, V.S., Dychkovskyi, R.O. & Lozynskyi, V.G. 2010. *Economical justification of effectiveness the sealing rockmass above the gas generator for borehole coal gasification*. Praze naukowe GIG, Gornictwo i srodowisko, kwartalnik, 3. Katowice: GIG: 51-59.

Improvement of technology of the gold- and diamond-contained ores concentration with the help of a new highly-effective disintegration and thin screening by using of dynamically active band sieves (DABS)

A. Bulat & V. Morus
M.S. Polyakov's Institute of Geotechnical Mechanics, Dnipropetrovs'k, Ukraine

ABSTRACT: Basing on many years' fundamental researches the N.S. Polyakov's Institute of Geotechnical Mechanics of National Academy of Science of Ukraine (IGTM of NAS of Ukraine) has created a new special technology of disintegration and thin screening of 0.3-5.0 mm size by using of rubber dynamically-active band sieves (DABS). Numorous experiments and our long-term practical experience in industries shows that technical and technological parameters of the technology helps to effectively recover fine-class gold and diamonds with 1.5-2 better productivity of the washing machine.

A.S. Polyakov's Institute of Geotechnical Mechanics of National Academy of Science of Ukraine (IGTM of NAS of Ukraine) works for many years in order to study, create, manufacture and widely introduce technologically highly-effective and long-lasting mineral processing equipment with working faces made of wear-resistant rubber and dynamically active band sieving surfaces (DABS) (Chervonenko 1997; Morus 1998). One of the directions of this research work that is currently intensively developed is to create a special highly effective equipment which could essentially improve technology of gold- and diamond-ore concentration. Keeping in mind, as our key objective, the task to increase output and, at the same time, to improve effectiveness of disintegration and thin screening by screening surfaces that operate with the streams of inputs containing great amount of lumpy ores we have developed scientific base for designing a new equipment for ore-washing and preparing plants in gold- and diamond-mining factories. The equipment is presented in two types of washing-and-classifying machines:

– scrubber-trommels with extended scrubbers for the difficult-to-disintegrate and difficult-to-wash materials;

– drum screens of a "dredging barrel" type that simultaneously disintegrate and separate materials by several size classes.

One of the examples of aggregates of the first type can be a scrubber-trommel СБР-100 (Figure 1) consisting of the following assemblies: scrubber, sludge separator, intermediate drum and classifying trommel and drive, supporting driving wheels, pressing wheels, stop wheel that are located on the platform.

Figure 1. General view of the СБР-100 scrubber-trommel.

The layout of the aggregates and assemblies should:

– disintegrate and wash off the crushed materials with size less than 100mm from the difficult-to-wash materials including kaolin clay;

– separate and discharge, during the flushing stage, clay sludge with size -0.5 mm;

– combine quartzite final washing with classification by size 5 mm.

The scrubber is a thick-wall cylinder drum edges of which are the flanges; there are bandages on the outer sides of the scrubber by which the scrubber rests against the wheels with rubber studs. One of the wheels is a traction one, three wheels are supporting and one wheel is stop. Besides, pressing wheels are installed on the hinged cross beams over the bandages. Pressing wheel installed over the driving bandage can be used to increase pulling force of the traction wheel in friction transmission in adverse weather conditions (rain, icing) and as a stopper for cross travels of the working member at the start moment. Pressing wheel located over the stop wheels serves to prevent tilting of the working member in case of center of gravity shifts towards discharge end when the load is re-transferred.

The scrubber has an edge wall in the charge sector of the scrubber, with round central hole in to which a feeding chute enters at a regulated slope angle. Entire inside surface of the scrubber is lined (Figure 2) by special rubber plates of different shapes and purposes and is equipped with a system of ring thresholds of different heights that form a line of reservoirs along the length of the working member.

Figure 2. Inside working faces of the СБР-100 scrubber-trommel.

Usage of exclusively rubber lining elements ensures efficient protection of the thin-wall cylinder drum and other key assemblies of the scrubber against dynamic and impact loads and abrasive wear, and also helps to solve a problem of maximal reduction of the working member weight and, consequently, cut operational costs. The lining plates are installed on the working face with straight and return lifters. The lifters and ring thresholds located inside the scrubber drum provide, at feeding a proper quantity of water, highly effective disintegration and clay washing off. All lining plates are fastened inside the scrubber by special clinching without any additional fastening items or glues.

Speed of the material run in the scrubber can be regulated by changing the scrubber longitudinal slope. Slope is regulated by changing tilt angel of the bearing platform with the help of special screw jack. Material after washing in the scrubber drum, together with the sludge, run over the first from the top ring threshold to the sludge separator. The sludge separator is flange-mounted in the discharge sector of the scrubber and is made in the form of a "squirrel's cage" inside surface of which is equipped with rubber DABS for screening 0.5 mm size class. Here, in the area with length 600mm, thin fractions of the dispersed clay are separated from material washed at the first stage; the washed mate-

rial is further classified and finally washed in the wet-screening section. Totally, the wet-screening sector of the СБР-type scrubber-trommels is made in the form of modular trommel system that could be formed and varied depending on required washing and classification parameters. At the QUARTZITE DM, Ltd., the sludge separator was followed by a blind drum (lined by the rubber) and two-section trommel with special rubber DABS-type sieving surfaces with the cells 5 x 30 mm.

The blind drum of this design option functions as a reserve section. If output of the sludge separation is not enough it is possible to install an additional section of the sludge-separating sieves. As another option, if it is necessary to get better separation by sizes an additional screening section can be installed instead of the blind drum.

Structure of the screening section is similar to the sludge separator – it is a drum with the length 2.2 m in the form of a "squirrel's cage". All ring elements of the drum – flanges, supports for the rubber sieves, parts of the band-holders – are made bent, with no additional mechanical treatment. On the inside surface of the screening section, rubber DABS-type sieves with multiturn transporting spirals are installed with the help of which material continuously travels along the sieve surface with a certain speed actually independently on the trommel axis titling. The rubber multiturn transporting spirals have one more very important function. Their top ribs (space between them is 80 mm) accept impact load from the big lumps and form a surface over which these fractions are transported to the discharge sector having no contact with the screening sieves for thin fractions. So, the thinnest rubber sieves are efficiently protected against the big lumps whose size multiply exceed the utmost screening size. The discharge sector of the trommel ends with a cap with special rubber-lined bars. The screening section can be designed with two sieves. We have already designed and manufactured first pilot two-sieve trommels with diameter of outside sections up to 3.3 m and DABS rubber working sieving faces for thin, fine and coarse screening of fractions with sizes from 0.5 mm to 300 mm. Trommel modules of such systems can include both existing proved sieving surfaces and new sieving elements with the cell profiles, protectors and special transporting spirals adapted to concrete industrial conditions. Each key working-member assembly and part of the scrubber and screening sections are made by bending method only, with no additional mechanical treatment providing easy maintenance and repair.

If compare technical and economic results of the СБР-100 operation observed during more than 6 years with operation of well-known equipment (Stepanenko; Vysotin 2006; Pyatokov) it is worth to note that usage of wear-resistant rubber in all disintegrating and sieving working faces in the scrubber-trommels helps to create machines with similar productivity but with 1.3-1.6 times less weight and less (up to 1.5 times) energy consumption. Weight of the assemblies and parts could be reduced to the extent when reliable operation of a new machine drive could be provided, in some cases, by a single electric engine against two engines in other similar existing machines. This aspect essentially simplifies design, reduces price of the scrubber-trommels and cut expenditures for their operation and repair.

These projects and technological and operational results, as well as our many-year's tests in the industries allowed us to apply the modular method in designing scrubbers with the widest range of application. To get effective dispersion of difficult-to-wash kaolin clays we recommend to use scrubbers with 6-10 m length. We worked out a concept and principles of designing any models of the scrubbers with diameters ranging between 1.5 m and 3.0 m and length ranging between 2.0 m and 10.0 m and with highly-elastic wear-resistant rubber lining. Some our scrubber-trommel models designed by individual orders of different enterprises are shown on the Figure 3. Mounting of classifying trommels with greater length on the large-size disintegration ag-

gregates causes no technological problems thanks to modular structure of the trommels. Trommels with sieving drum length 4-5 m provide the highest technological effectiveness of screening sizes from 0.5 mm up to 300.0 mm. Very important is the fact that our DABS sieving elements for fine screening the smallest cells (0.5-3.0 mm) are designed for contacting with the lumps sizes of which are 500-800 times bigger than utmost screening size. For example, basing on our 5-year operational experience in industries we modernized our special DABS sieving elements for screening 0.5; 1.0; 2.0; 3.0 and 5.0 mm size classes that can form working faces of the trommels for interacting with the lump size 300-400 mm. It should be noted that reliability of the sieve fastening, strength and wear-resistance is designed for not less than one season of operation. This is especially important for solving such pressing problem as obtaining a high effectiveness of thin productive class separation at the head of technological processes and, in particular, for treating material from the placers and man-caused deposits that includes thin mineral inclusions and productive classes with particle minimal sizes 0.5-1.0 mm. This problem is mostly characteristic for new technological concentration schemes of diamond- and gold-mining industries.

Our scrubber-trommel СБР 2.2 x 8/2 (Figure 4) is an example of an aggregate with longer disintegration sector for dispersing difficult-to-wash clays and trommel with rubber sieving elements for screening sizes 80 mm. This scrubber-trommel is designed to treat minerals with clay content up to 90% and to prepare diamond-contained inputs from placers for further concentration. This scrubber-trommel was installed in a flushing complex of URAL-ALMAZ Company.

Aggregates of our another design line are drum screens of the "dredging barrel" type that are widely used in processing less difficult-to-wash and easy washable minerals. The design is based on our experience in creating special highly-productive machines for dry separation of fine and thin classes from mined and supplied for further concentration ordinary coals with natural humidity between 6% and 12%. It is well known that screening of coal with such humidity content is always a very difficult, actually unsolvable problem. An example of our first successful solving of this problem is our design of a drum screen ГБК (Figure 5) with the DABS-type rubber sieves used for dry separation of the fine -3 mm classes at the head of technological line in Kievskaya Factory (Mine named after A.F. Zasyadko) with output capacity up to 400 t / h by base supply.

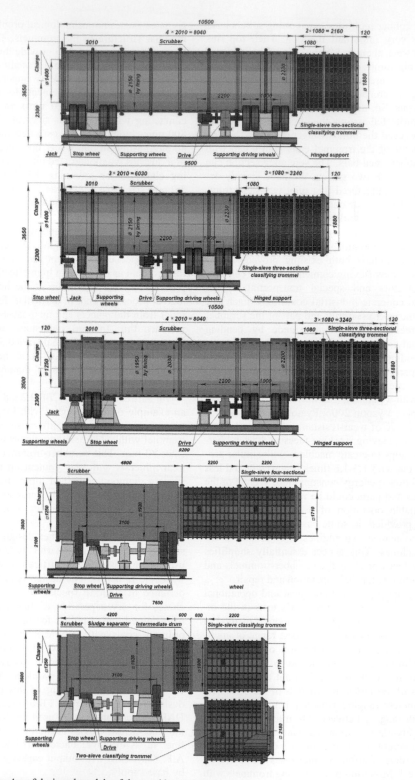

Figure 3. Examples of designed models of the scrubber-trommels.

Technical characteristics of scrubber-trommel СБР 2.2 × 8/2.

Type – scrubber-trommel, modular, 4 – sectional scrubber and 2 – sectional trommel, rubber lifter-threshold lining, with DABS sieves;		Drive type – electromechanical, friction	
		Electric engine – asynchronous; 220/380 V; 50 Hz	
Design feeding output, t / h	up to 400	Installed power, kW	2 x 30.0
Number of sieving elements	2	Frequency of the engine shaft rotation, min^{-1}	735
DABS cells size of the trommel, mm	80 x 80	Reducer type	cylinder
Diameter of the sieving surface, mm	1710	Driving wheels – rubber, big, highly elastic tyres (МВЭ)	
Total area of the sieve, m^2	16.8	Dimensions L x W x H, mm	7350 x 3290 x 2875
Angle of the drum axis slop, to horizon, degrees	2-10	Drum weight, t	13.5
Frequency of the drum rotation, min^{-1}	11-15	Weight, t	28.9

Figure 4. СБР 2.2 x 8/2 scrubber-trommel for flushing complex of URAL-ALMAZ.

Figure 5. General view of the ГБК screen at the bench tests.

Our seven-year experience of industrial testing and exploiting of the ГБК in technological line of the Kievskaya Factory helped us to improve our ГБК and create a dimension series of the drum screens with improved structures of the sieving sections and the drum itself and of assemblies and aggregates of the drive (Figure 6). According to the designers, this dimension series should include machines with maximal design capacity of 100 t / h; 200 t / h; 400 t / h and 600 t / h. These parameters are achievable through the properly specified length and diameter of the drum. Dimension series shown on the Figure 6 is

extended by screens with the following working diameters of the drum: 1500 mm; 2000 mm; 2300 mm; 2500 mm and 300 mm.

However, in order to solve specific problems of the main screening in other factories it is possible to apply other variants of the key structural and technological parameters and other individually designed models of the drum screens with taking into account in-situ specific peculiarities of transporting and loading facilities and different building objects.

An example of design with "two-arm" drive and two electric engines is our screen ГБА (Figure 7). It was designed on the basis of the ГБК screen for flushing complexes used in the new technological practices for fine natural gold recovery and preparation.

Additionally to the above features of the support platform with drive, design of this screen differs by 1) longer charging sector that, thanks to special rubber lifting liner, functions as a short receiving scrubber; and 2) specific layout of discharge sector and interacting assembly between drum and stop cross arm. DABS sieves for the thin screening of sizes 2.0 mm, 3.0 and 5.0 mm with slotted drum cells are oriented across the drum rotation axis. Critical requirements to the sieve design for such technologies is their high effectiveness by separation of undersized classes, reliable fastening and their long life at heavy loads caused by feeding material contained up to 40-80% of clay and lumps (300-500 mm and bigger).

The ГБА screen is in operation in technological line of the fine and thin gold recovery in the PIKAN deposit (Amur oblast, Russia) of the AMUR-DORE Company since 2009. General view of the main technological aggregate of mobile flushing machine with ГБА screen is shown on the Figure 8. The screen installed on the sledge platform provides highly effective disintegration, flushing and extraction of gold-contained minerals with size -2.0 + 2.0 mm; -0.3 + 2.0 mm; and -0.5 + 3.0 mm that are supplied from the undersized fraction receiving hopper to the lock sector for fine material where natural gold of appropriate sizes is extracted. Productivity of the machine is up to 140 t / h, and water consumption is about 400 m³ / h. It was the first experience of the Company in executing a practice of highly effective extraction of thin-class gold in the mobile plants with the drum screens of the dredging-barrel type. This experience has demonstrated that employing of highly effective facilities for the thin screening in the gold separating-and-flushing machines reduces amount of lost fine mineral particles and, in particular, particles of the flaky form giving a good chance to increase output of the flushing machines and complexes by 1.5-2 times.

Following the review of these results we managed to put our new technology in to operation in four other production units of this Company and in technological lines of other gold- and diamond-mining companies.

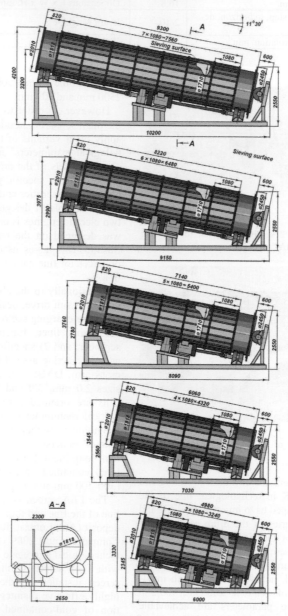

Figure 6. Dimension series of the drum screens with two-arm drives, one electric engine and DABS working surfaces.

Figure 7. Screen ГБА for the gold-recovery technologies.

Figure 8. ГБА screen in the flushing plant for mining thin-size gold.

REFERENCES

Chervonenko, A.G. & Morus, V.L. 1997. *Wear-Resistant Dynamically Active Band Sieving Elastomeric Surfaces For Separating Loose Materials and Pulps.* Proceedings of the II International Symposium on Mechanics of Elastomers. Dnipropetrovs'k. V.1: 296-309.

Morus, V.L. & Nikutov, A.V. 1998. *New Wear-Resistant Rubber Working Surfaces for Drum-Type Screens,* Regularity of Material Movements inside of Cylinders with Multiturn Transporting Spirals. In the book "Geotechnical Mechanics". Interorganization Collected Scientific Papers. Edition 7. Dnipropetrovs'k: 125-132.

Stepanenko, A.I. *Up-to-date Disintegrating Equipment.* Novosibirsk: httm: //gmexp.ru/about/.

Vysotin, A.V. & Stepanenko, A.I. 2006. *Concentration of Glass-Making Sand.* Novosibirsk: httm://gmexp.ru/about/.

Pyatokov, Vl.G (IrGTU) & Pyatakov, Vik.G. (Irgiredmet). *Light-Weight Scrubber Aggregate.* Mining Magazine, 2.

Development of gas hydrates in the Black sea

V. Bondarenko, K. Ganushevych, K. Sai & A. Tyshchenko
National Mining University, Dnipropetrovs'k, Ukraine

ABSTRACT: Given work presents urgency of the alternative sources of energy development, in particular, gas hydrates. Evaluation is given to the Black sea's gas hydrates amount and volume of methane in them. Also gas hydrates formation conditions are scrutinized. Leading technologies of gas recovery from gas hydrates are considered. Special attention is paid to CH_4 into CO_2 exchange within the gas hydrate deposit. Calculations for substantiation of the gases exchange parameters are presented with following conclusions.

1 INTRODUCTION

Standard of life of the most countries is directly proportional to their energy consumption (Guliyants 2010). This is substantiated by the fact that at modern consumption rates of standard energy carriers such as oil, coal and gas, according to the leading scientists, their amount will be enough for 150-300 years of world use. Thus, prospects of such minerals development lose their relevance with every year. In connection with that, all developed countries understanding hopelessness of the situation, begin to pay a lot of attention to development of alternative sources of energy. At present, the main and most perspective of such sources is gas hydrates. Scientific and practical interest for gas hydrates development i.e. extraction of natural gas from them has significantly increased during the last decade.

Such increased attention to gas hydrates is substantiated by their wide distribution in the oceans and seas that wash coasts of leading countries-importers of natural gas. Also, very interesting fact is that hydrates possess a very high specific concentration of gas – up to 200 m³ / m³ and are deposited relatively not deep (starting from the depth of 300-500 m under the sea bottom). Understanding all importance of the given prospect, the leading scientists of Ukraine do not set themselves aside from this trend, especially considering the fact that Ukraine has a great source of gas hydrates – the Black sea. The Black sea represents an ideal place for gas hydrates research because its annual average temperature allows to conduct studies practically all year round. In 1974 exactly from the bottom sediments of the Black sea the first samples of gas hydrates were received for the first time ever (Yefremova & Zhizchenko 1974).

2 NUMBER OF GAS HYDRATES IN THE BLACK SEA

In 2002 the studies conducted by Bulgarian scientists showed that the average depth at which the hydrates begin to form is 620 m embracing territory of about 288100 km² that presents near 68.5% of the Black sea total area (Vassilev & Dimitrov 2002).

Based upon the research of the same scientists, thickness of the hydrate stability zone (HSZ) in the Black sea reaches 160 m at depth of 1000 m. At depth of 1500 m the layer thickness makes up 260 m in average (from 110 to 650 m) and at 2000 m-depth – 350 m. If to calculate the amount of gas contained in gas hydrates of the Black sea (volume of gas hydrates makes up about 0.35 ×10¹² m³) then this number would make 40-50×10¹² m³ of methane! (Korsakov 1989).

Based upon the expeditions conducted in the 90's in the USSR, the amount of methane in the whole Black sea resulting from drilling and lifting samples of the sea bottom soil in more than 400 cores is not less than 100 trillion m³ (Solov'yov 2003).

The amount of natural gas in the Black sea's hydrates really impresses and based on the calculations of Ukrainian specialists in this field, given amount of gas will be enough for Ukraine for 1500-3000 years (http://www.goodvin.info/news/biznes/13432 v_ukraine_naydeno_novoe_mestorozhdenie_gaza_m etana.html). With that, regions of the Sea of Azov still remain to be unexplored.

Map of the Black sea showing gas hydrates forming places is presented on Figure 1.

As seen on Figure 1, basic mass of gas hydrates is located in Ukraine and Romania, less in Turkey, Bulgaria and Russia, there are deposits in Abkhazia and Georgia.

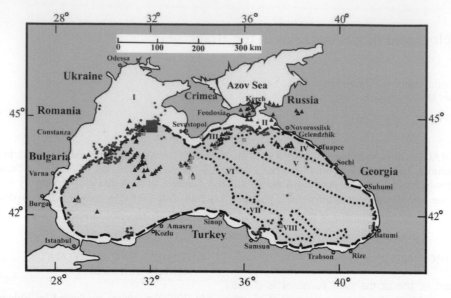

Figure 1. Location of methane hydrates in the Black sea (Starostenko 2010).

3 FORMATION CONDITIONS OF GAS HYDRATES IN THE BLACK SEA

When talking about formation conditions of gas hydrates the attention should be paid to composition of gas forming the hydrate, temperature and pressure of the forming media, sediment porosity and other.

Usually, hydrates form under temperature below +30 °C and high pressure. For example, at 0 °C methane hydrate forms at pressure of 3 MPa, and carbon dioxide at 1 MPa. If the temperature is +25 °C, methane hydrate forms at pressure of 40 MPa. Density of the Black sea's gas hydrates is within 0.9-1.1 g / cm³.

Water temperature in the Black sea below the seasonal temperature variations increases with depth and consequently the water surface temperature grows from 8.7 to 8.8 °C at sea depth being 400-500 m; and at sea depth of 2100-2200 m – to 9.05-9.1 °C. It allows to make a conclusion that in bottom sediments at sea depth of 620-650 m, favorable thermobaric conditions exist all over the place for formation and stable existence of methane hydrates (Kutas 2005).

The Black sea water, especially in shallow layers is far less salty than the oceans water. In average, 1000 grams of the Black sea water contains 18 grams of salt. Whereas, in Atlantic ocean 1000 grams of water contains 35 grams of salt, Red sea water contains 39 grams of salt. So, exactly relatively low content of salt in the Black sea has a favorable influence on gas hydrates formation (http: //www.crimeazoo.narod.ru /crimea/text3_1.htm).

4 EXTRACTION TECHNOLOGIES

If to consider methods of gas hydrates extraction, then, at present, there are several technologies in the world using which it is possible to receive gas from gas hydrates. Among them are the following ones: artificial lowering of pressure in the gas hydrate layer, pumping vapor or warm water into the gas hydrate deposit, pumping various inhibitors and salts, electromagnetic influence and other.

The most perspective method, according to the leading scientists of the world, is an exchange of methane to carbon dioxide captured from industrial enterprises. Such technology will solve three problems: 1. Methane production and its further industrial utilization; 2. Sequestration of carbon dioxide within the sea bottom "cementing" it by means of formation of more stable hydrate – carbon dioxide hydrate (this hydrate represents by itself more stable compound since it is less subjected to temperature and pressure variations); 3. Reduction of global warming as carbon dioxide captured from the enterprises does not go into the atmosphere but gets deposited under the bottom.

5 STUDIES IN THE NATIONAL MINING UNIVERSITY OF UKRAINE

A lot of attention to development of this technology is paid by the research staff of the Underground Mining Department. The second, modernized unit for gas hydrates receive was created this year at the department which allows to receive video surveillance during formation of hydrates created by various gases.

In 2011 the scholars of this department submitted patent "Method of gas methane extraction from sea gas hydrate deposits" that aims at improvement of gas recovery technology from gas hydrate deposits.

The essence of the technology consists of carbon dioxide delivery in gaseous state with temperature of +10 °C (±2 °C) along the inner pipeline with diameter of 200 mm (5) from the reservoir (2) located on the platform (3) to the depth of 100 m below the bottom of the sea that makes 600 m below the water surface, substitution of methane by carbon dioxide and formation of carbon dioxide hydrate with simultaneous methane recovery into the reservoir (1).

Methane is pumped out along the outer pipe with diameter of 400 mm (4) (Figure 2). Exchange process is conducted layer-wise (6) from the bottom to the top: radius from the borehole wall and height of the dissociation zone of each layer make up 3.5 m after 100 days that has been confirmed by the researches. After formation of the first CO_2-hydrate layer the pipe is raised up on the height of 3.5 m and a new layer development begins. Following layers are worked out by the same way up to the upper boundary of the hydrate deposit (3). After development of all the layers is over, the next borehole is drilled at a distance of 7 m from the worked-out area and methane recovery process begins by analogy.

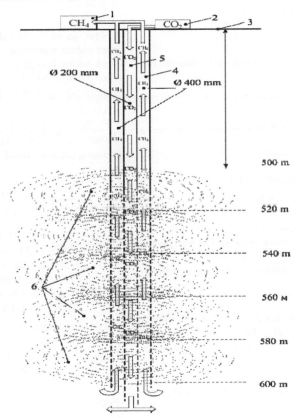

Figure 2. Scheme of CH_4 into CO_2 exchange in natural gas hydrates.

Phase transition temperature of the hydrate at 600m-depth was calculated to substantiate the above-mentioned technology:

$$T_{p.t.} = 9.75 \cdot \ln(P) - 0.7 \qquad (1)$$

Where pressure (P) is taken to be equal to 4 MPa, since critical pressure for gas methane is 4.641 MPa

(Chukhareva 2010). (Critical pressure is a pressure at which gas does not yet transfer into liquid state regardless the temperature).

$$T_{f.t.} = 9.75 \cdot \ln 4 - 0.7 = 9.75 \cdot 1.39 - 0.7 = 12.82 \ ^\circ C \ (2)$$

Given temperature contributes to prevention of CH_4 transfer into liquid (4 MPa < 4.6 MPa). That is, if the temperature exceeds 12.82 °C then CH_4 will transfer into liquid. Based on these data, the pressure in natural gas hydrate is artificially lowered down to 4 MPa and the temperature of hydrate deposit does not raise above 12.82 °C. This is done in order to control the process of exchange.

There is an interesting fact that during methane hydrate decomposition, 2.2 kJ / mole of heat releases at temperature of 0 °C, whereas during carbon dioxide hydrate formation the amount of 1.1 kJ / mole of heat gets absorbed (Moelwyn-Hughes 1975). Based on this, it can be concluded that CH_4 into CO_2 exchange will be conducted with ratio of 1:2. Hence, the following calculations can be made:

1 mole of $CH_4 = 16$ kg of CH_4;

1 mole of $CO_2 = 44$ kg of CO_2.

In terms of kilograms, the ratio will have the following look:

16 kg of CH_4 :(44 × 2)kg of CO_2;

16 kg of CH_4 :88 kg of CO_2;

1 kg of CH_4 :5.5 kg of CO_2. $\qquad$ (3)

That is, in order to pump out 1 kg of CH_4 from gas hydrate it is required to pump down 5.5 kg of CO_2 into this hydrate that undoubtedly expresses very important positive aspect of given technology: deposition of a big volume of green gas and, consequently, fight against global warming.

Further, using Mendeleev's-Clapeyron equation – $P \cdot V = \dfrac{m}{M} \cdot RT$, the mass of injected CH_4 can be calculated for the pipe height of 600 m with diameter of 400 mm. Initially, the volume of injected CO_2 for the depth of 600 m from the sea surface is calculated by the following equation:

$$V = h \cdot S = h \cdot \pi \cdot r^2 = 600 \cdot 3.14 \cdot 0.1^2 = 18.84 \, \text{m}^3 \quad (4)$$

Having expressed the mass "m" from the Mendeleev's-Clapeyron equation and substituting the volume "V" value, we will receive:

$$m = \frac{P \cdot V \cdot M}{R \cdot T} = \frac{4 \cdot 10^6 \cdot 5.5 \cdot 18.84 \cdot 44}{8.31 \cdot 283} = 7750 \, \text{т} \quad (5)$$

Based on the above-stated gases exchange equation (3), the volume of recovered CH_4 will make up 7750 t/5.5=1409 t.

The next stage is calculation of speed of CO_2 injection and CH_4 pumping-out. These values are calculated by the Reynolds number:

$$R_e = \frac{D \cdot U \cdot \rho}{\mu}, \quad (6)$$

where D – pipe diameter, m; U – flow speed, m / s; μ – dynamic viscosity of medium, Pa·s; ρ – density of medium, kg / s.

Having expressed the value of flow speed "U" from the given equation, we will receive the following:

$$U = \frac{R_e \cdot \mu}{D \cdot \rho}. \quad (7)$$

Now, in order to calculate the "U" value, the other values contained in the equation must be defined, taking into account that flow mode is laminar (such flow at which liquid or gas move by layers without mixing with each other). Thus, R_e value is accepted to be 2000.

Further, in order to calculate gases viscosity change "μ" depending on the temperature and pressure, the Sutherland's equation is used:

$$\mu = \mu_0 \cdot (273 + C) \cdot \frac{\left(\dfrac{T}{273} \right)^{1.5}}{T + C}, \quad (8)$$

where μ_0 – gas viscosity under normal conditions, Pa·s. According to the conducted studies (Chukhareva 2010), gases viscosity does not depend on pressure when its value is not greater than 5-6 MPa. Hence, the following gases viscosity values are accepted: for CO_2 – $137 \cdot 10^{-7}$ Pa·s, for CH_4 – $104 \cdot 10^{-7}$ Pa·s; C – Sutherland's coefficient, for CO_2 – 254, for CH_4 – 162.

Having substituted these values, it is possible to calculate viscosity of these gases:

$$\mu_{CO_2} = 137 \cdot 10^{-7} \left(273 + 254 \right) \times$$

$$\times \frac{\left(\dfrac{283}{273} \right)^{1.5}}{283 + 254} = 141.9 \cdot 10^{-7} \, \text{Pa·s}$$

$$\mu_{CH_4} = 104 \cdot 10^{-7} \left(273 + 162 \right) \times$$

$$\times \frac{\left(\dfrac{283}{273} \right)^{1.5}}{283 + 162} = 107.3 \cdot 10^{-7} \, \text{Pa·s}$$

Densities of the gases depending on pressure and temperature can be defined by the following equation:

$$\rho = \frac{M \cdot P}{R \cdot T} \quad . \tag{8}$$

Substituting needed values, we will receive the following:

$$\rho_{CO_2} = \frac{44 \cdot 10^{-3} \cdot 4 \cdot 10^6}{8.31 \cdot 283} = 74.84 \ \text{kg} / \text{m}^3$$

$$\rho_{CH_4} = \frac{16 \cdot 10^{-3} \cdot 4 \cdot 10^6}{8.31 \cdot 283} = 27.21 \ \text{kg} / \text{m}^3$$

Now, CO_2 injection speed can be defined:

$$U_{CO_2} = \frac{2000 \cdot 141.9 \cdot 10^{-7}}{0.2 \cdot 74.84} = 0.002 \ \text{m} / \text{s}$$

CH_4 pumping-out speed will be as follows:

$$U_{CH_4} = \frac{2000 \cdot 107.3 \cdot 10^{-7}}{0.2 \cdot 27.21} = \frac{0.02146}{5.442} = 0.004 \ \text{m} / \text{s}$$

As it is seen by the last two equations, methane pumping-out speed is 2 times greater than carbon dioxide injection speed.

Having summarized the calculations results, it can be concluded that using given technology consisting of CH_4 into CO_2 exchange with formation of carbon dioxide hydrate, the mass of CO_2 injected into the natural hydrate is 5.5 times greater than the mass of pumped-out CH_4 and methane pumping-out rate is 2 times higher than injection rate of carbon dioxide.

6 CONCLUSIONS

Extraction of natural gas from the Black sea's gas hydrates is a real task which has the place in modern scientific society.

Technology allowing to not only recover methane from gas hydrates of the Black sea but also to sequester captured from industrial enterprises carbon dioxide into the sea bottom is the most perspective and reasonable one because it provides methane re-

covery, formation of a more stable gas hydrate and reduction of global warming by means of CO_2 emissions reduction into the atmosphere;

The calculations result have shown that CH_4 recovery rate from the hydrate is 2 times higher than CO_2 injection rate and, with that, mass of the injected gas is 5.5 times greater than that of the recovered one.

Methane recovery is controlled by variation of pressure and temperature in hydrate deposit that prevents spontaneous abrupt dissociation of the hydrate leading to a sudden blowup. That is why, the following studies should be directed to improvement of this technology, and substantiation that this method is safe in terms of the environment influence.

REFERENCES

Guliyants, S.T., Egorova, G.I. & Aksent'yev, A.A. 2010. *Physico-chemical features of chemical hydrates.* Tumen, 2010.

Yefremova & Zhizchenko. 1974. Academy of Sciences. Volume 214. #5:1179-1181.

Vassilev, A. & Dimitrov, L. 2002. *Model evaluation of the Black sea gas hydrates.* Volume 56.

Korsakov, O., Dyakov, U. & Stupak, S. 1989. *Sov. Geologia,* 12:3-10.

Solov'yov, V.A. 2003. *Natural gas hydrates as potential mineral.* Russian chemical magazine. Volume 48:59-69.

http://www.goodvin.info/news/biznes/13432v_ukraine_na ydeno_novoe_mestorozhdenie_gaza_metana.html

Starostenko, V.I., Rusakov, O.M., Shnyukov, E.F., Kobolev, V.P. & Kutas, R.I. 2010. *Methane in the northern Black Sea: characterization of its geomorphological and geological environments.* The Geological Society of London.

Kutas, R.I., Kravchyuk, O.P. & Bevzyuk M.I. 2005. *Diagnosis of gas hydrates in near-bottom layer of sediments as the results of their heat conductivity measurement in situ.* Kiev: Institute of geophysics NAS of Ukraine.

http://www.crimeazoo.narod.ru/crimea/text3_1.htm

Chukhareva, N.V., Rudachenko, A.V. & Polyakov, V.A. 2010. *Definition of quantitative characteristics of oil and gas in a system of main pipelines: tutorial.* Tomsk polytechnic university: 311.

Moelwyn-Hughes, E.A. 1975. *The kinetics of reactions in solution. Moscow*: Chemistry.

The influence of performance funding strategy on capital cost of mining enterprises in Poland

M. Turek
Central Mining Institute, Katowice, Poland

A. Michalak
University of Technology, Faculty of Organization and Management, Zabrze, Poland

ABSTRACT: The problem of performance funding is very complex and multi-thread, especially regarding such specific companies as mining enterprises. They perform in a situation which is called crisis situation in the industry. In the hereby study the structure of funding two mining enterprises is presented. The structure is not typical for market conditions, it mainly bases on outside funding sources and results in alarming level of many financial ratios. Therefore, the proposal of building new funding structures of mining enterprises performance is described concerning various strategies of finance management and studying their influence on capital cost.

1 INTRODUCTION

Business functioning in the market requires adapting and conducting a specified funding strategy. It consists of a set of rules that in an ordered way determine assets structure and funding sources in order to allow company gain its aimed goals (Wawryszuk-Misztal 2007).

In the literature of finance discipline three kinds of performance funding strategy are distinguished (Tokarski 2006):

- aggressive strategy;
- conservative strategy;
- moderate strategy.

Aggressive strategy bases on assumption that floating and fixed[1] current assets and a part of fixed assets are funded by short-term outside capital. Other fixed assets are funded by fixed capital. There is a negative level of current net capital in this situation that means a low level of assets funded by fixed capital. Aggressive strategy aims to maximize income in relation to equity capital and accepts a higher financial level risk and also gives opportunity of using significant effects of financial leverage and

tax cover (Kołosowska et al. 2006). According to aforementioned strategy, financial liquidity is low which may indicate danger for company debts settling (Krzemińska 2005).

Conservative strategy deems that fixed capital funds involves not only fixed assets but also a fixed part of current assets. A conservative strategy is the opposite of aggressive funding performance strategy. Then a positive amount of current net capital appears. The financial risk is low in this case but nevertheless, considering a big share of long-term capital in company funding, including bank loans, funding costs are high. Financial liquidity is maintained on a high level. On the other hand, equity capital profitability is low due to high involvement of costly fixed assets in current performance (Krzemińska 2005).

The company conducting a **moderate strategy** funds its fixed assets by fixed capital and total current assets by short-term sources. A moderate strategy is an indirect strategy between aggressive and conservative strategy (Kołosowska et al. 2006). Current net capital floats around zero.

---

[1] Businesses observation indicates that the amount of current assets demand changes in time. This is caused by strong seasonal production and stock fluctuations. According to this, current assets may be differed into fixed not prone to changes and floating in the periods of high season. See: Z. Dobosiewicz, *Wprowadzenie do finansów i bankowości*, Wydawnictwo Naukowe PWN, Warszawa, 2005, pp.105.

2 CONTEMPORARY FUNDING STRATEGIES OF MINING ENTERPRISES IN POLAND AND THEIR FINANCIAL RESULTS

There were researched two out of five mining enterprises performing in Poland (determined in the article as I and II). They were created as a result of consolidation process in mining industry in the previous decade and consist in their structures of several to dozens of hard coal mines. The companies are sole stock corporations of National Treasury. The financial condition under which the researched mining enterprises perform may be considered as critical. The research conducted for the period of 2003-2009 indicates that capital structures in researched enterprises are consistent with aggressive strategy. There are presented contemporary funding structure strategies of researched enterprises in a simplified way in Table 1 and 2.

Table 1. Funding structure of mining enterprise I [%].

The kind of capital	years						
	2003	2004	2005	2006	2007	2008	2009
Own equity	3.93	12.12	17.48	14.67	14.63	14.44	14.24
Provisions	15.47	21.96	25.24	31.17	36.15	41.52	42.27
Long-term liabilities	22.30	18.12	15.69	14.53	10.31	10.43	9.27
Short-term liabilities	33.33	28.71	25.56	29.32	33.45	32.15	33.17
Accruals	24.97	19.09	16.03	10.31	5.46	1.46	1.05

Source: own study.

Table 2. Funding structure of mining enterprise II [%].

The kind of capital	years						
	2003	2004	2005	2006	2007	2008	2009
Own equity	53.10	39.95	40.44	36.19	34.38	32.66	29.10
Provisions	9.82	27.33	31.55	35.01	33.04	26.47	25.59
Long-term liabilities	1.54	0.74	0.38	0.40	0.38	2.94	3.66
Short-term liabilities	34.88	31.47	26.98	27.87	31.72	37.55	41.26
Accruals	0.66	0.51	0.64	0.53	0.47	0.37	0.39

Source: own study.

Funding structures presented above break most of funding rules. For example, in the first of researched enterprises own equity does not exceed the level of 20% in total liabilities which implicates the lack of gold balance rule fulfillment. A very high financial risk is also proved by the level of equity capital debt. In 2003 liabilities and provisions exceed the value of equity capital 24 times. In the subsequent years the value of this rate gains a considerable decrease to the level of 5-6, however, it is still high in relation to safety norm which is value of 1. This financial pathology is deteriorated by the fact that among outside capital short-term liabilities dominate.

A bad financial situation in coal industry also stresses the financial analysis of a second researched enterprise. In this case, the share of equity capital in funding structure oscillates around 40% and in addition, it presents a decreasing dynamics. Similarly to the first researched enterprise, short-term liabilities are on a significant position in outside capital structure.

When analyzing this untypical financial structure for market conditions additionally, attention should be aimed at the fact that analyzed enterprises, regarding their performance profile, characterize a highly stranded assets of a big amount that is practically impossible to cash (buildings and underground objects, excavation pits, professional mining machines etc.). Such assets should be funded by equity capital. Then in researched mining enterprises a negative net capital is observed proving that a considerable part of fixed assets is founded by outside capital and this situation is even more disturbing if, similarly to analyzed enterprises, it is a short-term outside capital. Such funding model, as we deal with in researched mining enterprises, results in an alarming level of most of financial ratios. Especially worrying are liquidity ratios of researched enterprises.

In the whole examined period, current financial liquidity of the first enterprise is on the much lower level than the normative value. Thus the enterprise in the analyzed period does not have the ability to regulate current liabilities by current assets in the light of adapted norms. Even when considering industry specificity, a systematic decrease of ability to settle current liabilities is alarming. In year 2006 it is the lowest in the researched period and allows only to settle 40% of current liabilities by current assets (Turek & Jonek-Kowalska 2008).

In the second of researched enterprises, where funding structure is better formed due to higher share of equity capital, situation is not satisfactory either. In the whole examined period this enterprise, similarly to previous one, does not have current financial liquidity. The ratios are much below standard norms. The problem of these two enterprises is a high level of current liabilities that exposes them to risk of financial liquidity loss and at the same time to settle these liabilities disability[2].

According to the results of analysis conducted it is possible to summarize that current strategies of mining enterprises funding cannot exist in the conditions of completely free market for mining activities. The financial situation of mining enterprises hereby described is a basic reason to build new funding strategies.

3 THE RULES OF BUILDING NEW FUNDING STRATEGIES OF MINING ENTERPRISES

When building funding strategies a basic assumption has been made that funding grounds on **dominant capital**. The share of dominant capital is the highest in funding structure. This feature do not possess other kinds of capital, called **supplementary** (Michalak 2007).

Dominant capital and supplementary capital(s) consist in some kind of funding structure. As there are many possible combinations of dominant and supplementary capital, there are also many options of funding strategies.

When considering financial situation of mining enterprises described above as **dominant capital** in funding performance of mining enterprises, it concerns equity capital. It may come from many sources, such as inner sources like income, amortization, rent etc. and outer sources like shares issue. On the other hand, **supplementary capital** is thought to be an outside capital coming from any source e.g. (Michalak 2007):
– bank loan (possibly consortium loans);

– capital coming from shares issue;
– support capital (e.g. from European Union funds);
– short-term securities;
– leasing;
– short-term liabilities and others.

There are assigned example supplementary capitals for a dominant capital in each strategy. Then it is possible to construct many options of performance funding strategy for mining enterprise, choosing various standards of dominant capital and various combinations of supplementary capitals.

In the conditions of mining enterprises functioning, when willing to build an optimal strategy of funding it has been assumed that in capital structure equity capital should dominate which derives from various sources and they are altogether combined for calculating the level of equity capital and its share in funding structure. It was estimated that the level of equity capital will not be lower than 50% of total capital. As fixed assets constitute about 80% of examined assets in mining enterprises, it is possible to envisage that regarding their unstable financial situation, these assets will be covered by fixed capital i.e. equity capital and current liabilities. Such funding structure is a demanded structure in mining enterprises, nevertheless, it is not possible to obtain in each enterprise and thus there will also be other options analyzed of performance funding in mining enterprises. The structure that will be established on the grounds of at least 50% share of equity capital and at least 80% share of fixed capital in capital structure will be the type consistent with **conservative** or **moderate strategy**. Some of mining enterprises that are in a better financial situation than in researched enterprises may adapt a more **aggressive** funding strategy of its assets, choosing funding structure resulting in negative net working capital, that is lower than 80% share of fixed capital in capital structure, however, the elementary assumption concerning minimum of 50% share of equity capital should be strictly retained.

While making analysis of contemporary funding structures in examined mining enterprises it should be stressed that the income significance in equity capital structure is not high and it cannot bear self-funding duty.

[2]More on this matter in: A. Michalak, M. Turek, Analiza struktury kapitału w kontekście źródeł finansowania przedsiębiorstw górniczych. Wyd. Sigmie PAN, Warszawa 2009, pp. 99-113; M.Turek, I.Jonek-Kowalska, Ocena płynności finansowej jako kryterium podejmowania decyzji zarządczych w przedsiębiorstwach górniczych, Wydawnictwo IGSMiE PAN, Warszawa 2009, pp. 115-125; M. Turek, A. Michalak, Uwarunkowania budowy modeli finansowania działalności operacyjnej przedsiębiorstw górniczych. Szkoła Ekonomiki i Zarządzania w Górnictwie, AGH Kraków, Przegląd Górniczy 2010, pp. 20-23.

It is necessary to complement equity capital from other sources. Current situation on coal market (restricted possibilities of generating income) and current public finances state (low probability of obtaining funds from current owner – National Treasury) should force the government to adopt activities leading to developing performance effectiveness.

One of the ways is privatization or additional shares issue. It is easy to notice that in both cases the share of equity capital in funding structure is low and in the second enterprise it is close to assumed value of 50% liabilities, however, its share in funding structure indicates a decreasing tendency. To retain proper functioning, this capital should be increased to the assumed level of 50% and that proves a need to issue shares. A remaining part of funding structure will constitute a suitable combination of long- and short-term outside capitals. These are so called supplementary capitals in separate funding models. Notwith-

standing, there are bank loans and shares issue among long-term outside capitals and short-term outside capitals that consist of non-interest short-term liabilities (purchase debts, pays due to be paid liabilities, tax liability etc.) and roll-over credit. According to the assumption of equity capital dominance, it may be stated that the level of outside capital should not exceed 50% of total capital. However, it is advisable to differentiate additionally outside capitals structure, especially regarding the relation of long- and short-term liabilities that influence the level of fixed capital.

Considering assumptions described above, example performance funding structures for mining enterprises may be offered, consistent with conservative, moderate and aggressive types. Examples of conservative performance funding structures for mining enterprises are presented in graphic form in Figure 1.

Conservative funding strategy – option 1 (data in %)					
fixed capital	EQUITY CAPITAL	Contemporary equity capital	25		80
		Additional shares issue	25		
	Long-term liabilities	Long-term loan	15	30	
		Bonds issue	15		
	Short-term liabilities	Non-interest liabilities	18	20	
		Loans and credits	2		
Conservative funding strategy – option 2 (data in %)					
fixed capital	EQUITY CAPITAL	Contemporary equity capital	25		80
		Additional shares issue	25		
	Long-term liabilities	Long-term loan	30		
	Short-term liabilities	Non-interest liabilities	18	20	
		Loans and credits	2		
Conservative funding strategy – option 3 (data in %)					
fixed capital	EQUITY CAPITAL	Contemporary equity capital	25		80
		Additional shares issue	25		
	Long-term liabilities	Bonds issue	30		
	Short-term liabilities	Non-interest liabilities	18	20	
		Loans and credits	2		

Figure 1. Example conservative operational activity funding strategies in mining enterprises Source: own study.

If the mining enterprise for which performance funding strategy is being built is in an unfavorable financial situation then it should adapt a **conservative** funding strategy. It means that equity capital in a certain model should cover fixed assets. It will provide a low financial risk level. Mining industry, as mentioned earlier, is specified by a high degree of assets blockage. In assets structure of mining enterprises fixed assets are dominating and a huge part of them is practically impossible to cash. The share of fixed assets in assets structure equals in the examined enterprises, a similar amount to the whole industry, about 80%. In such conditions fixed capital that consists of equity capital and long-term liabilities should constitute about 80% of all funding sources. If such thing is supposed that the share of equity capital should equal at least 50%, then the share of long-term liabilities in funding structure should oscillate about 30%. An optimistic forecast has been made in such models that the equity capital

coming from incomes gained will obtain a share of 7% in total capital structure (it is an average value based on historic data of two researched mining enterprises). On the other hand, other company capital reaches the level of almost 18% of total liabilities. A contemporary equity capital includes about 25% of total capital. In order to obtain the assumed level of 50% liabilities for equity capital, additional company capital should constitute 25% of total liabilities. According to aforementioned, it may be assumed that mining enterprises will gain an additional equity capital by shares issue. Other implications regarding outside long-term capital indicate that in a conservative model should equal 30% of total capital. In the first option it has been supposed that 15% of total capital derives from long-term loan and 15% from bonds issue and in the following options it has been assumed that the total needed long-term outside capital comes from one or another source.

Aggressive funding strategy – option 4 (data in %)					
fixed capital	EQUITY CAPITAL	Contemporary equity capital	25		60
		Additional shares issue	25		
	Long-term liabilities	Long-term loan	5	10	
		Bonds issue	5		
	Short-term liabilities	Non-interest liabilities	25	40	
		Loans and credits	15		
Aggressive funding strategy – option 5 (data in %)					
fixed capital	EQUITY CAPITAL	Contemporary equity capital	25		60
		Additional shares issue	25		
	Long-term liabilities	Long-term loan	10		
	Short-term liabilities	Non-interest liabilities	25	40	
		Loans and credits	15		
Aggressive funding strategy – option 6 (data in %)					
fixed capital	EQUITY CAPITAL	Contemporary equity capital	25		60
		Additional shares issue	25		
	Long-term liabilities	Bonds issue	10		
	Short-term liabilities	Non-interest liabilities	25	40	
		Loans and credits	15		

Figure 2. Example aggressive structures of mining enterprises operational activity funding. Source: own study.

If the mining enterprise which has performance funding strategy being built and uses and **aggressive** funding strategy then it may be estimated that in funding structure equity capital will still dominate (on the level of approx. 50% liabilities), however, complementary long-term liabilities will have a lower share than 30%. In this way fixed capital will not cover assumed 80% of assets in conservative models that consists of fixed assets. Example aggressive performance funding structures of mining enterprises are presented in Figure 2.

A **moderate** strategy will result in an intermediary situation between aforementioned strategy types. Example structure options consistent with this strategy are shown in Figure 3.

Moderate funding strategy – option 7 (data in %)					
fixed capital	EQUITY CAPITAL	Contemporary equity capital	25		70
		Additional shares issue	25		
	Long-term liabilities	Long-term loan	10	20	
		Bonds issue	10		
	Short-term liabilities	Non-interest liabilities	20	30	
		Loans and credits	10		

Moderate funding strategy – option 8 (data in %)					
fixed capital	EQUITY CAPITAL	Contemporary equity capital	25		70
		Additional shares issue	25		
	Long-term liabilities	Long-term loan	20		
	Short-term liabilities	Non-interest liabilities	20	30	
		Loans and credits	10		

Moderate funding strategy – option 9 (data in %)					
fixed capital	EQUITY CAPITAL	Contemporary equity capital	25		70
		Additional shares issue	25		
	Long-term liabilities	Bonds issue	20		
	Short-term liabilities	Non-interest liabilities	20	30	
		Loans and credits	10		

Figure 3. Example moderate performance funding structures of mining enterprises Source: own study.

4 CAPITAL COST IN SEPARATE FUNDING STRATEGIES OF MINING ENTERPRISES

Each of presented example funding structures in mining enterprises is specified by a different weighted average cost of capital (WACC). It is calculated by the following formula (Turek & Jonek-Kowalska 2009):

$$WACC = \sum_{i=1}^{n} w_i K_i ,$$

where w_i – the share of subsequent sources in investment funding structure, K_i – the cost of capital deriving from subsequent sources; n – the number of capital sources in investment funding structure.

In order to indicate weighted average cost of capital there should be known capital costs from several sources establishing funding structure. To carry analysis, average market capital cost values were adapted from several sources[3]. The calculations of WACC for funding structures of conservative type are presented in tables 3-5.

Table 3. WACC for a conservative strategy – option 1.

FUNDING SOURCES	SHARE IN THE STRUCTURE [%]	CAPITAL COST [%]	WACC COMPONENTS	WACC [%]
Income	7	10.28	0.07*0.1028	
Shares issue	25	14.98	0.25*0.1498	
Other company capital	18	10.28	0.18*0.1028	
Long-term loans	15	8.99	0.15*0.0899	10.41
Bonds issue	15	16.54	0.15*0.1654	
Non-interest liabilities	18	0	0.18*0	
Short-term loans and credits	2	13.52	0.02*0.1352	

Table 4. WACC for a conservative strategy – option 2.

FUNDING SOURCES	SHARE IN THE STRUCTURE [%]	CAPITAL COST [%]	WACC COMPONENTS	WACC [%]
Income	7	10.28	0.07*0.1028	
Shares issue	25	14.98	0.25*0.1498	
Other company capital	18	10.28	0.18*0.1028	
Long-term loans	30	8.99	0.30*0.0899	9.28
Non-interest liabilities	18	0	0.18*0	
Short-term loans and credits	2	13.52	0.02*0.1352	

Table 5. WACC for a conservative strategy – option 3

FUNDING SOURCES	SHARE IN THE STRUCTURE [%]	CAPITAL COST [%]	WACC COMPONENTS	WACC [%]
Income	7	10.28	0.07*0.1028	
Shares issue	25	14.98	0.25*0.1498	
Other company capital	18	10.28	0.18*0.1028	
Bonds issue	30	16.54	0.30*0.1654	11.55
Non-interest liabilities	18	0	0.18*0	
Short-term loans and credits	2	13.52	0.02*0.1352	

Table 6. WACC for an aggressive strategy – option 4.

FUNDING SOURCES	SHARE IN THE STRUCTURE [%]	CAPITAL COST [%]	WACC COMPONENTS	WACC [%]
Income	7	10.28	0.07*0.1028	
Shares issue	25	14.98	0.25*0.1498	
Other company capital	18	10.28	0.18*0.1028	
Long-term loans	5	8.99	0.05*0.0899	9.61
Bonds issue	5	16.54	0.05*0.1654	
Non-interest liabilities	25	0	0.25*0	
Short-term loans and credits	15	13.52	0.15*0.1352	

[3] Values from 2010, more on that matter in: M. Turek, A. Michalak, Uwarunkowania budowy modeli finansowania działalności operacyjnej przedsiębiorstw górniczych, Przegląd Górniczy 2010, pp. 20-23.

The most beneficial conservative strategy among the presented examples is option 2. It is characteristic for the lowest weighted average cost of capital which in this case equals 9.28%.

The calculations of WACC for an aggressive type of strategy is presented in Tables 6-8.

The most gainful option among the example aggressive options of funding structures is number 5.

This option shows the lowest weighted average cost of capital which equals 9.24%.

The calculations of WACC for a moderate strategy is presented in tables 9-11.

The most profitable option among the example moderate strategies is option 8. This model is characteristic for the lowest weighted average cost of capital which equals 9.46.

Table 7. WACC for an aggressive strategy – option 5.

FUNDING SOURCES	SHARE IN THE STRUCTURE [%]	CAPITAL COST [%]	WACC COMPONENTS	WACC [%]
Income	7	10.28	0.07*0.1028	
Shares issue	25	14.98	0.25*0.1498	
Other company capital	18	10.28	0.18*0.1028	
Long-term loans	10	8.99	0.10*0.0899	9.24
Non-interest liabilities	25	0	0.25*0	
Short-term loans and credits	15	13.52	0.15*0.1352	

Table 8. WACC for an aggressive strategy – option 6.

FUNDING SOURCES	SHARE IN THE STRUCTURE [%]	CAPITAL COST [%]	WACC COMPONENTS	WACC [%]
Income	7	10.28	0.07*0.1028	
Shares issue	25	14.98	0.25*0.1498	
Other company capital	18	10.28	0.18*0.1028	
Bonds issue	10	16.54	0.10*0.1654	9.99
Non-interest liabilities	25	0	0.25*0	
Short-term loans and credits	15	13.52	0.15*0.1352	

Table 9. WACC for a moderate strategy – option 7.

FUNDING SOURCES	SHARE IN THE STRUCTURE [%]	CAPITAL COST [%]	WACC COMPONENTS	WACC [%]
Income	7	10.28	0.07*0.1028	
Shares issue	25	14.98	0.25*0.1498	
Other company capital	18	10.28	0.18*0.1028	
Long-term loans	10	8.99	0.10*0.0899	10.22
Bonds issue	10	16.54	0.10*0.1654	
Non-interest liabilities	20	0	0.20*0	
Short-term loans and credits	10	13.52	0.10*0.1352	

Table 10. WACC for a moderate strategy – option 8.

FUNDING SOURCES	SHARE IN THE STRUCTURE [%]	CAPITAL COST [%]	WACC COMPONENTS	WACC [%]
Income	7	10.28	0.07*0.1028	
Shares issue	25	14.98	0.25*0.1498	
Other company capital	18	10.28	0.18*0.1028	
Long-term loans	20	8.99	0.20*0.0899	9.46
Non-interest liabilities	20	0	0.20*0	
Short-term loans and credits	10	13.52	0.10*0.1352	

Table 11. WACC for a moderate strategy – option 9.

FUNDING SOURCES	SHARE IN THE STRUCTURE [%]	CAPITAL COST [%]	WACC COMPONENTS	WACC [%]
Income	7	10.28	0.07*0.1028	
Shares issue	25	14.98	0.25*0.1498	
Other company capital	18	10.28	0.18*0.1028	10.97
Bonds issue	20	16.54	0.20*0.1654	
Non-interest liabilities	20	0	0.20*0	
Short-term loans and credits	10	13.52	0.10*0.1352	

6 CONCLUSIONS

Basing on the analysis of example options of conservative, aggressive and moderate strategy it is visible that the most profitable funding sources models are combinations that do not include shares issue. Among most gainful models of separate funding structure types a long-term loan becomes the only source of long-term outside capital. In general, the most profitable are aggressive models from the point of view regarding capital cost decrease. Their fixed capital consists of equity capital and capital from a long-term loan. These models are specific for a negative working capital. Apart from low capital cost, their characteristic feature is high financial risk triggering threat of financial liquidity loss.

The presented research was conducted in the frames of the project by Ministry of Science and Higher Education entitled *Investments in coal mining industry in terms of their financing* (N N524 464836) conducted by Silesian University of Technology.

REFERENCES

Dobosiewicz, Z. 2005. *Wprowadzenie do finansów i bankowości*. Warszawa: Wydawnictwo Naukowe PWN.
Kołosowska, B., Tokarski, A., Tokarski, M. & Chojnacka, E. 2006. *Strategie finansowania działalności przedsiębiorstw*. Kraków: Oficyna Ekonomiczna.
Krzemińska, D. 2005. *Finanse przedsiębiorstwa*. Poznań: Wydawnictwo Wyższej Szkoły Bankowej.
Michalak, A. 2007. *Finansowanie inwestycji w teorii i praktyce*. Warszawa: Wydawnictwo Naukowe PWN.
Michalak, A. & Turek, M. 2009. *Analiza struktury kapitału w kontekście źródeł finansowania przedsiębiorstw górniczych*. Warszawa: Wydawnictwo: Sigmie PAN.
Tokarski, A. 2006. *Strategie finansowania działalności przedsiębiorstw produkcyjnych*. Toruń Wydawnictwo Adam Marszałek.
Turek, M. & Jonek-Kowalska, I. 2009. *Dylematy kalkulacji kosztu kapitału w przedsiębiorstwie górniczym*. Szkoła Ekonomiki i Zarządzania w Górnictwie, Przegląd Górniczy, 9/20.
Turek, M. & Jonek-Kowalska, I. 2009. *Ocena efektywności finansowej jako kryterium podejmowania decyzji zarządczych w przedsiębiorstwach górniczych*. Kraków: Szkoła Eksploatacji Podziemnej 2009, Instytut Gospodarki Surowcami Mineralnymi i Energią Polskiej Akademii Nauk, Sympozja i Konferencje, 74.
Turek, M. & Jonek-Kowalska, I. 2009. *Ocena płynności finansowej jako kryterium podejmowania decyzji zarządczych w przedsiębiorstwach górniczych*. Warszawa: Wydawnictwo IGSMiE PAN.
Turek, M. & Michalak, A. 2010. *Uwarunkowania budowy modeli finansowania działalności operacyjnej przedsiębiorstw górniczych*. Kraków: Szkoła Ekonomiki i Zarządzania w Górnictwie, AGH, Przegląd Górniczy.
Wawryszuk-Misztal, A. 2007. *Strategie zarządzania kapitałem obrotowym netto w przedsiębiorstwach*. Lublin: Wydawnictwo Uniwersytetu M.Curie-Skłodowskie.

Analysis of combined support behavior of development openings with criteria of resource-saving technologies

I. Kovalevska & A. Laguta
National Mining University, Dnipropetrovs'k, Ukraine

O. Vivcharenko
OJSK "Pavlogradugol", Pavlograd, Ukraine

O. Koval
GP "Sverdlovantracite", Sverdlovsk, Ukraine

ABSTRACT: The mechanism of interaction of main load-bearing elements of various combined support constructions (a frame and rock bolts) is examined. Due to the analysis of support behaviors, computing experiments are based on finite element method and mine researches, the technology of development opening maintenance by frame-bolt supports is justified. This technology is more fitted to resource-saving criteria. The described support has spatial flexible mechanical connections of frame-legs and bolts as a one load-bearing system, with ability to redistribute loads on bearing elements while the process of its loading.

1 INTRODUCTION

By present time the huge experience of opening's exploitation, maintaining by combined supports has been accumulated. This supports include both positive and as negative evaluation of technical and economical aspects of its application in different mine-geological and mine-technical conditions. That is why we pay attention to development workings, which are situated in layered rock mass of Western Donbass weak rocks. There is high rate of rock pressure, which can be seen in the cross-section of an operation and along its length. Therefore, the spatial irregularity of force massive interaction with any type of support (even with a combined support) is taken place. So, the analysis of effectiveness of behavior of various constructions of such supports (combining pliable frames and bolts) is performed with evaluation of adaptation degree of its designs to spatial manifestations of rock pressure in development openings. From other side, a resource-saving factor of the rock mass near the contour involved in resistance to rock pressure is also important. That is required to evaluate the effectiveness of neighboring rocks to form something like a load-bearing construction around the opening, with high rebuff reaction. Thirdly, combined support installation and different elements, which are included in this support, must be worked together to resist to the displacement of a coal-containing rock mass. Fourthly, the strength balance of a construction directly influences on the resource-saving that is impor-

tant when considerable load variation occurs in space and time. Therefore the evaluation of combined support ability to self-regulate loads in proportional to bearing capacity of main load-bearing elements is put to the fore.

Thus, taking attention to resource-saving factor the analysis of effectiveness of present constructions of the frame support, combined with rock bolts, is done with following criteria of low cost maintaining of development openings in the weak coal-containing rock mass:

– the highest level of a rock mass near the contour involved to resist to rock pressure manifestations;

– the maximum adaptation degree of such combined support to the character of rock pressure manifestations in time and space;

– the coordination level (or level of synchronism) of main bearing element behavior of a combined support;

– the maximum strength balance degree of a combined support in longitudinal section and cross-section of an operation due to the self-regulation of loading on main bearing elements.

There are constructive and technological solutions of development operation maintaining of coal mines. And it is reasonable to divide the combination of a flexible frame with rock bolts, having analysis of its behavior, into three major groups:

– the combination of the pliable frame support (the metal three-link support is made from specific profile SVP, MFFS-3L ("metallic flexible three-linked frame support") and PFHS ("prolonged flexible hip-roof support") series are more widespread used) and the rock bolt systems, which are installed in order to opening shape. Rock bolt systems and the frame are not linked with each other (Bulat 2002);

– a frame and a rock bolt (all the bolt set in cross section of an operation or just one part of them) are linked with each other by the hard link that does not allow sufficient moving of main support elements (a frame, a rock bolt) from one another (Vygodin 1989);

– a frame and rock bolts are linked by flexible links, that allows to move together during sufficient rock pressure manifestations as in cross section and as in longitudinal section of maintaining operation (Kovalevska 1995).

Preliminary, to make it clear "frame-rock bolt support" term is objectively conformed only to the second and the third construction groups. There is the mechanical link of two main load-bearing elements – the frame and rock bolts. Constructions of the first group are just combination of two different types of support.

World and domestic practice of combined support application to maintain mine workings are well-know (Bulat 2002). The major aim is to decrease loading on a frame (that promotes the decrease of its metal-content) due to strengthening some volumes of massive near the contour, and to limit the volume of unstable rocks, perceived by a frame support. The functions of a frame and a bolt supports are considerably different from each other here, although can realize one task – ensure the stability of an opening. These constructions of first group are more widespread, even in Western Donbass mines, where combination of PFHS with polymeric-resin rock bolts is used to maintain development openings.

2 PRELIMINARY RESULTS

From the viewpoint of the resource-saving criterion the maximum inclusion level of rocks to resist to rock pressure manifestations is realized by the analyzed constructions of the first group (and even two other groups). The modeling of rock bolt's strengthening action shows that limited area is formed around bolts, where the bolt influence aims to avoid rock separation due to armature stretching and influencing reactive efforts of compression on a rock along the bolt axis. These compression efforts increase concerned stability of rock volumes due to conventional Coulomb-Morh strength theory (Borisov 1980). If the distance between rock bolts is short, joining of strengthened areas occurs and the load-bearing construction (a perceiving part) is formed. With increasing distance between bolts the load-bearing construction is not happened, however,

the volumes of strengthened rocks around bolts limit the volumes of unstable rocks, forming load on a frame support. That is why, increasing the density of bolting provides for reduction of load on a frame. Therefore, it contributes decrease of frame's metal-content. However, the material-content of bolt support and labor-content of its installation are simultaneously raised. So, it is necessary to find the compromise to provide for required stability of an operation with minimum cost of its installing and maintaining.

Due to considerations of geomechanical features of the loading process and relatively uniform support resistance along an opening, more preferable installation of rock bolts is in the middle between the frames (as it occurs in the first and third construction groups of combined supports). Then unstrengthened rocks load mainly on a frame, but not on the barrier between the frames, which can be maintained by a wire mesh. The influence step of reaction maximums of a support construction along a mine working is also reduced. According to investigations (Kovalevska 2005), it aligns the diagram of loading on a support and creates more favorable conditions for maintaining operations. Constructions of second group do not have mentioned advantages, because rock bolts are situated in cross-section of frame installation, and the interval between them is strengthened by a wire mesh only. Significant ununiform reaction of a support along a mine working, generating (Kovalevska 2005) considerable fluctuations of displacements of the rock contour of ultimate balance zone that contradicts bearing capacity of main load-bearing elements (a frame, a rock bolt and a wire mesh), occurs.

From the viewpoint of the second resource-saving condition of the maximum adaptation level of rock pressure manifestations, the constructions of all three groups have some differences. If in the cross-section of a mine working the adaptation level of construction to the diagram of predictable pressure can be controlled by the density of bolting, then in the longitudinal section of a mine working the problems of ununiform loading on a support construction are happened, by means of the rock mass irregularity and local manifestations, weakening its factors. Such rock pressure manifestations are forecasted difficulty. That is why support parameters are changed sometimes, especially in certain areas of an opening without taking into account local fluctuations of load and displacements of the rock contour.

For visualization of the analysis of support constructions' adaptive properties to local resistance to rock pressure manifestations along a mine working the scheme is given on Figure 1. It explains the resistance mechanism of each of three combined support groups. The first construction group (Figure 1b) is characterized by absence of mechanical links of frames and bolts, which are separately deformed in the area of resistance to $G(Z)$ loading. In the area of increased displacement of rock contour the redundant movement U causes additional bolt tension that increases its reaction of resistance of certain maximum Q_1^{max} in specific conditions, i.e. rock bolts reinforce resistance to increased rock pressure. Neighboring frame is independently deformed and due to its structural flexibility the frame resistance is little changed in this mode (Kovalevska 2007). And absence of the link with bolts does not allow increasing rebuff reaction of a frame. In result, in the area of increased rock pressure increase the rebuff of a support is only realized by rising resistance of some bolts and partly frames, being caught in this local area. Next to this area rock bolts and frames are not taken part in resistance to local increased loading. That is why, adaptation degree to the character of rock pressure manifestations of the first support group should be evaluated as low degree.

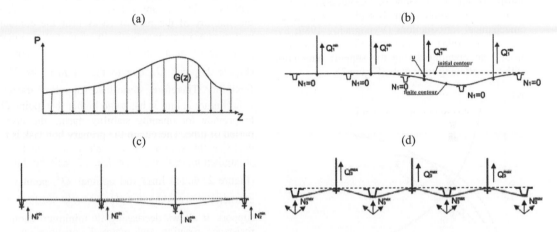

Figure 1. Diagram of $G(Z)$ loading along a mine working (a) and schemes of deforming support constructions are divided into groups of link between a frame and a bolt: (b) without link; (c) rigid link; (d) yielding link.

A frame-bolt support with hard link (Figure 1c) also increases its resistance in the area of local rock pressure manifestations. And the growth of the rebuff reaction is conditional by two components – increasing Q_2^{max} reaction of bolts, a part of which N_2^{max} is transferred to a frame; increasing the bearing capacity of a frame due to appearing additional supports (around the mine working) from the N_2^{max} influence of rock bolts. Given construction group has higher adaptive properties, relatively to previous group, as in the work of resistance to increased rock pressure two major elements – the frame and rock bolts – are included as united load-bearing system. Here the main defect is that changing parameter of force interaction of "rock mass-support of opening" system in the local area of increased rock pressure relates to support constructions only, situated in given area. Beyond it, neither frames nor bolts take part to resist to increased loadings, so the constructive links, between linking elements of supports along a working, are absent.

The third construction group, where bolts and frames are tied by pliable links along an operation, does not have mentioned defects (Figure 1d). There is the highest adaption degree of frame-bolt support to the character of rock pressure manifestations. In the area of higher loading the increased Q_3^{max} reaction of bolts is partly transferred to the frame as N_3^{max} efforts, which promote increase of its bearing capacity. But, all the bars of frame are tied together by pliable links along an opening with possibility of its longitudinal movements relatively to bolts and framelegs. Then in an area of increased rock pressure enhanced tension of pliable links is automatically transferred by tie strips behind the limits of local manifestation's area. And sufficient group of frame-bolt supports along a mine working is involved in the work of resistance to increase of local loading that leads to essential reduction of support's loading level at zone of increased pressure.

Therefore, following to the factor of adaptive properties the constructions of frame-bolt supports with pliable yielding links along an opening are preferable.

The third resource-saving condition of mine working maintenance by means of frame-bolt supports is formulated as the maximum possible degree of co-ordination of bolts and frames' behavior. It allows "moving away" increased rock pressure from one side, and resisting to pressure manifestations with the maximum rebuff reaction as a single load-bearing system, redistributing load to bearing elements during the process of its unloading, from another side. First of all, let us justify the reason-ability (from point view of resource-saving) of frame-bolting combination based on existing ge-omechanical introductions (Vynogradov 1989; Symanovych 2005) of rock pressure manifestations around an opening during the support interaction with the softening rock mass (Figure 2).

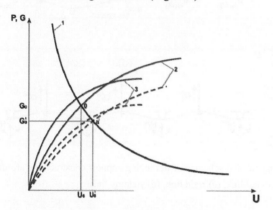

Figure 2. Substantiation of decrease the load on a frame support with combination of bolting: 1 – strain-load cha-racteristic of softening rock mass; 2 – loading of unstable rock's weight; 3 – strain-load characteristic of support; – – – frame support, ——— combination of frame support with bolts.

It is generally known that the increase of softening rock in volume causes U displacements of opera-tion's rock contour and further loading on a support by G efforts. If a support has enough constructive pliability it "moves away" increased G loadings that predetermine all-round distribution of pliable supports for unstable rock conditions. Link of G loading on the support with its U pliability (Figure 2, line 1) was proven numerous analytical and ex-perimental researches. Accordingly, G loading is decreased to minimum during sufficient construc-tive support pliability. But, the "1" dependence characterizes one side of the process only, linked with loosening (exfoliating) of rocks near the con-tour, which are increased in volume, but saving the

stability. Definitely, if not to limit rock contour displacement the resistance reaction to them is not practically happened. However, scientists and spe-cialists point to the second process's side of con-tained rock mass displacement (Figure 2, line 2). It characterizes growth of unstable rock volumes due to increase of pliability support. These rocks create G loading on a support by its weight. Therefore it is impossible to avoid loading on a support, but other way the value of G loading is increased. Thus, there are two opposite geomechanical tenden-cies of load formation on a support: 1) softening with loosening and unraveling rock mass; 2) loosing the stability of the part of rocks near the contour with further its caving. Such situation predetermines existence of minimum G_O loading on a support (Figure 2, point O) due to the value of its U_O flexibility. Therefore, strain-load support character-istic (Figure 2, line 3) has to go higher then point O to provide for opening stability during the given period of time. Thereupon the primary bolt task is to decrease the volume of unstable rocks due to its strengthening action. Then Line 2 will be fallen (Figure 2, dotted line) and optimal O^I point will characterize considerably lower G_O^I loading on support. It allows decreasing its minimum enough resistance reaction and material consumption ac-cordingly.

Described mechanism of minimum load formation supposes the condition that is a frame and bolts have to provide for required pliability of support con-struction and to interactively (synchronously) change resistance reaction during the moving proc-ess of opening's rock contour. Obviously, simple combination (without constructive links) of frames and bolts does not allow to realize such condition, because of: 1) a frame support possesses certain flexibility according to its constructive features; 2) a rock bolt forms enough rigid rock load-bearing con-struction, having lower flexibility. Therefore the majority of loadings will be received by more rigid element. It can lead to breaking element and suffi-cient loss of general bearing capacity of support construction. If rock bolts do not form rock load-bearing construction, its separate work with a frame does not also suppose any consistency, but it leads to limitation of unstable rock volume according to line 2 in Figure 2.

In our view, the coordination of behaviors of a frame support and bolts is possible due to combin-ing them by constructive links. In this case, it is necessary to analyze the work of a frame-bolt sup-port as a single construction for both rigid and pliable links that was realized by means of its strain-load characteristics. During combining by the rigid link in the initial loading period of a frame-bolt support, the components of frames and bolts' reaction are summed up and its total rebuff is increased due to rising bearing capacity of a frame with additional supports (around an operation) from rigid link influence (see Figure 1c). Further, with increasing of rock contour displacement, the resistance of a ferro-polymeric bolt (Samorodov 2001), even in other constructions of rock bolts without expansion ends, is reduced by means of the destruction of the part of the contact of bolt armature with rock bore walls. This process is specified by little flexibility of rock bolts (more spreading in the practice of mine working maintenance), which based on many experiments, does not usually exceed several tens of millimeters. Therefore very limited pliability of a bolt (in few times less, then a frame support has) the total reaction of frame-bolt support is decreased with increase of rock contour displacement. And further, it is stabilized to the level of combination of frame resistance in the flexible mode, plus the action of bolts' reaction during big movements by means of residual friction of the armature to rock bore's walls (Bondarenko 2005).

Potential possibilities of a frame-bolt support (i.e. increasing its resistance reaction in the flexible mode) are fully opened when spatial pliable links of frames and bolts occur. There is also the effect of increase of frame bearing capacity while creating additional supports. But it is increased due to features of flexible link behavior. The essence of the process consists in following. During displacement of rock contour deforming the frame transfers forces to a neighboring rock bolt through pliable (flexible) link. Forces tend to "pull out" this bolt from a borehole. It is called "pulling out" forces. In the given area of the growth of bolt resistance flexible properties of its construction are added by flexibility of pliable link, thus the total value of flexibility is increased and the tension of pliable link becomes weaker. Its N_3^{max} reaction (see Figure 1c) on a frame-leg becomes lower accordingly. But, other rock bolts, which are not reached to "the top" of its resistance reaction, try to prevent to make weaker the tension of pliable link. That is why, the self-regulation of "pulling out" forces between a whole group of bolts, connected by one pliable link, occurs along an opening. Decreasing a part of resistance reaction of separate rock bolt in the flexible behavior is immediately compensated by neighboring bolts and a frame-bolt support generally reaches effective flexible behavior of constant high resistance, exceeding such one for support constructions of previous groups. In result, with the third resource-saving

criterion the frame-bolt supports with spatial pliable links have unquestionable advantages.

The fourth criterion of resource saving, concerning the achievement level of the balanced strength condition of main bearing elements, is closely linked with the previous criterion. The essence of the balanced strength condition consists in the redistribution of frame-bolt support's forces in proportion to bearing capacity of every bearing element. And it is reasonable that this redistribution occurs automatically in the process of resistance to rock pressure manifestations. The balanced strength of construction will always have advantage in "bearing capacity-material content" relation. There is sense to compare only two groups of frame-bolt supports – with rigid and flexible links. The rigid link of frames and bolts does not admit any redistribution of forces due to their features. For example, due to small displacements of the rock contour the limited volume of unstable rocks is formed and the frame becomes underloaded. In the same time the given value of displacements is enough to destroy the part of contact of bolt armature with rock walls of a borehole and the rock bolt looses a part of its resistance reaction. So, the one element of construction is underloaded, and the second one is overloaded with partial destruction. From other side, due to increased displacements of the rock contour the frame works in flexible behavior and it is full loaded. Due to the rigid link with rock bolts the frame "pulls out" an armature from the borehole, leading to the bolt destruction partially or fully, that is also undesirable. Therefore, during the whole period of force interaction of the bolt with a frame-leg the maintenance of the process of the self-regulation of forces is required in proportion to the bearing capacity of frames and bolts. It is realized by the creation of spatial pliable links between them. Really, the loading on a separate bolt, in different stages of approaching to the value of its bearing capacity, is redistributed by means of pliable links on neighboring bolts. And this process occurs automatically due to the slipping of connecting element relatively to the tail part of a rock bolt. From other side, the frame, transformed to the flexible behavior, does not "pull out" neighboring bolts due to the stock of the flexibility of pliable links in direction to the displacement of the concerned frame's area. Thus, the analysis of constructions and behaviors of the frame-bolt support confirms the greatest fitness to resource-saving criteria of development opening maintenance by frame-bolt constructions with spatial pliable links, which providing for the self-regulation of behaviors of interaction of main load-bearing elements of "rock mass-opening's support" system.

3 RESULTS

The analysis of computer modeling results of the research of the field of the current σ stresses (Figure 3) has been fixed that the character of distribution σ for the frame-bolt support with pliable links of rib bolts with frame-legs has principal differences from σ field for basic version of the combined support (combination of frames with the bolts without pliable tie strips). So, in the upper part of a support practical constant field σ is installed around an arch, from its U-bolt to the U-bolt of the flexibility of the frame. This field is conformed to the action of even separate bending moment: in the central part of SVP cross-section, nearby its neutral axis, current stresses are minimal ($\sigma = 10...25$ MPa); σ stresses are increased in the upper and the lower part of SVP cross-section, achieving the maximum on the surface of SVP, which equals to $80...100$ MPa. Though stresses are in few times less, then the evaluated limit of pliable flow steel St.5 ($\sigma_T = 270$ MPa), and therefore the upper frame part is under quite stable condition.

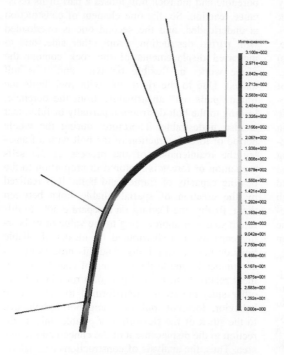

Figure 3. Diagram of current σ stresses in the frame-bolt support with spatial-flexible tie strips.

In the U-bolt of flexibility some changes of the unloading area $\sigma = 10...25$ MPa is observed (it goes to the upper part of SVP cross-section). And behind the limits of this area, up to the SVP surface, the value of current stresses is $\sigma = 25...40$ MPa. That is not enough for achieving the flexibility behavior of a frame and it generally saves the cross-section of the mine working till the ground setting.

In the curvilinear part of a frame-leg, beginning from the flexible U-bolt, the small area ($150...200$ mm of width) with constant reduced stresses $\sigma = 30...50$ MPa remains. It points to the practical absence of the given area of bending moment. It is fully explained by the neighborhood of the flexible U-bolt, having properties of a quasi-plastic joint. During movement to the upper rib bolt, in the curvilinear part of the frame-leg σ stress is undergone significant changes: 1) current stresses are increased along all area of SVP cross-section, but basically in the upper and lower parts, it points to bending moment action; 2) when stress comes to the upper rib bolt on 200 mm distance, in the surface part of SVP cross-section the σ component reaches $250...275$ MPa, that is conformed to the evaluated limit of pliable flow steel St.5. Lower the place of the installation of upper rib bolt the σ stress with local areas of manifestation of pliable steel flow St.5 appears practically and symmetrically. However, areas of pliable condition are quite limited and conditional by contact stresses during the tension of a flexible tie strip and its influence on SVP. Repeated reduction of area measurements of flexible condition is obvious (in comparing with basic version of support) due to only one connection of bolts and frame-legs by the flexible tie strip.

In the curvilinear part of a frame-leg the current stresses are decreased along the height of SVP cross-section: $\sigma = 25...40$ MPa – in the central part; $\sigma = 70...100$ MPa – in the marginal area. During the movement to the lower rib bolt direction the σ stress is changed again, as in the area of the upper rib bolt. The σ component is increased in the all area of SVP cross-section and achieves the $250...275$ MPa maximum in the external and internal parts of a cross-section. The quite limited area of pliable steel condition St.5 appears here, which is also conditional by contact stresses of the tie strip and the bottom part of SVP. Downwards along the frame-leg, in the area of its support, σ stresses are decreased and noticeably made even along the height of SVP cross-section up to $40...110$ MPa level, except the local area (up to 70 mm) of a frame-leg end. That is also made by contact stresses of the frame-leg with a thrust block. However the σ stresses do not reach the level of pliable condition in the pointed area.

It is also worth that in the area of upper rib bolt, the district of flexible St.5 condition is expanded along the frame-leg length up to 170 mm and along the height of SVP cross-section up to 35 mm. In the similar area of the basic support version the area's length of flexible SVP condition reaches up to 1120 mm in the frame-leg and the height is up to 90 mm. That is why the sizes of flexible condition of the area of a frame-leg are decreased in 2.6…6.6 times in the district of upper rib bolting, transforming into the local zones of contact stresses, which do not influence on the stability of frame-legs and whole support. In the lower rib bolt area the area length of flexible condition of a frame-leg is not more then 180 mm, and along the height of SVP it is expanded up to 40 mm. In the basic support these sizes are 400 mm and 110 mm accordingly, i.e. are increased in 2.6…2.8 times.

4 CONCLUSIONS

In whole, the major conclusion of the analysis of frame's stress-strain state (SSS) of advanced frame-bolt support consists in assigning the fact of multiple size reduction of pliable condition areas along the leg length till the level of local zones of contact's stresses, situated in the bottom part of SVP. It provides for required stability of the opening's frame. Also the reasonability of technical solution is confirmed in the part of bolts and frames, strengthening by pliable tie strips, with possibility of spatial flexibility. These tie strips provide for original additional support for frame-legs, making them stronger to resist to rock mass displacements inside the opening from the mine working walls. And it does the functions of loading redistribution on bearing elements of a support as in cross-sections, and as in longitudinal-sections of a mine working.

REFERENCES

Bulat, A.F. & Vynogradov, V.V. 2002. *Supporting bolting of the mine openings of coal mines.* Dnipropetrovs'k: Vyl'po: 372.

Vygodin, M.A. & Evtushenko, V.V. 1989. *Increasing methods of mine working stability in the Western Donbass mines.* Mine building, 5: 11-14.

Kovalevska, I.A. 1995. *Rock bolt and combined support interaction with rock mass and development of calculation method of its rational parameters.* Dnipropetrovs'k: National mining academy of Ukraine: 200.

Borisov A.A. 1980. *Mechanic of rocks and rock mass.* Nedra: 360.

Kovalevska, I.A. 2005. *Calculation of parameters of stability management of subsystem "strengthened rocks-support of underground openings".* Dnipropetrovs'k: System technologies: 113.

Kovalevska, I.A. 2007. *General scheme of calculation of loading on subsystem "strengthened rocks-support".* Dnipropetrovsk: Scientific bulletin, 1: 11-12.

Vynogradov, V.V. 1989. *Geomechanics of mass condition management near mine working.* Kyiv. Scientific idea: 192.

Symanovych, G.A. 2005. *Stability of underground mine workings.* Dnipropetrovs'k: System technologies: 164.

Samorodov, B.N., Marysuk, V.P. & Nagovitsyn, U.N. 2001. *Experience of mine working support by ferro-concrete bolts.* Mining journal, 4: 29-31.

Bondarenko, V.I., Kovalevska, I.A., Symanovych, G.A. & Porotnikov, V.V. 2005. *Theory and practice of tubular bolt application.* Dnipropetrovs'k: System technologies: 321.

Geophysical prospecting of gas hydrate

S. Sunjay
Geophysics, BHU, Varanasi-221005, India

ABSTRACT: Clathrate Gas-Methane Hydrate(White Gold) Crystal Fuel is a crystalline substance composed of water and gas, in which solid lattices of water molecules trap gas molecules in a cage-like structure or clathrate, present in permafrost regions and beneath the sea in outer continental margins. Gas Hydrate are usually inferred on seismic profiles by large amplitude bottom simulating reflector (BSR), which occur near the sea floor; Cutting across underlying dipping strata, The velocity inversion at the BSR caused by moving from high velocity hydrate cemented sediments to low velocity water or gas filled sediments below; The polarity reversal of the BSR with respect to sea floor. The occurrence of a BSR in seismic reflection data is the most important indicator of hydrates in marine sediments. However, hydrates can exist without creating a BSR if there is no underlying free gas or if the hydrates do not appreciably stiffen the sediment matrix. The weak reflectivity observed above the BSR (blanking) acoustic blanking indicates the absence of any signal because of increased transmission and obliteration of sediment impedance structures owing to the general replacement of pore water by hydrate, therefore, the zone with acoustic blanking characteristics is also referred to as the hydrate stability zone which is defined as the sedimentary package which contains the gas hydrates; the degree of blanking is proportional to the amount of hydrate in the pore space. The BSR is conspicuously underlain by transparent zones which are totally devoid of reflections and called as wipeouts. If hydrate layers are thin, tuning effects can occur and make it more difficult to interpret the gas hydrate or determine whether gas hydrates are thin or thick. Therefore, a special approach is required to identify thick high concentration hydrate layers by integrating rock physics modeling, amplitude analysis, and spectral decomposition.

1 INTRODUCTION

Gas Hydrates are detected on seismic section, primarily through identifying an anomalous reflector know as bottom simulating reflector (BSR) that mimics the shape of the sea floor, cuts across the dipping strata and has opposite polarity with respect to the sea-floor event. Third is important manifestation in identifying the BSR is the *polarity reversal* of the wavelet. The abrupt change in the velocity pattern from high velocity zone associated with the gas-hydrates to lower velocity water saturated or gas filled sediments just below the BSR produces a polarity reversal. The velocity configuration produces characteristic reduction in amplitude in the reflector above the BSR. Wave attenuation is an important seismic attribute that contains significant information about physical rock Properties. Seismic reflection techniques are the most important tool for locating gas hydrate zones. Hydrates have very strong effect on the acoustic reflection because of high acoustic impedance contrast, since the cementation of grains by hydrates produces a high velocity velocity (3.3 km / sec), deposit. Sediments below the hydrate-cemented zone if saturated with water will have low velocity and if gas is trapped in these se-

diments the velocity of the underlying layer will be still lower (1.5-1.7 km / sec). Because the strength of the reflected signal is proportional to the change in acoustic impedance, the base of hydrate-cemented zone produces strong reflections. The degree to which amplitudes are diminished depends on the amount of hydrates present. The phenomenon of amplitude reduction is termed as '*blanking*' –a characteristic feature for identifying gas-hydrates or BSR. It is to be mentioned here that sometimes the *transparency* is caused by homogenization of lithology. Three main manifestations (BSR, Blanking zone, Polarity reversal) have been used to recognize the presence of hydrates in the seismic section. The velocity build-up above the BSR can be used to quantify the amount of hydrates present above BSRs.

2 GEOPHYSICAL EVIDENCES

Gas Hydrate has been recognized in drilled cores but their presence over large areas can be detected more effectively by seismic reflection methods. Gas Hydrate are usually inferred on seismic profiles by Large amplitude bottom simulating reflector (BSR), which occur near the sea floor, Cutting across underlying dipping strata, The velocity inversion at the BSR caused by moving from high velocity hydrate cemented sediments to low velocity water or gas filled sediments below, The polarity reversal of the BSR with respect to sea floor, Blanking zone above the BSR due to addition of gas hydrates into pore fluids. The occurrence of a BSR in seismic reflection data is the most important indicator of hydrates in marine sediments. However, hydrates can exist without creating a BSR if there is no underlying free gas or if the hydrates do not appreciably stiffen the sediment matrix. The weak reflectivity observed above the BSR (blanking) =Acoustic blanking indicates the absence of any signal because of increased transmission and obliteration of sediment impedance structures owing to the general replacement of pore water by hydrate, therefore, the zone with acoustic blanking characteristics is also referred to as the hydrate stability zone which is defined as the sedimentary package which contains the gas hydrates. Some of the blanking is not obvious, because the acoustic blanking is related to the hydrate cementation in the sediments; the degree of blanking is proportional to the amount of hydrate in the pore space. The BSR is conspicuously underlain by transparent zones which are totally devoid of reflections and called as wipeouts.

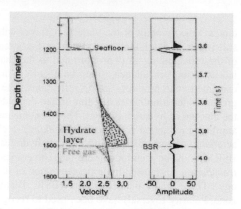

Figure 1. Polarity reversals at the BSR, lower velocities than seawater beneath the BSR perhaps indicating free gas. (Courtesy: http://gsc.nrcan.gc.ca/gashydrates).

In thin-bed interpretation of reflection seismic data, the term stratigraphic interpretation is synonymous with waveform analysis. The objective is to derive from the reflected wavelet as much relevant information as possible about geological formations with thicknesses that are below seismic resolution. For this purpose, the instantaneous amplitude is the least useful, since it is the envelope of the wavelet and masks all subtle waveform changes.

Nevertheless, it offers the advantage of being phase-independent, and is a useful tool for studying the total energy contained in a wavelet or in a wave packet. Because complex attributes separate amplitude information from phase information, any low-amplitude subtle waveform changes can be detected more clearly by instantaneous phase and instantaneous frequency than is evident in the raw data. The complex attributes are useful in that the instantaneous amplitude and the instantaneous frequency both show tuning effects, which indicate the halfway event represents a composite reflection from a thin bed. The instantaneous phase is also useful, although it is slightly less sensitive than the instantaneous frequency. The most unique property of instantaneous frequency is that it can be abnormally high compared to its Fourier frequency components of the wavelet, and it can also be negative. Some of these abnormal values occur around inflection phase anomalies and low-amplitude troughs. Thus, instantaneous frequency and instantaneous phase can be very useful tools to delineate subtle waveform character changes due to thin-bed wavelet interference. Instantaneous frequency is similar to instantaneous phase in that they both illustrate lateral continuity of waveform character and are independent of amplitude. Hence, they are also useful tools for delineating geological features such as pinch outs, angular unconformities, onlaps, faults, and channels etc. Since instantaneous frequency is the derivative of the instantaneous phase, so its section tends to appear slightly noisier than the corresponding phase section due to the high-frequency enhancement property of differentiation. In particular, frequency tuning effects should be investigated further as to how and when they occur. For complex attributes, research should be directed towards linking the qualitative observations to the geological changes.

The reflection amplitude increases as gas hydrate saturation increases. This positive amplitude change suggests phase reversals can occur in these sands if gas hydrate saturation changes laterally. The reflection amplitude is close to zero and between 20% and 40% gas hydrate saturation in these data, which would correspond with what has been described as a "blanking zone" in other gas hydrate systems For free-gas charged sands and hydrate-over-free-gas sands, the dominant reflection amplitudes decrease as gas hydrate or free gas saturation increase. Different concentrations of free gas and gas hydrate over free gas can produce the same reflection coefficient. For example, a given reflection amplitude of -0.35 could be caused by 20% free gas, 32% gas hydrate over 20% free gas, or 52% gas hydrate over 10% free gas (Figure 1). So BSR amplitude interpretation is typically ambiguous for a gas hydrate layer. Complex Trace Analysis Amplitude analysis with respect to variation of gas hydrate saturation is important.

Seismic properties of thin layers: The normal incidence properties of one layer and two layers are studied in terms of amplitude, frequency, and complex attributes of the composite wavelet. The offset-dependent properties of one layer are also studied. The amplitude results for one-layer models indicate that, as the thickness increases from zero to the (l/8) λ_d value, the amplitude changes quadratically. However, if the two reflection coefficients have equal magnitudes and opposite polarities, the amplitude increases linearly. At (1/4) λ_d thickness, all four models show tuning effect. The amplitude results for two-layered models show that the amplitude changes quadratically as the thickness of one of the two layers increases from zero to the (1/8) λ_d value, with tuning effect occurring at close to the (1/4) λ_d thickness. However, the model with alternating polarities for the three reflection coefficients exhibits a minimum at approximately the (1/16) λ_d thickness, and a maximum at close to the (l/4) λ_d thickness. These properties do not change appreciably as the thickness of one of the two layers increases within a range of five fold. In the frequency study, the results indicate that, as the thickness increases, the peak frequencies of the composite reflections decrease slowly. However, for the one-layered model whose reflection coefficients have unequal magnitude and opposite polarities and the two-layered model whose reflection coefficients have alternating polarities, the peak frequencies increase as the thickness increases from zero to the (1/16) λ_d . Value, and then decrease as the thickness increases further. The complex attributes study indicates that the instantaneous frequency is useful for studying wavelet interference. Amplitude tuning effect combined with frequency tuning effect appears to be a good indicator of the existence of thin layers. However, the use of complex attributes remains largely empirical and a pattern recognition tool. The results of the offset-dependent study show that tuning effect can change drastically the effect of lateral changes in Poisson's ratio in terms of amplitude, peak frequency, and complex attributes. To interpret AVO effect properly in thin-bed interpretation, the effect of offset-dependent tuning must be accounted for.

Tuning effect: A phenomenon of constructive or destructive interference of waves from closely spaced events or reflections. At a spacing of less than one-quarter of the wavelength, reflections undergo constructive interference and produce a single event of high amplitude. At spacing greater than that, the event begins to be resolvable as two separate events. The tuning thickness is the bed thickness at which two events become indistinguishable in time, and knowing this thickness is important to seismic interpreters who wish to study thin reservoirs. The tuning thickness can be expressed by the following formula: $Z = V_I / 2.8 f_{max}$, where Z = tuning thickness of a bed, equal to 1/4 of the wavelength, V_I = interval velocity of the target, f_{max} = maximum frequency in the seismic section. The equation assumes that the interfering wavelets are identical in frequency content and are zero-phase and is useful when planning a survey to determine the maximum frequency needed to resolve a given thickness. Spatial and temporal sampling requirements can then be established for the survey. Tuning thicknesses for both zero-phase and minimum-phase data are slightly less than the Rayleigh resolution limit. Event amplitudes can be better measured from minimum-phase data than from zero-phase data. Amplitude detuning is probably not required for minimum-phase data for bed thicknesses greater than about one-half of the Rayleigh resolution limit. Because event amplitudes in zero-phase data are significantly affected by tuning, amplitude interpretations based on zero-phase data should be calibrated or detuned for correct amplitude analysis.

Resolution: The ability to distinguish between separate points or objects, such as sedimentary sequences in a seismic section. High frequency and short wavelengths provide better vertical and lateral resolution. Seismic processing can greatly affect resolution: deconvolution can improve vertical resolution by producing a broad bandwidth with high frequencies and a relatively compressed wavelet. Migration can improve lateral resolution by reducing the size of the Fresnel zone. Nonlinear seismic imaging enables the end-user to retain the conventional linear seismic images and provides additional nonlinear seismic images that identify the porous and fractured reservoir rocks. In areas where the current seismic fails to map the stratigraphic or fractured hydrocarbon traps, nonlinear seismic technology can provide the useful reservoir information.

Nonlinear Seismic Imaging: In a nonlinear elastic system, the principle of superposition does not hold and the frequency mixing, harmonic generation, and spectral broadening takes place. These changes that add new frequencies to the frequency spectrum provide us with a means of measuring the elastic nonlinearity parameter of the reservoir rocks. This elastic nonlinearity parameter is unique, and can be effectively used as a seismic attribute to map the rock properties of the reservoirs for improving the results of the exploration and exploitation efforts. The sensitivity of the nonlinear response to the porosity, fracturing, and pore fluids of the reservoir rocks is relatively larger than the linear measurements being used today. Industry needs to take advantage of this additional seismic attribute to reduce the ambiguity of the seismic-based geologic interpretation. Nonlinear seismic imaging enables the end-user to retain the conventional linear seismic images and provides additional nonlinear seismic images that identify the porous and fractured reservoir rocks. In areas where the current seismic fails to map the stratigraphic or fractured hydrocarbon traps, nonlinear seismic technology can provide the useful reservoir information.

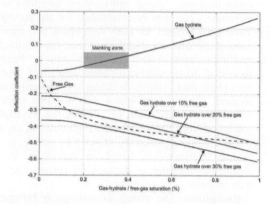

Figure 2. Normal incidence reflection coefficients versus gas hydrate and free gas saturation. The gas hydrate and free gas curves are in sands overlain by clay. The gas hydrate over free gas impedance curves are within a sand. (Courtesy: Zijian Zhang and Dan McConnell, AOA Geophysics).

Here, we focus more on interface responses at gas hydrate and gas contacts using rock physics models based on grain contact theories. To observe the effect of amplitude on different thicknesses of gas hydrate, two sand wedge models, (a) and (b), with various gas hydrate and free gas saturation levels (Figure 3). In model (a), high amplitude occurs with high hydrate saturation in hydrate-bearing sands with the maximum amplitude occurring at $\lambda/5$, where λ is the wavelength. The amplitude starts to decrease at $\lambda/5$ and significantly decreases as the layer thins below $\lambda/10$ (Figures 4a and 4b). Figures 4c and 4d show the results for a variably thick gas.

The strong amplitude can be seen at the interface between gas hydrate and free gas, a type of reflec-

tion that could form BSRs. layer of gas hydrate is significantly weaker than at the bottom of the gas hydrate layer (Figure 4c). The weak amplitude could easily be missed by the interpreter as the top of a thick layer of gas hydrate-saturated sand.

Hydrate zone with 30% gas beneath it for model (b).

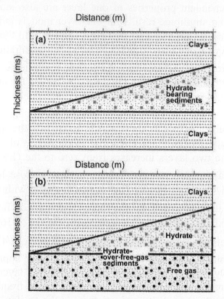

Figure 3. Wedge models for hydrate-bearing sediment (a) and hydrate-over-free-gas sediments (b): (a) three-layer model of a gas hydrate layer in clay; (b) gas hydrate layer with a clay cap and free gas bottom. The thickness of wedge varies from 0 to 30 ms. (Courtesy: Zijian Zhang and Dan McConnell, AOA Geophysics).

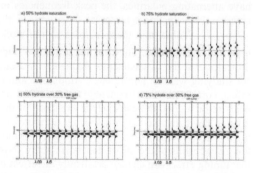

Figure 4. Synthetic seismic from the wedge models for hydrated sediments and hydrate-over-gas sediments. A Ricker wavelet of 50 Hz dominant frequency was chosen to generate zero-offset synthetic seismic data. The P-wave velocity of the clay is 1810 ms and the density is 2.06 g / cc. The properties of wet sand, hydrated sand and gas sand were derived from the well log data and the rock physics model. (Courtesy: Zijian Zhang and Dan McConnell, AOA Geophysics).

Spectral Decomposition. Conventional thickness analysis by picking horizons cannot be used if the peak-trough time separation is less than the tuning thickness (Partyka 1999). The spectral decomposition method, however, is a valuable tool with the ability to map thin beds (Partyka *et al.* 1999; Castagna *et al.* 2003). Partyka (1999) indicates a robust approach to seismic thickness estimation for thin beds showing that thickness can be derived from amplitudes at appropriately low discrete frequencies. The technique may be especially useful for identifying gas hydrate deposits and determining their thickness.

3 CONCLUSIONS

With a view to energy security of the world ,unconventional energy resources: Methane Gas Hydrate, coalbed methane , shale gas, tight gas, oil shale, Basin Centred Gas and heavy oil-exploration and exploitation is pertinent task before geoscientist. Wavelet analysis, known as a mathematical microscope, has scope to cope with nonstationary signal to delve deep into geophysical seismic signal processing and interpretation for hydrocarbon exploration and exploitation. Non-Stationary statistical Geophysical Seismic Signal Processing (GSSP) is of paramount importance for imaging underground geological structures and is being used all over the world to search for petroleum deposits and to probe the deeper portions of the earth. **SeisLab for Matlab** Software for the Analysis of Seismic and Well-Log Data.

REFERENCES

www.journalseek.net.
Fire in the ice http://www.netl.doe.gov/technologies/oil-gas/FutureSupply/MethaneHydrates/newsletter/newsletter.htm
Realizing the energy potential of methane hydrate for the united states, ISBN: 0-309-14890-1, http://www.nap.edu/catalog/12831.html
http://www.mms.gov/revaldiv/GasHydrateAssessment.htm
http://www.netl.doe.gov/technologies/oil-gas/publications/Hydrates/pdf/MethaneHydrate_2007Brochure.pdf; the JIP is managed by Chevron.
http://www.netl.doe.gov/technologies/oil-gas/FutureSupply/MethaneHydrates/projects/DOEProjects/Alaska-41332.html; this cooperative agreement is managed by BP Exploration Alaska,Inc. (BPXA).
www-odp.tamu.edu /publications/citations/cite164.html.
http://iodp.tamu.edu/scienceops/expeditions/exp311.html.
http://energy.usgs.gov/other/gashydrates/india.html.
Castagna, J.P., Sun, S. & Siegfried, R.W. 2003. *Instantaneous spectral analysis. Detection of low-frequency shadows associated with hydrocarbons.* The Leading Edge, 22: 120-127.
Lee, M.W., Collett, T.S. & Inks T.L. 2009. *Seismic-attribute analysis for gas-hydrate and free-gas prospects on the North Slope of Alaska.* Natural gas hydrates – Energy resource potential and associated geologic hazards: AAPG Memoir, 89: 541-554.
Lu, S. & McMechan, G.A., 2002. *Estimation of gas hydrate and free gas saturation, concentration, and distribution from seismic data.* Geophysics, 67: 582-593.
Mavko, G., Mukerji, T., & Dvorkin J. 1998. *The Rock Physics Handbook.* Cambridge, UK: Cambridge University Press.
Partyka, G.A., Gridley, J.A., & Lopez J.A. 1999. *Interpretational aspects of spectral decomposition in reservoir characterization.* The Leading Edge, 18: 353-360.

Technical and Geoinformational Systems in Mining – Pivnyak, Bondarenko & Kovalevs'ka (eds)
© 2011 Taylor & Francis Group, London, ISBN 978-0-415-68877-2

Justification of design parameters of compact load-haul dumper to mine narrow vein heavy pitching deposits

L. Shirin, Y. Korovyaka & L. Tokar
National Mining University, Dnipropetrovs'k, Ukraine

ABSTRACT: Consideration of loading unit performance within the system "rock mass-load-haul equipment-stowing mass" justification of its design parameters covers to mine narrow vein heavy pitching deposits. To intensify the machine performance it is proposed to use vibratory loader while entering ladle into rock mass, and while unloading. Under the same geometrical characteristics of ladle it will help to improve rate of load-haul machine work.

1 INTRODUCTION

From the viewpoint of efficient use of resources the paper uses systematic approach of justification of design parameters of compact load-haul dumper. Considering loading unit performance within "rock mass-loading and hauling equipment-stowing mass" approach justification of parameters of recommended loading equipment is run by means of the three stages.

In the context of method (Korovyaka 2003) stage one determines design dimensions of the loading equipment across the width of working excavation and throughout its height (minimum width of working excavation is $m_{min} = 1.2$ m, unevenness of wall of stope is $b_l = b_h = 0.2 \div 0.25$ m, shift of dynamic axis of working is $\alpha° = 15$ deg., and face movement is $a_1 = 2.0$ m) (Figure 1).

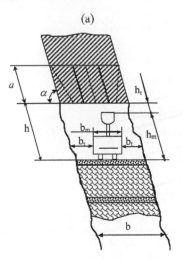
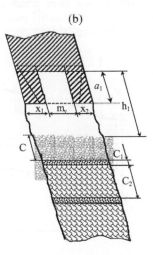

(a) (b)

Figure 1. Design model of face space parameters control: b – width of stope, m; a – depth of holes, m; a_1 – face movement, m; h – distance from stowing layer surface up to the top drilled, m; h_1 – height of free working space over shot rock, m; α – vein's dip, deg.; m_v – thickness of vein, m; $x = x_1 + x_2$ – width of enclosing roof and floor out-of-seam dilution, m; C – height of shot rock, m; C_1 – a lift of rapid-hardening stowing, m; C_2 – a lift of dry stowing, m; b_t and h_t – manufacturing clearances of width and height according to Safety Rules, m; b_m – minimum constructive width of loading equipment, m; h_m – maximum of loading equipment ladle raise, m.

While ore hauling by means of load-haul dumper making maximum allowable thickness of stowing mass is important process parameter of goaf stowing. It depends on the fact that under raise bench working procedures concerning drilling and charging as well as shot ore loading are performed from consolidating stowing. According to DBO certificate, standard height of face space should be permanent between top and filling mass surface. Under the height, manufacturing clearance h_t being minimum allowable on Safety Rules distance between top and vertical position of loader shovel is kept. Manufacturing clearance h_t is required for free ranging movement of equipment with vertical position of shovel within block (Figure 1a). Manufacturing clearance b_t across the width of face space is the parameter which determines loading equipment adaptability while hauling ore within block.

Taking into account real hypsometry of face space walls and shift of dynamic axis of working obtained as a result of simulation, stage two determines adaptive capacity of loading equipment. That is, minimum length of the equipment is determined for minimum width defined during stage one.

Semitheoretical problem of justification of design parameters of compact load-haul dumper can be solved by means of simulation of its design adaptive capacity within narrow stope which stochastically changes its direction and spatial outline as a result of drilling and blasting, and requirement to follow ore body.

There was performed structural analysis of configuration of side wall and bottom wall of working excavation formed under controlled drilling. Then approximating surfaces of stope were built up, and they had m mining height, $A = 0.2 \div 0.25$ m amplitude of benches and basins, curvatures r_l of side wall and r_h of bottom wall reflecting shift of trend azimuth and vein's dip $\Delta\beta = 15°$ within 5 m segment.

Solving the problem of determining load-haul dumper parameters for its operability assurance in stoping faces which can vary greatly per unit of working's length has been realized in terms of *Poisk bℓ* program (Shirin & Korovyaka 1998).

To describe algorithm and the program of the equipment parameters determination state description of the problem solved with a computer. It is required to find extreme length l of rectangle with fixed width b inscribed in zone D_i limited from left and right by vertical lines $X = x_0$ and $X = x_N$. From above and from below they are limited by certain curves $f_u(X)$ and $f_e(X)$ accordingly (Figure 2):

$$D = \{(X,Y), X \geq X_0; X \leq X_N; Y \leq f_u(X); Y \geq f_e(X)\}.$$

The minimum value is chosen among all l_X values obtained within $[x_0, x_N]$ area of search:

$$\ell = min_{X \in [X_0, X_N]}\{\ell_X\}.$$

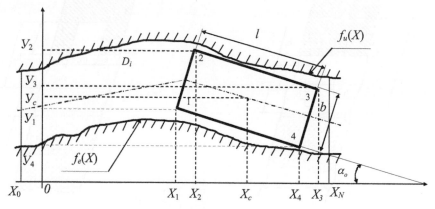

Figure 2. Adaptation of load-haul dumper to shifts of dynamical axis and hypsometry of stope walls: $[x_0, x_N]$ is area of loading equipment optimum length search; X_c, Y_c are coordinates of center of rectangle describing load-haul dumper; (x_1, y_1), (x_2, y_2), (x_3, y_3), (x_4, y_4) are coordinates of rectangle's vertexes; $f_u(X)$ and $f_e(X)$ are certain curves limiting width of stope.

The value is taken as a length of rectangle (load-haul dumper under study).

Stage three defines constructive parameters of compact load-haul dumper working element under its minimum structural dimensions (length is $l = 3.0$ m, width is $b = 0.8$ m, height with raised ladle is $h = 1.8$ m).

2 PROBLEM DEFINITION

For compact batch overhead loaders working within narrow stope with variable dynamical axes loading equipment duty cycling is one of the key characteristics of rock mass loading process. Loading equipment (having reduced clearance) with downdip ladle moving from draw hole to rock mass pile at the expense of motional energy being proportional to equipment weight and its velocity squared as well as travel mechanism moving force introduces ladle at a depth of L' into a pile. At the expense of the lift after ladle has been introduced it turns vertically before leaving pile, and dips some rock. Then lifting drive is broken, ladle stops, and equipment moves to draw hole where lift is connected again. With it ladle rises up to maximum upward position, unloads at the expense of handle on buffer, falls, and loading equipment starts its new cycle.

Theoretically, duration of working cycle T_c of batch ladleman loading equipment as it is shown in (Bartashevski, Strashko, Shirin & Shumrikov, 2001) is additive quantity covering timing to perform a number of serial operations:

$$T_c = \sum_{i=1}^{n} t_i, \text{ where } i = 1,2,3,... \qquad (1)$$

where t_0 – time to shift a handle (as a rule, 1 to 2 seconds); t_1 – time for equipment to move from a draw hole to a pile; t_2 – time to bring ladle into rock mass pile; t_3 – time for ladle to draw rock mass; t_4 – time for equipment to leave with loaded ladle; t_5 – time for ladle lift to be unloaded; t_6 – time for ladle to be unloaded; t_7 – time to lower ladle into origin return.

As cycle time is a quantity inversely proportional to output then its cutting is connected with increase in theoretical output which reflects loading equipment feasibility. Besides, as (Yevnevich 1975) informs such parameters or factors effect cycle duration: power and mechanical characteristic of drives, reduction ratio of carrier and lift of ladle, equipment weight, shape and dimensions of ladle, physical and mechanical properties of rock mass, and distance from pile to draw hole.

3 ANALYSIS OF THE PROBLEM STATE

Stage one considers factors effecting initial charge of ladle under separate ladling from rock mass pile. Figure 3 shows kinematic chain of ladling rock mass which density is ρ_0. The Figure demonstrates possible motion trajectories of front ladle edge within the pile form. As it is seen, types of the trajectories depends on ladling method, value R_0, rotational center "0" position, and ratio of lift speed and pressure of ladle introduction into rock mass (Poluyanski 1981).

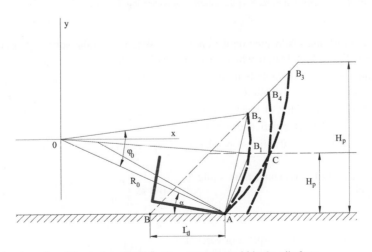

Figure 3. Specified trajectories of front edge of ladle bottom movement within the pile form.

Symbolize introduction depth as L'_d, height of a pile as H_p, and square determining initial loading a ladle as F'. Trajectory of ladle front edge under sequential operation of introduction and loading (when ladle winds) is AB_1 and AB_2 curves. With it (as Figure 4 shows) trajectory may cross form of pile either on horizontal face at B_1 point or on its aslope at B_2 point depending on height of a pile H_p. Theoretically, it can be shown in terms of the inequations:

$H_p \leq R_0\left(sin\alpha + sin(\varphi_0 - \alpha)\right)$ for the first variation, and $H_p > R_0\left(sin\alpha + sin(\varphi_0 - \alpha)\right)$ for the second variation. (2)

If it is referred to combined ladling method when lift of ladle introduced into some depth keeps pace with continuous introduction then trajectory of ladle front edge will go with AB_3 and AB_4 curves.

Thus, front edge of ladle bottom under its vertical winding separates some volume of rock mass from pile. The volume of rock mass is proportional to involved square F'. When ladle leaves pile the volume of rock mass (separated from the pile), and values of the pile height H_p and introduction depth L'_d are different, the square of polygon $BCDEK$ is separated (Figure 4) and ladle is partially loaded. Theoretically, the figure square is close to square F' value. With it, vertex $D(D')$ lies on perpendicular erected to ladle bottom plane form point B (that is front edge).

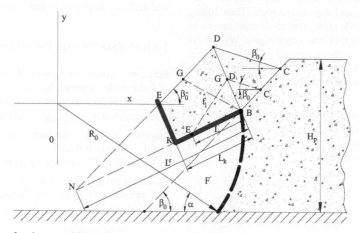

Figure 4. Location of rock mass within ladle at the moment of its leaving the pile.

Inside the ladle rock mass is located on the angle of natural levee β_0 with l' length which in turn may be both more and less than ladle bottom length l_k depending upon introduction depth L'_d.

Rock mass facing a pile with CD slope has slope angle to β'_0 level. Most of all, practical calculations takes the angle as that equal to friction slope, that is $\beta_0 = \beta'_0$.

With it, volume of rock mass taken from a pile in one working cycle is:

$V_\kappa = f_1 \cdot B_1$, (3)

where B_1 – ladle width, m; f_1 – square in vertical plane limited by $BGEK$ contour.

While defining square f_1 consider that height is $H_p > R_0\left(sin\alpha + sin(\varphi_0 - \alpha)\right)$, and introduction depth L_d is equal to ladle bottom length that is $L'_d = l_k$. For the case, (Figure 5) square of $BGEK$ figure is:

$f_1 = \frac{1}{2}\left[l_k(h + h_2) - h_1 \cdot l_0\right]$, (4)

But as $\begin{cases} h_2 = h - l_\kappa \cdot tg(\beta_0 - \tau) \\ h_1 = h \cdot cos(\beta_0 - \tau) \end{cases}$, then

$l_0 = |DG| = |BC|$ will be equal to: (5)

$l_0 = \frac{cos(\beta_0 + \tau)}{sin 2\beta_0}$, where $\tau = \varphi_0 - \alpha_0$ (6)

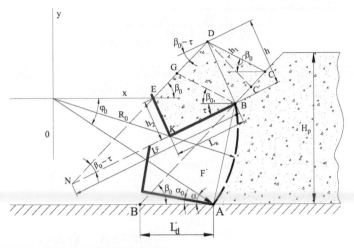

Figure 5. Design diagram for loading ladle with rock mass.

Formula (4) will be:

$$f_1 = \frac{1}{2}\left[2 \cdot h \cdot l_k - \frac{h^2 \cdot cos(\beta_0 + \tau) \cdot cos(\beta_0 - \tau)}{sin\, 2\beta_0} - l_\kappa^2 \cdot tg(\beta_0 - \tau)\right]. \qquad (7)$$

In the formula h is undetermined value. It characterizes location of rock mass at the moment of ladle leaving a pile. While determining it (to simplify the problem) specify approximate equation of squares F' and F'' ($BCDEK$ figure). According to Figure 5:

$$F' = \frac{1}{2}\left[L_d' \frac{1}{ctg\beta_0 - tg\left(\alpha - \dfrac{\varphi_0}{2}\right)} + \right.$$

$$\left. + R_0^2 \cdot \left(\frac{\pi}{180^0}\varphi_0 - sin\,\varphi_0\right)\right]. \qquad (8)$$

Taking into account

$$R_0 = \frac{L_d' \cdot sin\beta_0}{2\,sin\dfrac{\varphi_0}{2} \cdot cos\left(\beta_0 + \alpha - \dfrac{\varphi_0}{2}\right)}, \qquad (9)$$

Expression (7) will be:

$$F' = \frac{1}{2}L_d'^2\left\{ \frac{1}{ctg\beta_0 - tg\left(\alpha - \dfrac{\varphi_0}{2}\right)} + \right.$$

$$\left. + \frac{\left(\dfrac{\pi}{180^0}\varphi_0 - sin\,\varphi_0\right)sin^2\,\beta_0}{\left[2\,sin\dfrac{\varphi_0}{2} \cdot cos\left(\beta_0 + \alpha - \dfrac{\varphi_0}{2}\right)\right]^2}\right\}. \qquad (10)$$

In turn, from Figure 5 we get:

$$F'' = \frac{1}{2}\left[l_k(h + h_2) + h_1 \cdot |BC|\right]. \qquad (11)$$

Substituting into (9) values h_1, h_2 and $|BC|$ from Formula (6) we get:

$$F'' = \frac{1}{2}\left[2 \cdot h \cdot l_k + \frac{h^2 \cdot cos(\beta_0 + \tau) \cdot cos(\beta_0 - \tau)}{sin\, 2\beta_0} - l_k^2 \cdot tg(\beta_0 - \tau)\right]. \qquad (12)$$

Equating $F' \approx F''$ if $l_k = L_d'$ determine depth h:

$$h = d_1 \cdot L_d', \qquad (13)$$

where

$$d_1 = \sqrt{\frac{sin\,2\beta_0}{cos(\beta_0+\tau)\cdot cos(\beta_0-\tau)}\left\{\cfrac{\cfrac{sin\,2\beta_0}{cos(\beta_0+\tau)\cdot cos(\beta_0-\tau)}+\cfrac{1}{ctg\,\beta_0-tg\left(\alpha-\frac{\varphi_0}{2}\right)}+}{\cfrac{\left(\frac{\pi}{180^0}\varphi_0-sin\,\varphi_0\right)sin^2\beta_0}{\left[2\,sin\frac{\varphi_0}{2}\cdot cos\left(\beta_0+\alpha-\frac{\varphi_0}{2}\right)\right]^2}-tg(\beta_0-\tau)}\right\}} - \frac{sin\,2\beta_0}{cos(\beta_0+\tau)\cdot cos(\beta_0-\tau)}.$$

Finally, substituting value h into Formula (7) and result obtained into (3) find ladle capacity of recommended compact loading equipment:

$$V_{\kappa} = k_1 \cdot B_{1,} \cdot L_d^2, \qquad (14)$$

where

$$k_1 = d_1 - 0.5 \cdot d_1 \frac{cos(\beta_0+\tau)\cdot cos(\beta_0-\tau)}{sin\,2\beta_0} -$$

$$-0.5 \cdot tg(\beta_0-\tau).$$

It results from Formula (14) that under otherwise conditions maximum value of ladle volume is $f\left(k_1 B_1 L_d^2\right)$ function, and $V_{\kappa\,max}$ if $\beta_0 = \frac{\pi}{2}$. On the other hand, value of ladle loading to the full extent depends on frequency of its introduction into rock mass (Semko 1960).

To determine relationship between ladle loading volume and current time apply directly-proportional dependence of loading time on rock mass being loaded under limiting factors. As a rule, maximum loading volume (Yevnevich 1975) is among them:

$$\frac{dV}{dt} = k \cdot V(b-V), \qquad (15)$$

where $\frac{dV}{dt}$ is time of ladle loading, m^3 / s; k is ladle loading factor; V is current value of ladle loading under one-time introduction, m^3; b is maximum ladle loading.

Reduce differential expression (15) to nondimensional values: $V = \frac{V_{min}}{V_{max}}$; $b = 1$ if $t = t_i$.

Then symbolize (15) as:

$$\int \frac{dV}{V(b-V)} = \int k \cdot dt, \qquad (16)$$

Separate integral in right member into the two addends, and integrate:

$$\frac{1}{b}\left[\int \frac{dV}{V} + \int \frac{dV}{b-V}\right] = \int k\,dt. \qquad (17)$$

Obtain:

$$\frac{1}{b}\left[ln\,V - ln(b-V)\right] = kt - C_1. \qquad (18)$$

With it, constant of integration C_1 is determined with the help of initial conditions $C_1 = -\frac{1}{b}ln\,a$, where a is starting volume of ladle loading.

(18) helps to get final output:

$$V_{(t)} = \frac{b}{1+a\cdot e^{-bkt}}. \qquad (19)$$

(19) shows that current value of ladle loading depends on values "a" and "k". Specify $\tilde{a} = k_1$, then rewrite (19) as:

$$V_{(t)} = \frac{1}{1+k_1\cdot e^{-kt}}. \qquad (20)$$

Values k_1 and k depend on different factors – physical and mechanical properties of rock, its hardness, lumpiness etc. as well on method of ladle introduction into rock mass.

4 RESULTS

To increase efficiency of ladle equipment a number of papers recommend different ways of loading process intensification (Yevnevich 1975 & Poluyanski 1981). For example, Institute of Geotechnical Mechanics of the National Academy of Sciences of Ukraine developed technique of ladle loading intensification with the use of vibration exciter as vibration loader. Hydraulic pulsators or pneumo pulsators with $10 \div 15$ Hz rhythm and $A = 3 \div 5$ mm of ladle loading edge oscillation amplitude. With it, factors k_1 and k increase up to $0.6 \div 0.8$.

Figure 6 shows dependence diagrams of ladle loading on its introduction into rock mass without applying vibration on ladle and with it. Graphical dependences are obtained with the help of *Mathcad* program.

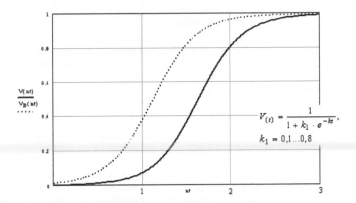

$$V_{(t)} = \frac{1}{1 + k_1 \cdot e^{-kt}},$$

$$k_1 = 0,1 \ldots 0,8$$

Figure 6. Dependence of ladle loading on its introduction into rock mass frequency: —— conventional ladle; ----- vibrating ladle; $nt = \dfrac{t_u}{t}$, where t_u – average time of cutting cycle; and t – current time of single cutting.

From the dependences it follows that under conventional loading the same volume needs more cuttings to compare with vibrating ladle. Thus, application of vibrating exciter both during introduction into rock mass and during unloading will help to increase efficiency of load-haul-dumper even if geometrical characteristics of ladle stay to be identical.

Hence, according to design model (Figure 1) volume of rock mass to be loaded (if length of semiblock is 25 m, working face movement is 1.8 m, degree of fragmentation is $k_p = 1.5$, and mining power is 1.2 m) will be 81 m³. Then time to haul gangue to draw hole under continuous operation of compact load-haul-dumper with fixed ladle is about 10 hours, and with vibrating ladle it is about 7 hours.

While determining dimensions of ladle on rolling handle following ratios are taken:

$$l_\kappa = 1.14 \sqrt[3]{V_\kappa}, B_\kappa = l_\kappa, h_b \approx 0.4 \cdot l_\kappa, H \approx 1.2 \cdot l_\kappa,$$

where l_κ – ladle bottom length; B_κ – ladle width; H – ladle height from the front; and h_b – ladle bottom height.

There are identified following design factors for compact load-haul-dumper ladle: $V_\kappa = 0.3$ m³, $l_\kappa = B_\kappa = 0.76$ m, $H = 0.91$ m, and $h_b = 0.3$ m.

Since cutting is performed by means of pressing loading equipment into rock mass pile then adhesion weight is determined as:

$$G_c = n \frac{P}{\Psi - z(W_m + W_\kappa - W_d)}, \text{H}$$

where n – reserve coefficient equal to $1.1 \div 1.15$; P_{BH} – rated force of a ladle introduction into a pile, H; Ψ – a coefficient of wheels adhesion with stowing mass; z – the relation between working weight of equipment and its adhesion weight; W_m – running resistance of equipment; W_κ – resistance of equipment on curves equal to $(0.25 \div 0.3) \cdot W_m$; $W_d = 0.7 \cdot v^2 / L_d$ – dynamic resistance; and v – stroke speed of equipment, m / s.

Usually, pressure of ladle into pile equal to rock mass reaction when lumpiness is no more than 400 mm is determined as:

$$P = 341 \cdot a \cdot L_d^{1.25} \cdot B_k \cdot k_h \cdot k_f, \text{ кH}$$

where a – a factor taking into account tightness and abrasive properties of rocks and mineral (at average, it is $0.17 \div 0.2$ for iron ore; 0.15 for sandstone and granite, and 0.12 for sandy shale); $k_h = (1.16 \div 1.57) \cdot (2 + lg\, H_p)$ – a factor taking into account influence of pile's height; and k_f – a ladle form factor (roughly, it is taken as that equal to $1.2 \div 2.0$).

91

5 CONCLUSIONS

From the viewpoint of mining mechanization the area is rather promising while mining ore bodies which thickness is >1.5 m. To cut qualitative losses of minerals while mining seams (to dilute them as a result of mixing with dead rocks) it is necessary to be geared to ore body selective mining and diluted rock walls. It depends on the fact that ore dilution and losses, and cripples economy greatly not only while mining but in the process of processing. Excessive mining dead rocks (connected with dilution) then separating in tails harms environment too as vast territories are required to place tailing pounds which negative influence is known.

The research are implemented in method of determining rational parameters of operation schedule to mine narrow vein heavy pitching deposits by means of compact load-haul dumpers, and in "Initial Standards" to design compact load-haul dumpers for mining narrow heavy pitching veins. The Standards are agreed with Institute of Geotechnical Mechanics of the National Academy of Sciences of Ukraine, approved and passed to the State Design Institute "Krivbassproject" to be applied.

REFERENCES

Korovyaka, E.A. 2003. *Control of Parameters of Face Space and Stowing Mass within Stopes of Narrow Heavy Pitching Veins.* Mining of Ore Deposits. Scientific and Technological Collection. Krivoy Rog, 82: 49-55.

Shirin, L.N., Korovyaka, E.A. & Shirin, A.L. 1998. *Model Analysis of Adaptive Capacity of Load-haul Dumpers to Mine Narrow Vein Deposits.* Inter-agency Collection of Scientific Papers of IGTM of the NAS of Ukraine, 6: 67-73.

Bartashevski, S.E., Strashko, V.A., Shirin, L.N. & Shumrikov, V.V. 2001. *Mathematical Models of Work-cycle Time of Ladleman Loading Machine.* Vibration in Methods and Technology, 3(19): 46-49.

Yevnevich, A.V. 1975. *Transport Machines and Complexes.* Moscow: Nedra: 415.

Poluyanski, S.A., Savitski, Yu.P., Strashko, V.A. & Voloshanyuk, S.N. 1981. *Calculations of Basic Technical Parameters and Efficiency of Mine Ladleman Loading Equipment.* – Kyiv.: Naukova Dumka: 76.

Semko, B.P. 1960. *Concerning Influence of Ladleman Loading Equipment on the Process of Bringing into Rock Pile.* Collection "Topics of Mine Transport": Gosgortechizdat, 4: 390-407.

Technical and Geoinformational Systems in Mining – Pivnyak, Bondarenko & Kovalevs'ka (eds)
© *2011 Taylor & Francis Group, London, ISBN 978-0-415-68877-2*

The top caving system with roof fall for excavation of thick coal seams

M. Lubryka
"Jas-Mos" Coal Mine

J. Lubryka
Elgor+ Hansen Company, Kopex

ABSTRACT: The top caving system with roof fall is designed to excavation of thick seams. The solution assumes that the extracted coal from the top layer is recovered by the lifted up and pivoted tail canopy onto the retractable rear Armoured Face conveyor (AFC). The rear AFC AFC is installed behind bases of powered roof support units and is pulled on the floor. Hydraulic cylinders and chains are used as pulling members to hide the AFC under the tail canopy when waste rock is encountered. The "JAS-MOS" coal mine intends to apply that extraction method to the 510/1-2 seam. The seam deposited at the depth of 750 to 900 m and its thickness ranges from 9.6 to 13.6m. The top caving method should be supported by a longwall shearer, a system of powered roof support units with electrohydraulic control as well as a power supply and control system for the set of the longwall mining machinery with very high degree of reliability. Such a set of equipment cannot work without an advanced visualization system with the possibility of parameterization.

1 INTRODUCTION

Jastrzębie Coal Company SA (JSW plc.) has a number of seams, where many parts of them are of sizeable thickness. It is why the Managing Board of the company made the decision that application of the highly efficient top caving system to extraction of thick seams is a matter of crucial importance for further development of the company.

"Jas-Mos" coal mine intends to extract coal from the seam 510/1-2 in is part W2. The seam is deposited at the depth ranging from 750 to 900 m with its thickness from 10 to 13 m. Aiming to extract the seam with the optimum efficiency and to observe the stringent safety rules, the mine decided to consider top caving as a very promising option of mining operations.

"Jas-Mos" coal mine has developed the plan to extract that part of the seam with application of the longwall system with roof fall and with subdivision of the entire seam into two layers with the thickness of 3.5 m each. Therefore the seam with the thickness of 7 m is to be extracted from the overall bed thickness that varies from 10 to 13 m, as it was mentioned before.

Having considered all the circumstances, possible obstacles and difficulties with subdivision of the extraction process into layers as well as the wish to achieve higher efficiency of the seam excavation with simultaneous observance of all the standards and conditions intended to guarantee safety of surface facilities, the idea was put forward to apply the system of top caving under the areas that are chiefly used for agricultural purposes. It is the system that has not yet been used in the Polish mining industry and consists in longwall extraction with roof fall and advanced progress of the roof-adjacent layer in ahead.

The experience of both Russian and Chinese mining industry demonstrates that application of that system makes it possible to extract about 75% of coal deposited within the borders of longwall parcels with simultaneous minimization of indispensable roadway drivage jobs. Therefore the assumption could be made that the seam height to be extracted within borders of longwalls where the top cavity system is applied shall amount to 8-9 m.

2 CHARACTERISTIC PARAMETERS OF THE SEAM

Application of the top cavity system with roof fall is anticipated in the part of the W2 panel encircled from the north by the line of eastern drifts and by the fault area ~ 11.0 m from the south, by the ventilation roadway from the west and the border of the mining area from the east. The identified industrial resources of coal within the mentioned region amount to ca. 9.7 m. tons where the extractable resources are estimated to 3.1 m. tons.

The considered part of the bed is made up of merely a single coal layer – the seam 510/1-2 with the thickness of 9.6-13.6 m. The coal is classified to the grades 35.2A, 35.2B and 37.1. Coal in that seam is brittle by its nature as its cohesion factor is estimated to $f \leq 0.5$.

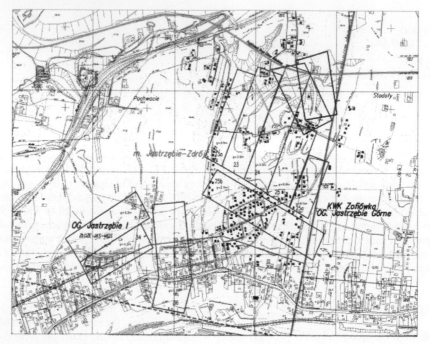

Figure 1. Part W2 of the seam 510/1-2.

The considered part of the bed represents the eastern, less steep wing of the geological anticline of Jastrzębie, with the central line sinking to the north-east direction and unduled. Inclination of coal strata in that wing averages to 16°/E.

The area features with many factors of both continuous tectonic effects (a tectonic fault and flexure) as well as non-continuous ones (normal and upthrown faults). The tectonic properties as well as formation and thickness of the seam are some of the essential factors that determine the seam development and extraction methods.

Major directions of the coal seam cleavage are nearly parallel to the central line of the anticline. The seam is directly covered by the layer of clay shale or mudstone, where the clay shale comprises local intrusions of sand. This roof-adjacent layer has the average thickness of 9.9 m and is loose and easily separable from the upper sandstone layer with the thickness of ca. 20 m. Upwards, within the neighbourhood of clay shale, the 508 seam can be found that approaches the 510/1-2 seam along the diagonal direction, where the both seam converge on the mining area of Jastrzębie Górne. In pace with

mutual approaching of the both seams the sandstone layer represents still thinner and thinner wedge and completely disappears. Thickness of the 508 seam ranges from 0.0 m to 2.3 m. Upwards, above the 508 seam, a thick layer of sandy and gravelled deposits can be found (sandstone and pudding stone alternately) with the thickness of 75 to 80 m and with local intrusion of clay shale formations.

The floor of the considered seam is made up of mudstone (arenaceous shale).

Compression strength of the individual rock grades amounts to:

– sandstone – $R_c = 62 - 74$ MPa;

– clay shale (claycy stale) – $R_{c\,av.} = 21.4$ MPa;

– arenaceous shale (mudstone) –

$R_{c\,av.} = 33.7 - 39$ MPa,

– coal – $R_c = 4.4 - 11.5$ MPa.

The slakeability factor is 0.8 for clay shale and 1.0 for arenaceous shale and sandstone.

The decline angle of strata within the considered region ranges from 5° to 25°/E and NE but locally the decline angle is as high as 30°/E.

Coal extraction used to be carried out from the

seam 505/1 that is deposited about 120 to 140 m above the considered seam 510/1-2. Nowadays the seam 505/2 is being extracted, about 110 to 130 m upwards.

The coal panel to be extracted may suffer from slips with the drops up to several meters and various directions of running. Rock in the regions of slips is brittle and cracked, conducive to falls or slides.

Hazards:

1st degree of water hazard.

4th category of methane hazard.

Class B of coal dust explosion hazard.

The seam in its W2 part is classified to the 1st degree of hazards due to rock bumps.

The seam is not rated to the class of rock that is conducive or hazardous in terms of methane or rock outbursts.

The southern part of that seam segment is built up with detached houses whilst the surface over central and northern areas is chiefly used for agricultural purposes.

For selection of the mining system that shall be applied to extraction of the seams 510/łłg, 510/1 and 510/1-2 in the part W-2 the following circumstances had to be considered:

– the need to conserve the land surface within the borders of the coal extraction area, in particular in the densely built-up southern part of the region;

– sophisticated geological conditions within the part of the seam that is to be extracted (substantial brittleness of coal, occurrence of considerable cleavage, presence of a loose shale layer with the thickness of several meters in the overlaying stratum of the seam), which present substantial obstacles to roadway driving operations;

– the need to achieve cost effectiveness of the mining operations with maximum possible extraction of the seam.

3 EXTRACTION SYSTEM

The roof fall system with advanced extraction of the roof-adjacent layer is dedicated to so called 'thick' seams. The solution assumes that the coal extracted from the top cavings is slid down on the partly lifted and pivoted tail canopy onto the rear Armoured Face Conveyor (AFC). The rear AFC is installed behind bases of powered roof support units and is pulled on the floor. Hydraulic cylinders and chains are used as pulling members to hide the AFC under the tail canopy when waste rock is excavated. When the powered roof support units are moved forward the rear AFC is left partly behind the tail canopy to enable more efficient loading of the extracted winning to the AFC. The hydraulically extended part of the tail canopy is provided with picks to disintegrate excessively large lumps of winning.

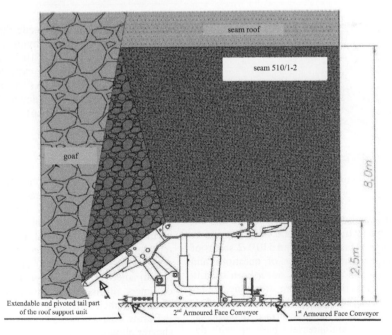

Figure 2. Essential components of the longwall appliances.

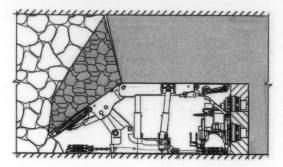

Figure 3. Use of a shearer to extraction of the floor layer of coal.

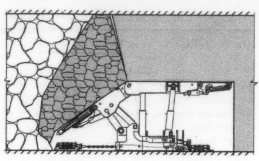

Figure 4. Advancing movement of the powered roof support unit with leaving the rear AFC immobile.

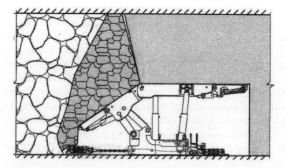

Figure 5. The first step associated with extraction and loading of the roof coal layer by controlling the extendable tail roof canopy.

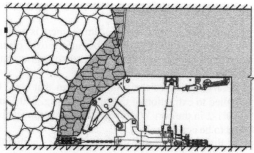

Figure 6. The second step associated with extraction and loading of the roof coal layer by controlling the pivoted part of the tail roof canopy.

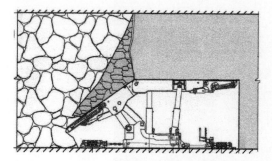

Figure 7. Final stage of extraction and loading of the roof coal layer.

Anticipated application of the innovative and experimental coal extraction method of top caving to excavate seams 510/łłg, 510/1 and 510/1-2 seams of the W-2 part entails a number of questions and wonderings. As the method has never been used in the Polish mining industry there are many doubts with regard to safety of the extraction method as well as to its effects to the land surface, in particular to these parts that are intensely built up with detached houses. Due to the aforementioned doubts

the "Jas-Mos" coal mine made the decision to prepare an option of the top caving extraction method, where the seams 510/łłg, 510/1 and 510/1-2 seams of the W-2 part would be extracted with use of a single longwall that would serve as an field test to enable clarification of all doubts before major extraction of the entire W-2 part.

4 SHEARER

Current trends towards optimization of the entire process related to coal extraction and eventual improvement of its efficiency impose the need to use more productive and reliable equipment that make up the entire set of longwall machinery. As shearers represent the basic machines of longwall sets, their productivity and reliability are crucial for overall performance of mining operations. However, the designers of modern longwall shearers put main stress onto their reliability (Kostka).

Beside the advanced diagnostic system the shearer should be equipped with a positioning system to enable location of the machine within the longwall as well as the functionality of the "pattern cut" with possibility of its automatic reproduction. Shearers should be designed is such a way that longwall extraction of coal should be automated as much as possible. In addition, the machine should be provided with a system of data transmission to the supervising system located in the main gate and on the coal mine surface.

The main stress is put to the reliability factor, which is achieved by application of equipment and components with the high MTBF (Mean Time Between Failures) factor.

The shearer operator should be enabled to select the source of control signals between the remote RF control and local wired control with use of dedicated devices. Selection of the RF control not only should enable remote control the shearer but also remote view of the monitored parameters and messages about emergency conditions. Such a solution absolves the operator from the need to walk each time to the central monitor, which is very important when overall length of shearers exceeds 14 m.

Worldwide manufacturers of shearers, such as ZZM from Zabrze, Poland, Joy or Eickoff offer the machines that demonstrate all the foregoing qualities.

5 CONTROL SYSTEM OF EQUIPMENT INCLUDED INTO THE LONGWALL EXTRACTION SET

Automatic control and mechanization are the most important factors of highly efficient coal extraction technology with use of longwall machinery and application of the top caving system with roof fall, where electrohydraulic control of powered roof support units is also a crucial component (Adamus-

iński 2010). The set of longwall machinery should be controlled from the operator station located in the main/tail gate. Operators of the control station are provided with visualization appliances to display current information about operational parameters of the longwall equipment and to adjust these parameters in the real time mode.

Innovative solutions for MV electric equipment of explosion-proof design have been developed by Elgor-Hansen to supply highly productive sets of extraction equipment. These solutions perfectly proved themselves for application in underground mining operations and made it possible to gain necessary experience. Simultaneous tracking of the most recent research and development solutions enabled the company in a short time to offer the equipment that not only guarantee the high degree of safety but also demonstrate the desired operational qualities. The systems form Elgor+Hansen are constantly improved and customized according to individual needs and requirements of customers.

The control system for the machinery of a highly productive extraction system with use of top caving and roof fall method is a really sophisticated and very complex solution intended to carry out a great number of functions, such as control, visualization, keeping records and diagnostics. Deployment of the controlled equipment in quite a large space suggests that the control system should be of the field type with use of industrial data transmission networks for local communication. The information shall be processed with use of the multi-level hierarchical structure, starting from the control level of individual components via parameterization of the equipment, diagnostic and record keeping functions, visualization and data storage up to the top level where the entire technological process is controlled.

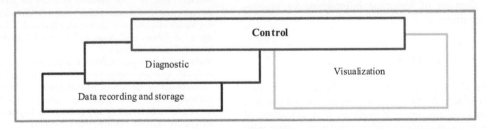

Figure 8. The tasks to be carried out by the control system for the set of extraction equipment with application of the top caving method with roof fall.

The control system for powered roof support units must be furnished with additional functionalities associated with application of the top caving system with roof fall. Such a system must be based on control modules installed on every unit of the longwall set. By appropriate configuration of the modules

they should adhere to the operation mode selected for the entire system. Each control module receives values of measured parameters from components of the system. Therefore the operators, depending on the required configuration, should be capable to initiate individual, separate functions or to run the system automatically.

In terms of operational applicability, the visualization system for the electric equipment of the longwall set should be structured into:

– local visualization;
– central visualization.

The local visualization shall cover only a single device, such as a compact power centre, a transformer unit or a single powered roof support unit. The visually presented information shall be restricted to a certain device, appropriately structured and organized to enable easy local diagnostics for the operator. When a need to adjust parameters of the device appears, the local visualization system should also make it possible. Additionally, on some cases the system of local visualization should enable local storage of operational information.

On contrary, the central visualization system shall cover all equipment incorporated into the set of extraction machinery. It will perform a large number of functions, including:

– visualization of the control system status;
– visualization of the shearer operation;
– visualization of the powered roof support system with regard to indications of the installed sensors and gauges;
– automatic advance of individual powered roof support units depending on indications of the installed sensors and gauges;
– adjustment of parameters that determine advancing operation of the longwall set;
– visualization of the equipment included into the train of electric appliances;
– visualization of the equipment included into the control system;
– parameterization of the control system;
– central data storage;
– viewing of historical information for maintenance purposes.

6 CONCLUSIONS

Extensive and detailed analyzes dedicated to extraction efficiency of 'thick' seams are carried out in a number of countries. Experience from extraction of 'thick' seams with use of the top caving system with roof fall acquired in Russia and China serve as a proof that the proposed technology is a cost-effective and technically justified method for efficient extraction of seam with the thickness ranging from 10 to 13 m. Additionally, the technology is supported by substantial economic benefits as it makes possible to extract 75% of resources deposited within boundaries of longwall panels with minimum amount of roadway drivage efforts. All in all, the technology offers the optimum extraction of seams with maintaining the most stringent safety requirements. The anticipated objectives can be achieved with use of highly efficient and reliable systems developed to control sets of longwall extraction machinery.

REFERENCES

Lubryka, M. 2003. *Assessment of financial efficiency achieved in investment projects at the "JAS-MOS" coal mine.* The School of Economy and Management in Mining Industry.

Lubryka, M. 2006. *Analysis of extraction opportunities for parts of seams with small sizes and irregular shapes with application of the "non-conventional" extraction methods.* Cracow: Proceedings of the School of Underground Mining. Symposiums and Conferences of PAN and AGH, 66.

Lubryka, M. 2007. *The determining aspects to seek for opportunities to sustain extraction of coking coal in the context of deposit exhaustion within the region of Jastrzębie town.* Power Engineering Policy. Vol. 10. Special Edition, 2.

Kostka, M. *The longwall shearer KSW-20008 in the aspect of modern power suppling, control and diagnostic technologies.* Zabrze.

Czechowski, A., Lubryka, J. & Lubryka, M. 2008. *The study of Control Software System and Visualization for Special Mining Machine.* 21st World Mining Congress & Expo 2008.

Lubryka, M. & Lubryka, J. 2010. *Control and power supply of highly efficient mining plow systems.* Advanced mining for sustainable development. International Mining Conference. Ha Long Vietnam.

Adamusiński, M., Morawiec, M., Jędruś, T. Lubryka, J. & Macierzyński, D. 2010. *Modern solutions for power supplying, control and automation system on the example of the highly efficient plow system supplied with 3.3 kV voltage at Zofiówka coal mine.* 13th National Conference of Mining Electric Systems.

Lubryka, M. & Lubryka, J. 2010. *Control and power supply of highly efficient mining plow systems.* Dnipropetrovs'k: Proceedings of the school of underground mining. New Techniques and Technologies in Mining – Bondarenko, Kovalevs'ka & Dychkov'kyy (eds) 2010 Taylor & Francis Group, London, ISBN 978-0-415-59864-4.

Lubryka, M. & Lubryka J. 2010. *Control and power supply of highly efficient mining plow systems.* International Forum and Contest for young scientists. The State Mining Institute of Sankt Petersburg: Journal of the Mining Institute.

Technical and Geoinformational Systems in Mining – Pivnyak, Bondarenko & Kovalevs'ka (eds)
© 2011 Taylor & Francis Group, London, ISBN 978-0-415-68877-2

Modeling of dynamic interaction of technological loading with elastic elements of sifting surfaces in mining and ore-dressing equipment

O. Dolgov
National Mining University, Dnipropetrovs'k, Ukraine

I. Dolgova
Prydneprovska academy of civil engineering and architecture, Dnipropetrovs'k, Ukraine

ABSTRACT: The subject matter is the research of dynamics of the rubber strings serving as the separation media in technological process. The rubber ribbon-strings used as the elements of sifting surfaces substantially differ from beams. Examining their motion as motion of the thin-walled bars, differential equations of their bending and torsion vibrations are developed. A mathematical problem is reduced to integration of the system of linear non-homogeneous differential equations at the given boundary conditions.

1 INTRODUCTION

The questions related to exploitation of self-cleaning sifting surfaces are closely associated with research of dynamic interaction of technological loading with elastic elements. Rubber ribbon-strings (RRS), used as such elements, on the geometrical parameters, behavior, constructional features essentially differ from beams. Many of them rather represent the thin-walled rods. Their structural features consist in possibility to sustain longitudinal elongations at torsion strains. Consequently, the longitudinal normal stresses, proportional to these strains, are reduced in each cross-section to system of balanced longitudinal forces. These, not examined in the theory of pure torsion additional stresses, arising owing to relative deplanation of cross-section, can reach rather great values.

The conditions of RRS work under the action of technological loading aggravate this circumstance and result in the necessity of account of these additional factors at theoretical research of their behavior for various operating regimes.

There is a fundamental concept of a bar centre line in the classic bending theory, based on the law of the plane sections. This is the line through the centroids of their transversal cross-sections. This basic concept of strength of materials, which in essence is Saint-Venant's principle, for thin-walled bars requires radical revision and specification.

If the bent element of a construction can be treated as a thin-walled rod, in considering of such element the concept of a line of bending takes on special significance. For the rods, which have in cross-section two axes of symmetry, this line coincides with centroids of cross-sections. For rods of any asymmetrical profile or a profile with one axis of symmetry, the line of centers of bending will not coincide with centroids of cross-sections. The teeth of RRS do not give possibility to consider a rod as an element with two axis of symmetry. Applying to such rods the classic theory of bending, which allows to consider only bending strains, it is necessary to accept for a rod axis not a line through centroids, but a line of centers of bending and to assume that the transversal loadings, causing a bending of a rod according to the law of plane sections, are applied at points of a line of the centers of bending. If transversal loading reduces to centroids such loading, except a bending, will cause as well torsion. The usual theory of a bending in this case becomes inapplicable (Vlasov 1959).

2 PROBLEM STATEMENT

The theory of thin-walled rods is based on the geometrical hypotheses; their main sense consists in the following:

– the rod in a plane of cross-section has a rigid (not deformable) profile;

– he shear strain of the middle surface, characterized by the right angle change between co-ordinate lines, is accepted equal to zero.

The sense of the first hypothesis consists in the fact that the elementary portion of the rod contained between its two cross-sections, is considered in the plane as absolutely rigid body. Then in a plane of cross-section, it possesses three degrees of freedom corresponding two linear and one angular displacement. At a deformation of an infinitesimal part from a plane of cross-section, it is considered as an elastic deformable body.

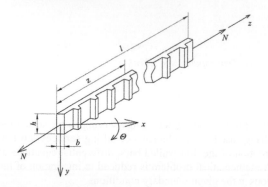

Figure 1. Design diagram of rubber ribbon-string.

The second hypothesis makes it possible to ignore right angle change between co-ordinate axes before and after deformation, i.e. to neglect right angle change between these axes as the geometrical factor, which does not have owing to smallness essential value for stresses in a rod.

Let us consider RRS, having unilateral teeth, and choose the co-ordinate system shown on the Figure 1. We will consider RRS under the action of technological loading and longitudinal eccentric force N. Let co-ordinates of a point of application of the force N are equal e_x and e_y.

Accepting aforementioned hypotheses and taking into account the quantities of corresponding components of disturbing forces, which are caused by the action of technological loading, we can obtain following differential equations of RRS vibrations:

$$EJ_y \frac{\partial^4 u}{\partial z^4} - \rho J_y \frac{\partial^4 u}{\partial z^2 \partial t^2} + \rho F \frac{\partial^2 u}{\partial t^2} - N \frac{\partial^2 u}{\partial z^2} + \rho F a_y \frac{\partial^2 \theta}{\partial t^2} + N(e_y - a_y) \frac{\partial^2 \theta}{\partial z^2} = Q_x(z,t),$$

$$EJ_x \frac{\partial^4 v}{\partial z^4} - \rho J_x \frac{\partial^4 v}{\partial z^2 \partial t^2} + \rho F \frac{\partial^2 v}{\partial t^2} - N \frac{\partial^2 v}{\partial z^2} - \rho F a_x \frac{\partial^2 \theta}{\partial t^2} - N(e_x - a_x) \frac{\partial^2 \theta}{\partial z^2} = Q_y(z,t),$$

$$\rho F a_y \frac{\partial^2 u}{\partial t^2} + N(e_y - a_y) \frac{\partial^2 u}{\partial z^2} - \rho F a_x \frac{\partial^2 v}{\partial t^2} - N(e_x - a_x) \frac{\partial^2 v}{\partial z^2} + EJ_\omega \frac{\partial^4 \theta}{\partial z^4} -$$

$$- \rho J_\omega \frac{\partial^4 \theta}{\partial z^2 \partial t^2} + \rho F r^2 \frac{\partial^2 \theta}{\partial t^2} - GJ_p \frac{\partial^2 \theta}{dz^2} - N(r^2 + 2\beta_x e_x + 2\beta_y e_y) \frac{\partial^2 \theta}{\partial z^2} = M(z,t),$$

(1)

where u, v – displacements of the line of the centers of bending along coordinate axes x and y respectively; θ – angle of rotation of the RRS cross-section in plane xy; N – longitudinal force applied with eccentricity e_x and e_y; $Q_x(z,t)$, $Q_y(z,t)$, $M(z,t)$ – components of disturbing forces (technological loading); a_x, a_y – coordinates of the centre of bending; E, G – Yuong's and shear modulus respectively; J_x, J_y, J_ω – axial and sectorial moments of inertia; ρ – density of material; b – thickness of the RRS base; F – area of the cross-section of the base;

$$r^2 = \frac{J_x + J_y}{F} + a_x^2 + a_y^2 \; ; \; \beta_x = \frac{U_y}{2J_y} - a_x \; ;$$

$$\beta_y = \frac{U_x}{2J_x} - a_y \; ; \; U_x = \int_F y^3 dF + \int x^2 y dF \; ;$$

$$U_y = \int_F x^3 dF + \int_F y^2 x dF \; .$$

It should be noted that at the action of the force N all equations in (1) are related, i.e., they define together with boundary conditions spatial bending and torsion vibrations.

Let us consider forced vibrations of ribbed RRS, which has in cross-section horizontal axis of symmetry ox, supported on diaphragms, rigid in xy plane and flexible in zy plane. Thus, mathematically the problem is reduced to the integration of simultaneous equations (1) at the given set boundary and initial conditions, assuming $a_x = 0$.

3 PROBLEM SOLUTION

Let us consider the next boundary conditions:
at $z = 0$, l, $u = v = \theta = 0$,

$$\frac{\partial^2 u}{\partial z^2} = \frac{\partial^2 v}{\partial z^2} = \frac{\partial^2 \theta}{\partial z^2} = 0. \tag{2}$$

Provided (2), solution can be represented in the form:

$$u(z,t) = \sum_{n=1}^{\infty} u_n(t) \sin \lambda_n z, \quad v(z,t) = \sum_{n=1}^{\infty} v_n(t) \sin \lambda_n z,$$

$$\theta(z,t) = \sum_{n=1}^{\infty} \theta_n(t) \sin \lambda_n z, \tag{3}$$

where $\lambda_n = \dfrac{n\pi}{l}$, n – positive integer, l – length

of RRS. Let us suppose that external loading can be expanded into Fourier's series:

$$Q_x(z,t) = f(t) \sum_{n=1}^{\infty} q_{xn} \sin \lambda_n z,$$

$$Q_y(z,t) = f(t) \sum_{n=1}^{\infty} q_{yn} \sin \lambda_n z, \tag{4}$$

$$M(z,t) = f(t) \sum_{n=1}^{\infty} m_{zn} \sin \lambda_n z,$$

where $f(t)$ – common function, which define the law of variation of external forces with time; q_{xn}, q_{yn}, m_{zn} – coefficients of Fourier's series.

By substituting expressions (3) and (4) into equations (1), we shall obtain for the n-th members:

$$EJ_y \lambda_n^4 u_n + \rho \left(J_y \lambda_n^2 + F \right) u_n'' + \rho a_y F \theta'' +$$

$$+ N\lambda_n^2 u_n - N\left(e_y - a_y\right)\lambda_n^2 \theta = q_{xn} f(t),$$

$$EJ_x \lambda_n^4 v_n + \rho \left(J_x \lambda_n^2 + F \right) v_n'' + N\lambda_n^2 v +$$

$$+ N e_x \lambda_n^2 \theta_n = q_{yn} f(t), \tag{5}$$

$$EJ_\omega \lambda_n^4 \theta_n + \rho \left(J_\omega \lambda_n^2 + Fr^2 \right) \theta_n'' +$$

$$+ \rho a_y F u_n'' - N\left(e_y - a_y\right)\lambda_n^2 u_n +$$

$$+ N e_x \lambda_n^2 v_n + \left[N\left(r^2 + 2\beta_x e_x + 2\beta_y e_y \right) + GJ_p \right] \times$$

$$\times \lambda_n^2 \theta = m_{zn} f(t).$$

The general solution of homogeneous equations corresponding to equations (5) can be obtained in the form of the simple harmonic vibrations:

$$u(t) = A \sin(k_n t + \alpha), \quad v(t) = B \sin(k_n t + \alpha),$$

$$\theta(t) = C \sin(k_n t + \alpha). \tag{6}$$

By substituting expressions (6) into homogeneous equations and setting determinant of the coefficients of A, B and C equal to zero, we develop expressions for determination of natural frequencies:

$$\begin{vmatrix} C_{11} & 0 & C_{13} \\ 0 & C_{22} & C_{23} \\ C_{31} & C_{32} & C_{33} \end{vmatrix} = 0, \tag{7}$$

where $C_{11} = J_{ny}\left(1 - \dfrac{k_n^2}{k_{ny}^2} \right) + N$,

$$C_{13} = -\rho F a_y \frac{k_n^2}{\lambda_n^2} - N\left(e_y - a_y\right),$$

$$C_{22} = J_{nx}\left(1 - \frac{k_n^2}{k_{nx}^2} \right) + N, \quad C_{23} = N e_x,$$

$$C_{31} = -\rho F a_y \frac{k_n^2}{\lambda_n^2} - N\left(e_y - a_y\right), \quad C_{32} = N e_x, \tag{8}$$

$$C_{33} = J_{n\omega}\left(1 - \frac{k_n^2}{k_{n\omega}^2} \right) r^2 + N\left(r^2 + 2\beta_x e_x + 2\beta_y e_y\right),$$

$$J_{nx} = EJ_x \lambda_n^2, \quad k_{nx}^2 = \frac{EJ_x \lambda_n^4}{\left(J_x \lambda_n^2 + F \right)\rho},$$

$$J_{ny} = EJ_y \lambda_n^2, \quad k_{ny}^2 = \frac{EJ_y \lambda_n^4}{\left(J_y \lambda_n^2 + F \right)\rho},$$

$$J_{n\omega} = \frac{1}{r^2}\left(EJ_\omega \lambda_n^2 + GJ_p \right), \quad k_{n\omega}^2 = \frac{EJ_\omega \lambda_n^4 + GJ_p \lambda_n^2}{\left(J_\omega \lambda_n^2 + Fr^2 \right)\rho}.$$

Each value $\lambda_n = \dfrac{n\pi}{l}$, which characterizes according to solution (3) the mode shape, corresponds to compound bending and torsion vibrations.

After determining the natural frequencies from equations (7) the general integral of the system can be represented in the following form:

$$u_n(t) = \sum_{j=1}^{3} A_j \sin(k_{nj}t + \alpha_j),$$

$$v_n(t) = \sum_{j=1}^{3} B_j \sin(k_{nj} + \alpha_j), \qquad (9)$$

$$\theta_n(t) = \sum_{j=1}^{3} C_j \sin(k_{nj}t + \alpha_j).$$

Let us express arbitrary constants A_j and B_j through C_j, making use equations (5) and taking into account formulas (8):

$$B_j = -\frac{C_{23}^j}{C_{22}^j} C_j, \quad A_j = -\frac{C_{13}^j}{C_{11}^j} C_j, \qquad (10)$$

where $C_{mn}^j = C_{mn}$ for each of three frequencies $k_{nj} (j = 1, 2, 3)$.

Substituting formulas (10) into solutions (9), one can obtain:

$$u_n(t) = \sum_{j=1}^{3} \rho_{1j} C_j \sin(k_{nj} + \alpha_j),$$

$$v_n(t) = \sum_{j=1}^{3} \rho_{2j} C_j \sin(k_{nj} + \alpha_j), \qquad (11)$$

$$\theta_n(t) = \sum_{j=1}^{3} C_j \sin(k_{nj} + \alpha_j),$$

where $\rho_{1j} = \dfrac{C_{13}^j}{C_{11}^j}; \quad \rho_{2j} = \dfrac{C_{23}^j}{C_{22}^j}.$

General integral (11) of the homogeneous system corresponding to equations (5) can be represented in other form:

$$u_n(t) = \sum_{j=1}^{3} \rho_{1j}(C_j \sin k_{nj}t + D_j \cos k_{nj}t),$$

$$v_n(t) = \sum_{j=1}^{n} \rho_{2j}(C_j \sin k_{nj}t + D_j \cos k_{nj}t), \qquad (12)$$

$$\theta_n(t) = \sum_{j=1}^{n} (C_j \sin k_{nj}t + D_j \cos k_{nj}t).$$

Expression for partial integral of nonhomogeneous equations (5) can be determined by the method of a variation of arbitrary constants or by method of initial parameters. Let us express six constants in expressions (12) in terms of displacements and speed at initial instant.

We have from formulas (12):

$$u_n(0) = \sum_{j=1}^{3} \rho_{1j} D_j, \quad v_n(0) = \sum_{j=1}^{3} \rho_{2j} D_j,$$

$$\theta_n(0) = \sum_{j=1}^{3} D_j,$$

$$u_n'(0) = -\sum_{j=1}^{3} \rho_{1j} k_{nj} C_j, \quad v_n'(0) = -\sum_{j=1}^{3} \rho_{2j} k_{nj} C_j, \qquad (13)$$

$$\theta_n'(0) = -\sum_{j=1}^{3} k_{nj} C_j.$$

By solving the system (13), one can obtain:

$$D_1 = \frac{S_3 S_6 - S_1}{S_1 S_4} u_n(0) - \frac{S_3}{S_1} v_n(0) +$$

$$+ \frac{S_1 S_2 S_5 - \rho_{11} S_1^2 - S_2 S_4}{S_4} \theta_n(0),$$

$$D_2 = -\frac{S_6}{S_1} u_n(0) - \frac{S_4}{S_1} v_n(0) + \frac{S_2}{S_1} \theta_n(0), \qquad (14)$$

$$D_3 = -\frac{S_5 S_6 - S_1}{S_1 S_4} u_n(0) +$$

$$\frac{S_5}{S_1} v_n(0) - \frac{S_2 S_5 - \rho_{11} S_1}{S_4} \theta_n(0),$$

where $S_1 = S_5 S_6 - S_4(\rho_{21} - \rho_{22})$,
$S_2 = \rho_{11} S_6 - \rho_{21} S_4$, $S_3 = 2\rho_{11} - \rho_{12} - \rho_{13}$,
$S_4 = \rho_{11} - \rho_{13}$, $S_5 = \rho_{11} - \rho_{12}$, $S_6 = \rho_{21} - \rho_{23}$.

By analogy coefficients $C_j (j = \overline{1.4})$ can be written:

$$C_1 = -\frac{S_3 S_6 - S_1}{k_{n1} S_1 S_4} u_n'(0) +$$

$$+ \frac{S_3}{k_{n1} S_1} v_n'(0) - \frac{S_1 S_2 S_5 - \rho_{11} S_1^2 - S_2 S_4}{k_{n1} S_4} \theta_n'(0),$$

$$C_2 = \frac{S_6}{k_{n2} S_1} u_n'(0) + \frac{S_4}{k_{n2} S_1} v_n'(0) - \frac{S_2}{k_{n2} S_1} \theta_n'(0), \qquad (15)$$

$$C_3 = \frac{S_5 S_6 - S_1}{k_{n3} S_1 S_4} u_n'(0) - \frac{S_5}{k_{n3} S_1} v_n'(0) +$$

$$+ \frac{S_2 S_5 - \rho_{11} S_1}{k_{n3} S_4} \theta_n'(0).$$

Using formulas (14) and (15), one can determine

functions (12), which are factors of series (3), if initial conditions are known.

Let the displacements and speed of any particle at initial moment of time will be equal to zero and at instant t_1 the speeds are $u'_n(t_1)$, $v'_n(t_1)$, $\theta'_n(t_1)$. Then the law of motion can be written:

$$u_n(t) = \sum_{j=1}^{3} \rho_{1j} C_{1j} \sin k_{nj}(t - t_1),$$

$$v_n(t) = \sum_{j=1}^{3} \rho_{nj} C_{1j} \sin k_{nj}(t - t_1), \qquad (16)$$

$$\theta_n(t) = \sum_{j=1}^{3} C_{1j} \sin k_{nj}(t - t_1),$$

where C_{1j} are determined from formulas (15) at $t = t_1$.

Imparting of these velocities to the RRS points is equivalent to the action of instantaneous force. Let us express the speeds and accelerations through this force, assuming that it is known. If the force, which components are equal to $q_{xn} f(t_1)$, $q_{yn} f(t_1)$ and $m_{zn} f(t_1)$, acts rather small time Δt_1, it imparts accelerations $u''_n(t)$, $v''_n(t)$ and $\theta''_n(t)$. These accelerations are determined from equations (5), if assume in them at $t = t_1$, $u_n(t_1) = v_n(t_1) = \theta_n(t_1) = 0$.

Thus, for accelerations we have:

$$u''_n(t_1) = \frac{a_y F m_{zn} - \left(J_\omega \lambda_n^2 + Fr^2\right) q_{xn}}{\rho\left[a_y^2 F^2 - \left(J_\omega \lambda_n^2 + Fr^2\right)\left(J_y \lambda_n^2 + F\right)\right]} f(t_1),$$

$$v''_n(t_1) = \frac{q_{yn}}{\rho\left(J_x \lambda_n^2 + F\right)} f(t_1), \qquad (17)$$

$$\theta''_n(t_1) = \frac{a_y F q_{xn} - \left(J_y \lambda_n^2 + F\right) m_{zn}}{\rho\left[a_y^2 F^2 - \left(J_\omega \lambda_n^2 + Fr^2\right)\left(J_y \lambda_n^2 + F\right)\right]} f(t_1).$$

Hence, the speeds $u'_n(t)$, $v'_n(t)$ and $\theta'_n(t)$ take the form of equations (17) if their right-hand parts multiply by Δt_1.

Substitution of the values of speeds into equations (15) and (16) yields the values of the functions $u_n(t)$, $v_n(t)$ and $\theta_n(t)$, which take part of expansion (3) for any moment of time after the action of a force.

If the series of impulses acts, the obtained results should be summed up. In the case of continuous action of impulses these sums turn into integrals. Thus, the gained expressions will be the solution of nonhomogeneous equations (5).

4 CONCLUSIONS

The proposed approach to research of dynamics of elastic elements of sifting surfaces allows choosing rational parameters and operating modes of the corresponding mining and ore-dressing equipment, in particular, rubber-string washing drums (Dolgov & Ravishin 1992). The described above algorithm allows calculating frequencies of RRS vibrations, forces of interaction of technological loading, contact time, etc.

The obtained results, in particular, show that the 10-percent relative extension substantially increases efficiency of a washing drum operating mode. The further increase in longitudinal strains leads to increase in efficiency factor of a mode, however, not so greatly. As one would expect, increase in axial stretching efforts, i.e., relative extension, causes the increase of the maximum force of interaction of technological loading, and decrease of a contact time. The analysis of the results, thus, leads to a conclusion that for used parameters a rational range of longitudinal strains of elastic elements of sifting surfaces of washing drum are 15-20%. As practice shows, the further increase in extension, and, as a result axial stretching efforts, leads to their intensive wear and falling out.

REFERENCES

Vlasov, V.Z. 1959. *Thin-walled elastic rods.* Moscow: Fizmatgyz: 508.
Dolgov, A.M. & Ravishin, V.P. 1992. *The rubber-string washing drum. Proceedings of Cracovian mineral processing conference.* Zacopane: 185-191

4. CONCLUSIONS

REFERENCES

The system of the air cooling of deep mines

I. Shayhlislamova & S. Alekseenko
National Mining University, Dnipropetrovs'k, Ukraine

ABSTRACT: The system of air-cooling has been represented. It includes the method and the installation for the regulating of the thermal rate of the deep mines by means of redistribution of heat-moisture potential of mine air and installation for the thermal relaxation of mineworkers' organisms. The system suggested can be used in mines where inadmissibly high air temperatures occur in the mining and development faces as well as the low air temperature in the slope bottom of the air feeding shaft.

1 INTRODUCTION

The economic crises in our country, the complication of the mine industry restructuring, high electricity charges demand to think about the perspective of the mineral resources mining in the depth down to 1500-2000 m, the temperature of this field will be 50-75 °C and more, the temperature in the working face will raise to 45-50 °C. While passing deeper levels, the temperature and the moisture saturation of mine air increase because of the warm flows caused by the coal massif, the work of mechanisms, the oxidization and other reasons. The main source of heat releasing in mines is the coal massif whose heat release is above 50%.

Adverse climatic conditions cause heat-strokes, serious diseases and miners' efficiency decrease.

The analysis of thermal shock rate among mineworkers (Valutsina 2001) has shown certain dependence of the rate on the age. This dependence has been noticed for both acute and chronic heat exhaustion. It has been specified that the average age of mineworkers with acute overheating (AO) is 27.5 years meanwhile that with chronic overheating (CO) is 38,4 years old. In spite of the treatment taken 56.7% of workers with the AO have been found incapacitated for their profession under conditions of heating microclimate, among them 84% of patients with average severity and all the patients with serious acute heat exhaust. 74.2% of the patients with chronic overheating have also been found incapacitated for their profession after medical examination; thus, after the Medical Sanitation Expert Committee's reference, they have been employed in spheres with normal microclimatic conditions.

In the near future technological and economic reasons will not make it possible to normalize thermal conditions in mining and development high-temperature faces of deep level mines. Therefore, the issue may become even more urgent.

Thus, there is a great demand both to find new methods and technical equipment and to improve the existing ones, to improve microclimate in deep mines by using alternative energy sources. The wide range of the techniques giving the possibility to improve considerably and sometimes to normalize the climate conditions during mining in the deep mines has been developed at the Department of Aerology and Labour protection at the National Mining University.

2 THE MAJOR PART

A method and installation of mine air conditioning with preliminary heat moisture saturation of mine air have been developed for the regulating of the thermal rate (Patent 53467, 2006). The matter of the offered method of regulating thermal rate of workings with preliminary heat moisture saturation is the following. The heating of the mine air while moving along the workings from air-feeding shaft to working faces and preparatory faces is known to be accompanied by intensive moistening of the mine air therefore the relative humidity in all workings raises, as a rule, up to 95-100%. As a result, only one third of the value of heat increasing of the mine air is connected with its heating, while two thirds of the whole value of heat increase of the mine air is caused by moistening.

Under the natural thermal conditions in mines and pits (Figures 1) in slope-bottom air feeding shaft (SAFS), there is a zone of relative low thermal potential of mine atmosphere (LTP), particularly: the temperature of the air is usually 5-15 °C under the humidity 95-100%.

The harmful heat inflows move to the air in main line workings ($Q_e = 30-50$ kJ / kg) and the moisture inflows move to the air ($m_e = 10-20$ g / kg), as far as there is a great thermal head and a head of the proportioned pressing between the air and surrounding coal massif.

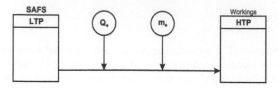

Figure 1. The scheme of the initial thermal potentials and inflows.

While getting saturated with heat and moisture in the main line workings, the air moves to the faces with inadmissibly high temperatures (35-40 °C) under the humidity 90-100%. As a result, a zone of high thermal moisture potential (HTP), unsuitable for safety and productivity of labour is formed.

The task is to reduce the harmful inflow of moisture to the air in the main line and field workings.

This effect can be achieved by "filling" that space in the air that will be taken by water vapour. It will be done with forced heating and moistening the air in the workings SAFS till the level that excludes the main part of moisture inflow to the air while moving to the working faces.

The technique solving this problem has been offered (Patent 53467, 2006). It is based on the forced reducing the temperature head and the head of the proportioned pressing of the water vapour between the mine air and surrounding coal massif, as well as the reducing the air temperature difference along the length of the working; as a result, the harmful inflows of heat and moisture will be reduced according to ventilation stream while moving from SAFS to faces.

To achieve this effect, one must remove the heat in quantity ΔQ out of the air which has reached the HTP zone (Figure 2) and transfer it to the LTP zone, filling the fresh ventilation stream with heat and moisture. The following results are to be achieved:

– the heat moisture potential of the HTP zone has been reduced and this space has been transformed into the space of RTP (reduced thermal potential), improving the microclimate of SAFS workings and faces;

– the heat moisture potential of the LTP zone has been increased and this zone is transformed into the ITP zone (increased thermal potential), improving the microclimate of SAFS workings;

– the temperature-moistening head between the mine air and surrounding coal massif in workings has been reduced, decreasing harmful inflows of heat ($Q_p \approx> 0$) and moisture ($m_p \approx> 0$) into the ventilation stream.

Thus, the positive effect of microclimate improvement in workings is to be achieved with heat massif exchange between the LTP and HTP zones.

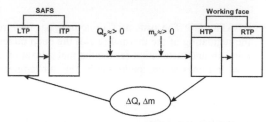

Figure 2. The scheme of exchanges of the heat potentials and inflows.

Figure 3 shows a quality temperature change of the mine air under the natural thermal mode (line B-C-C^I) and under the conditioning of the mine air with the method proposed (line $1-1^I-2-2^I-2^{II}$).

Point 1 (B) indicates the natural air temperature in workings of the slope bottom shaft before air-cooling. Point C characterizes the natural air temperature at the entrance of the mining, and the point C^I characterizes the natural air temperature at the exit of the faces.

The segment $1-1^I$ means the heating and moistening of the air in the air cooling in the slope-bottom workings. The segment 1^I-2 characterizes the air temperature change in main line workings under dispersal cooling.

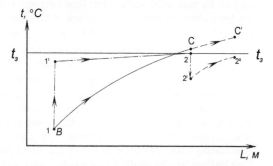

Figure 3. Qualitative change of mine air temperature.

Point 2 shows the air temperature in front of the air cooler. Point 2^I characterizes the temperature of conditioned air when leaving the air cooler. Temperature difference between C and 2^I points indicates air cooling in mining fields workings of such type. Point 2^{II} indicates the temperature of conditioned air when leaving faces.

Under the natural thermal mode one can see intensive air temperature increase when it moves along workings from air feeding shaft to faces (*B-C-C^I* line) and provided the air is being conditioned according to the given method, air heating in main line workings occurs less intensive ($1^I - 2$ line) than under the natural mode.

Due to the air preheating and humidification in a water cooler of evaporative type situated in slope bottom of air feeding shaft, there is certain decrease of the temperature head and the head of partial pressures of steam between ventilation stream and surrounding rock mass in workings which lead to mine sections. This results in reduction of heat and moisture inflow to the air mowing along workings.

Figure 4 shows the simplified scheme of a plant placement to control the heat mode in a sloping face of mine.

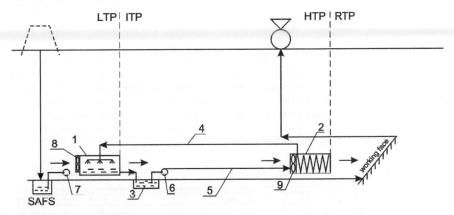

Figure 4. Simplified scheme of a plant placement in a mine.

The underground plant includes a hydraulically connected jet water cooler placed in the workings of pit-bottom (in the area of LTP) – 1, an air cooler of surface type – 2 situated near active working face (in the area of HTP), an accumulator of cooled water – 3 connected with the sump of air feeding shaft, circulating pipe-lines of heat-transfer agent – 4, 5 and pumps – 6, 7, fans – 8 and 9, and regulators of water and air consumption.

In water cooler 1two cross flows interact – the cold air and sprayed warm water – which results in the air being heated and moistened and water being cooled.

The cooled water from water cooler1 moves to the tank 3 (an accumulator of cooled water) then by means of pump 6 along pipe-line 5 it is fed into air cooler 2 where it is warmed due to the heat exchange with the air in the HTP zone. Cooled water from air cooler 2 along pipe-line 4 moves to water cooler 1 where it is cooled.

The cooled air leaving air cooler 1 is mixed with fresh ventilation flow, then it moves along main workings and gets into air cooler 2 where it is cooled and dried and after that moves to active working faces.

The plant provides continuous regulation of air heating and moistening dynamics in workings and faces within given limits without refrigerating machines. Due to this, costs as well as overheat and overcooling of miners reduce.

Cooling effect of the plant for mine air conditioning according to the given method is defined by two key factors:

– lower quantity of dangerous heat inflows and water steams to the ventilation stream from pit-bottom workings of air feeding shaft to the workings of mine sections;

– heat quantity carried from circulating water in the air cooler.

The overheating of miners occurs because heat production of their bodies is higher than emission of heat into the environment (cooling). It is known that human heat emission can be the result of convection, thermal radiation, moisture evaporation and conduction. This is convective heating of miners' bodies not cooling that occurs in high-temperature faces as the air temperature is higher than that of their skin. Radiant heat exchange of miners depends on the temperature and blackness degree of the surrounding surfaces, radiant flux geometry, skin temperature and external surface of working clothes. High radiating capacity of the working area is explained not only by high temperature and blackness degree of the surrounding surfaces, but also by the peak value of radiation angular coefficient as excavations are closed radiating systems. The heat radiation sources also include fresh rock and coal which have the temperature higher than the average temperature of the surrounding surfaces. As to the evaporation cooling of minors bodies, it is hampered because the air has a high saturation rate (about 90%) under these conditions, the sweat hardly evaporates but forms drops and a thin film which cover the skin and worsen convective heat exchange.

The analysis of possible heat exchange ways (heat production, convection, evaporation, radiation) does not provide the proper cooling of miners' bodies in high-temperature working faces which leads to the overheating and heat shock. Therefore it is necessary to look for the way out based on a partial (but acceptable) overheating of miners' bodies.

To cool miners in the development working faces from time to time, the *Relaxator* plant has been worked out (Patent 70653, 2007). The given techni-

cal solution can also be used to rescue workers overtaken by accidents in mine opening and having no possibility to go to the save zone, to protect mine workers temporarily and give the first aid.

The plant for protecting miners from overheating (Figure 5) includes: a screen in the form of a closed chamber – 1, an air quench system in the form of a perforated pipe – 2, a cold generator as a pneumatic turbine – 3 connected with a pipeline the external compressed air source, a catcher – 4 with a tray – 5 and an opening for the dried air exit – 6 which is connected with the wall passage and the chamber ceiling, and through them with the air quench system. In the front part of the chamber there are two openings for regulating condensate and discharge air emission. Moreover, the chamber has an independent compressed air source – 7, and the chamber itself is equipped with a hermetic door and is set on a chassis to make it possible to move. The chamber also has seats and sections for keeping insulating self-rescuer, respirators medical apparatus for providing the injured with the first aid.

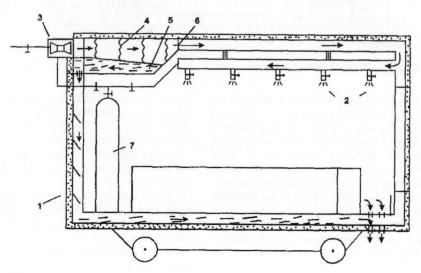

Figure 5. The *Relaxator* scheme.

The *Relaxator* works in the following way. The turbine 3 is connected to the pipe of the external source of the compressed air which enters the turbine through the high pressure pipeline. In the turbine 3 the compressed air dilates doing effective work of the wheel rotation. According to the first Thermodynamics Law, while advancing through the turbine the compressed air power budget decreases by the quantity of the work retracted from the turbine. This mechanical power is retracted from the turbine with the air flow in a different way. The decrease of the com-

pressed air power budget is observed in its pressure and temperature decrease, relative humidity increase and condensation of part of steam. Moreover, the condensate is of a low temperature. The air used in the turbine 3 together with condensate drops goes to the catcher 4 where the condensate separates from the air. As a result, the air becomes dry saturated and through Exit 6 moves to the ceiling air passage and the chamber side walls while the condensate moves to the tray 5 of the catcher 4. While moving along the air passage the air cools the internal wall surface and the chamber ceiling. At the same time the air is partly warmed while its relative humidity decreases considerably (down to 40-50% and lower). Therefore, the warm and dry air enters to the air quench system 2 and through the pipe branches it gets to the recreation area.

The condensate from the tray 5 of the catcher 4 gets to the back wall passage of the chamber. The turnback plates direct the condensate to the wall which is oriented to the middle of the chamber and is cooled by the condensate streaming down. After that the condensate gets to the floor passages, cools the floor and flows outside through the openings.

The auxiliary independent compressed air source 7 is used in the following cases: under absence or damage of the external source; under necessity of air consumption increase in the air quench system; while using the chamber as a shelter during accidents. For this purpose, the chamber has a section where extra self-rescuers are stored.

Therefore, the heat relaxation of mine workers in the chamber 1 is implemented in a complex way:

– by convection while blowing on the body with the air whose temperature is lower than the body surface temperature;

– by evaporating sweat form the external body surface and moisture from the respiratory tract internal surface while blowing on the body with dry air;

– by radiant heat exchange between the cold internal surface of the chamber and mineworkers' bodies (radiative cooling).

The use of all the main ways of the heat exchange of mineworkers' bodies in the *Relaxator* plant provides quick decrease of overheating, heat balance normalization which excludes heat shock risks, promotes the renewal of mineworkers' efficiency. Moreover, the *Relaxator* plant can provide efficient protection of mineworkers in case of emergency including high temperature protection during mine working fires.

The application of radiative cooling installations in high-temperature working faces allows decreasing the reduced radiant temperature by 10-15 °C and reduce mineworkers' radiant heating by 100-150 Watt. This radiative cooling effect can be compared to the convective cooling effect increase of the air flow provided the air is cooled down by 10-12 °C.

3 CONCLUSIONS

As for different mining conditions of deep mines the problem of thermal labour conditions normalizing in breakage and development faces must be solved in complex taking into consideration all means and components to improve microclimate. It will permit raising labour efficiency and reducing morbidity rate of miners.

The advantage of the method and mine air conditioning installation is absence of refrigerating machines and air conditioners using Freon. It increases safety of the plant, reduces costs connected with operation and service of underground devices, and simplifies considerably the servicing of the mine air conditioning plant. The method and plant for mine air conditioning provide the improvement of thermal labour conditions not only in workings and faces of mining sections but in pit-bottom workings where miners are always present.

The advantage of *Relaxator* plant is the ability to provide mineworkers' thermal relaxation in severe microclimatic conditions as well as efficient protection of mineworkers in case of emergency in production units including protection from the irrespirable atmosphere and heat shock protection.

REFERENCES

Valutsina, V.M., Tkachenko, L.N. & Ladariya, E.G. 2001. *The Frequency of thermal lesions of deep coal mine workers.* Journal of Hygiene and Epidemiology: Volume 5, #1: 50-52.

Patent 53467 Ukraine, International Patent Classification 7 E21 F 3/00. *Mine air conditioning – the method and plant* / Muraveinik, V.I. Alekseenko, S.A., Shayhlislamova, I.A. and others; the patentee – the National Mining University. – #2002064680; applied 07.06.2002; published 15.12.2006, Bull. #12.

Patent 70653 Ukraine, International Patent Classification 2006 E21 F 3/00, E21 F 11/00. *Overheating protection installation* / Muraveinik, V.I., Alekseenko, S.A., Shayhlislamova ,I.A. & Korol, V.I.; the patentee – the National Mining University. – #20031211992; applied 22.12.2003; published 25.06.2007, Bull. #9.

Technical and Geoinformational Systems in Mining – Pivnyak, Bondarenko & Kovalevs'ka (eds)
© *2011 Taylor & Francis Group, London, ISBN 978-0-415-68877-2*

Identifying method for abnormal values of methane release in mining level blocks

V. Okalelov, L. Podlipenskaya & Y. Bubunets
Donbass State Technical University, Alchevsk, Ukraine

ABSTRACT: We study the problem of identifying outliers in the time series of methane in the excavation sites of coal mines. The classification of outliers and methods for their diagnosis are designed. An algorithm that transforms a number of outliers in the modified series was described, which can be used to describe a longwall-analog and, accordingly, the forecast of methane release on a projected excavation area.

One of the most well-known in the theory and practice of various events forecasting is the analog approach. It is widely used in the prediction of methane release in the longwall. In this case we are talking about the forecast of only average values of methane release without regard to their pronounced dynamic character. In this connection there arose the necessity to improve existing methods of prediction to assess the dynamics of methane release in the longwall projected based on the data of longwall-analog (Okalelov 2008).

While solving this scientific and technical problem there arose the necessity to develop a method for analyzing time series of methane release in the active longwalls, aimed at establishing of causal relationships between forecasting performance and influencing factors (Podlipenskaya 2007).

The first stage of this analysis is to assess the reliability of the source data, which can be performed on the basis of a priori method comprising the following steps:

– assessment of the homogeneity of the study population;

– analysis of the distribution of the studied population characteristics;

– to identify logically significant causal relationships between the characters and phenomena.

Assessment of the homogeneity of the source data that characterize the dynamic process is recommended in the following sequence:

– identification and analysis of outliers;

– determination of the homogeneity degree of the totality by one or more essential characteristics;

– choosing the optimal selection variant of homogeneous populations.

In the statistical theory and practice there have been developed different approaches to assessing the degree of homogeneity (Sadovnikova 2007).

The most difficult and controversial is the question of ways and criteria for the selection of homogeneous groups of objects within the original population.

Any researched population, along with the values of attributes, formed under the influence of factors directly characteristic of the object may contain attributes and values obtained under the influence of other factors that are not characteristic to the studied object. These values stand out sharply and, consequently, using this methodology of statistical analysis without studying these observations will result in serious errors.

Usually, outliers can be detected visually using a graphical representation of time series, but before you tweak the values found in this way, they should be subjected to quantitative and qualitative analysis.

Under the outliers of time series we understand the observation, which differs significantly from the rest of the observations set and localized in time (or other discrete variable number). We suggest classifying the outliers in relation to the symptoms associated with the processes of methane release at the excavation sites, operating at highly gas-bearing strata, as follows:

1) outliers of the first kind;

2) outliers of the second kind;

3) points of a structural shift.

The anomalous first-order (Figure 1) appears as a form of a strong change in the level indicator (jump or decline), followed by an approximate reconstruction of the previous level. Abnormality of this type is usually observed in a very narrow neighborhood of certain points of time axis.

At Figure 1, which characterizes methane release within a month in 17-th Orlovskaya longwall at "Molodogvardeiskaya" mine PC "Krasnodonugol", the outlier of the first kind can occur in the point which corresponds to the 8-th day.

Outlier of the second kind (Figure 2) is an atypical behavior of the indices at a visible interval of time axis.

At Figure 2, which characterizes methane release within 100 days the possible interval of outliers of the second kind is between 50-th and 60-th days.

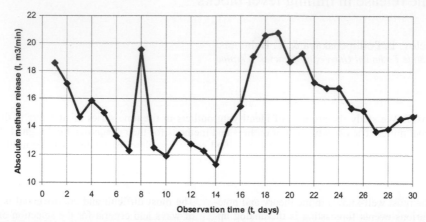

Figure 1. Methane release within a month in 17-th Orlovskaya longwall at "Molodogvardeiskaya" mine PC "Krasnodonugol".

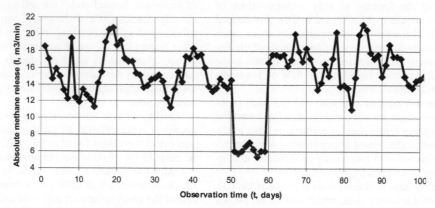

Figure 2. Methane release within 100 days in 17-th Orlovskaya longwall at "Molodogvardeiskaya" mine PC "Krasnodonugol".

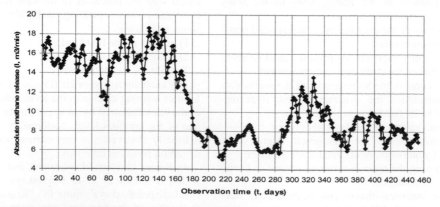

Figure 3. Methane release in 26-th Orlovskaya longwall at "Molodogvardeiskaya" mine PC "Krasnodonugol".

Points of the structural shift (Lukashin 2003) are those values of time series, in which there is an abrupt change in the behavior o whole time series. A similar case occurred in the 26th Orlovskaya longwall at "Molodogvardeiskaya" mine PC "Krasnodonugol" (Figure 3), where on the 162-th day of operation the ventilation scheme of excavation site has changed. Thus, $t = 162$ corresponds to the point of structural change.

Investigation of time series from the standpoint of causality shows that anomalously may be due to various reasons, which must be considered during accepting the decision on the rejection or correction of each individual outlier.

There are anomalous values, reflecting the objective process development, but very different from the general trend, as they show their extreme effects rarely. They do not always have to be excluded from the time series and may even be useful at the stage of studying the causal mechanism of the phenomenon. The presence of the peak values for the same time in different time series indicates, as a rule, the causal connection between the respective indices.

Some anomalous values appear due to single changes in the terms of the production process, such as changing the ventilation scheme. These values should not be excluded from consideration, and taken for "turning" (threshold) from which should be verified a mathematical model of time series for methane release. In the case of a single change of operational technological parameters it's rational to divide the original observations into several series.

Anomalous values arising from errors in measuring the indicator during recording and transmission the information, as well as the values associated with various catastrophic events do not influence the further course of events, the aggregation and disaggregation of indicators, etc., should be excluded from consideration anyway, as they distort the perception of the phenomenon nature and may have a significant impact on the conclusions obtained from analysis of a series containing such misrepresentations.

For the diagnosis of outliers of time series there were developed different criteria, for example, Irwin's method (Fedoseyev 1999). The idea of this method is that for all types or only for anticipated outliers value λ_t is calculated:

$$\lambda_t = \frac{y_t - y_{t-1}}{S_y}, \tag{1}$$

where y_t – current value of time series of methane release; y_{t-1} – previous value of time series of methane release; S_y – standard deviation of methane release.

$$S_y = \sqrt{\frac{\sum\limits_{t=1}^{n}\left(y_t - \overline{y}\right)^2}{n-1}}. \tag{2}$$

$$\overline{y} = \frac{1}{n}\sum\limits_{t=1}^{n} y_t. \tag{3}$$

If the module of calculated value λ_t exceeds the table level (Fedoseyev 1999), then level is considered abnormal and is replaced in the series by the appropriate calculated level.

In the introduced classification of types of anomaly time series values we propose a modification of the Irwin's criterion, which can be used both to diagnose outliers and identify the kind of anomaly.

Let the original dynamic range of indicators of methane excavation site is:

$$Y = \{y_1, y_2, \ldots, y_{t-1}, y_t, y_{t+1}, \ldots, y_n\}. \tag{4}$$

Let's consider an algorithm for recognizing outliers:

1. Calculation of numerical characteristics for the whole series on formulas (2-3).

2. Sequential computation of values, starting with $t > 2$:

$$\lambda_t = \frac{y_t - y_{t-1}}{S_y}. \tag{5}$$

3. Check a condition:

$$|\lambda_t| > \lambda_{kp}. \tag{6}$$

If condition (6) is not satisfied, then the point is not considered abnormal, and proceed to checking the next point (item 2).

If condition (6) holds, then the point y_t is declared abnormal, and proceeds to specify the kind of anomaly (item 4).

4. It is calculated $\lambda_{t+1} = \frac{y_{t+1} - y_t}{S_y}$ and also checked similar to (6) the condition:

$$|\lambda_{t+1}| > \lambda_{kp}. \tag{7}$$

If it holds, the composition is found:

$$\mu_t = \lambda_t \cdot \lambda_{t+1}. \tag{8}$$

Further variants are possible:
a) $\mu_t < 0$. Consequently, the point is anomalous

of the first order. It is taken away from a series, replacing by the arithmetic value of the close points. For more reliable replacement of anomalous value you can use more complex interpolation methods (Bahvalov 2007). Extracted anomalous point y_{t_k} together with its number t_k is put into the auxiliary unequidistant series of the first kind abnormal points $A_1 = \{y_{t_k}, k = 1, ..., m\}$, where m – the number of abnormal points of the first kind. Next, move to step 1, using the corrected value of the current point y_t.

b) $\mu_t > 0$. Hence, the point y_t may be anomalous of the second order, or is the beginning of a structural shift of the series. To determine the nature of the point go to the next item 5.

If condition (7) is not satisfied, then the conclusion is similar to case b) – the point y_t may be anomalous point of the second kind, or is the beginning of a structural shift of the series. Go to next item 5.

For convenience of presentation of test results on the anomaly, we introduce vector-indicator of outliers $\Psi = \{\psi_1, \psi_2, \psi_3\}$, whose components take values in a binary encoding 0 or 1 (Table 1).

Table 1. Values of vector-indicator of outliers.

Conditions	Values of indicators		
	Condition is satisfied	Condition is not satisfied	ψ_t
$\|\lambda_t\| > \lambda_{kp}$	1	0	ψ_1
$\|\lambda_{t+1}\| > \lambda_{kp}$	1	0	ψ_2
$\mu_t = \lambda_t \cdot \lambda_{t+1} < 0$	1	0	ψ_3

Then we can identify the points of the original series in the following combinations of indicators:

a) $\psi_t = \{0, \psi_2, \psi_3\}$ – means that the point is not anomalous (second and third components of the vector ψ_t can take values 0 or 1), in this case go to item 2 and check the next point;

b) $\psi_t = \{1, 1, 1\}$ – shows that the point is anomalous of the first kind;

c) indicators $\{1, 1, 0\}$, $\{1, 0, 1\}$, $\{1, 0, 0\}$ signal the possibility of appearing of the second kind anomaly or structural shift, then go to item 5.

5. Diagnosis of abnormality of the second kind, as well as a structural shift is performed online. Visualization of data in the form of the graphs allows you to check the hypothesis about the tested observation as the starting point of the anomalous data interval or the turning point to changing the structure of the series. Then, series is divided into specific intervals and through the variance analysis it is established a statistical difference between the intervals and the conclusion is made about the type of detected anomalies.

For further analysis if a point of structural shift is revealed the researches are made for each characteristic parts separately, but if there is an abnormality of the second kind, then to replace the outliers interval it is necessary to study the whole series (without anomalous points of the first kind), which will clarify the causes of the outliers interval, to establish the regularity of dynamic range in the absence of these causes and to make adequate replacement of points of anomalous interval.

Thus, we can conclude that the time series of methane release at the excavation site may contain outliers of various kinds. The proposed approach to identify the outliers in dynamic series describing the processes of methane release at the excavation site of a coal mine can be used to reconstruct the original series of methane release, and to establish cause-effect relationship between methane release and the influencing factors.

REFERENCES

Okalelov, V.N., Podlipenskaya, L.Ye., Bubunets, Y.V. & Dolgopyatenko, S.I. 2008. *Prognosis and control for dynamics of methane release in the breakage face.* Coal of Ukraine, 7: 21-24.

Podlipenskaya, L.Ye. & Bubunets, Y.V. 2007. *Dynamics investigation of methane release at the excavation site.* Alchevsk: DonSTU. Issue, 23: 56-66.

Sadovnikova, N.A. & Shmoilova, R.A. 2007. *Analysis of time series and prognosing.* Moscow: Issue 3: Studying-practical handbook. Publishing centre ЕАОИ: 272.

Lukashin, Yu. P. 2003. *Adaptive short-range forecasting methods of time series.* Moscow: Finances and Statistics: 416.

Fedoseyev, V.V., Garmash, A.N., Daiyitbegov, D.M., Orlova, I.V. & Polovnikov V.A. 1999. *Economic-mathematical methods and applied models.* Moscow: studying handbook for universities. UNITY: 391.

Bahvalov, N.S., Zhydkov, N.P. & Kobelkov, G.M. 2007. *Numeric methods. Moscow: studying handbook for students of phys.-math.* Specialized universities – 5-th edition. BIMOM. Knowledge laboratory: 637.

Technical and Geoinformational Systems in Mining – Pivnyak, Bondarenko & Kovalevs'ka (eds)
© 2011 Taylor & Francis Group, London, ISBN 978-0-415-68877-2

Research of dynamic processes in the deep-water pumping hydrohoists lifting two-phase fluid

Y. Kyrychenko, V. Kyrychenko & A. Romanyukov
National Mining University, Dnipropetrovs'k, Ukraine

ABSTRACT: A comprehensive methodology for the calculation of dynamics of the two-phase flows has been first developed. The methodology allows studying the whole spectrum of transient processes in the deep-water pump-based installations and provides the precision level needed for this class of problems. On the basis of the developed methodology a special HydroWorks 2p software is developed, allowing to define the parameters of transient processes in the deep-water pumping installations. Using the software it is defined that pressure oscillations and ensuing dynamic stresses often reach critical values, which may pose a risk in terms of efficiency of installation and violation of its integrity.

1 INTRODUCTION

Nowadays in Ukraine there is a shortage of some strategic non-ferrous metals extracted from the continental deposits in traditional way. In this regard, the further growth of mineral resources base in Ukraine is closely linked with the development of ore deposits of the World Ocean.

One of the most promising methods of transportation of solid minerals from the seabed is a pump-based hydrohoist (Fox 1981).

Decision of the of National Security and Defense Council on May 16, 2008 "On measures to ensure Ukraine's development as a maritime state" powered by Presidential Decree № 463/2008 of 20 May 2008, provides for the development of a new "National Programme for research and use of Azov-Black Sea and other regions' of the oceans resources in the years 2009-2034". Thus, development of technical means of lifting minerals from the seabed is one of the priority areas of research. This article focuses on the urgent problem of development of mineral potential of the World Ocean, the solution of which is directly linked to the development of effective methods of regulation and management of deep-water pump-based hydrohoists.

The lifting process of mineral raw materials on a basic watercraft associates with the solution of tasks of calculating the dynamics of two-phase (water and solids) flow, which is due to many transient processes accompanying the work of the pumping unit. Deep-water pumping hydrohoists (DPH) usually operate in non-stationary or quasi-stationary modes because of the long hydraulic paths and specific exploitation characteristics.

The existing methods of calculating the DPH mostly base on the idea of the mixture as a homogeneous fluid (Nigmatulin 1987). The authors of these methods tend to focus on the problems of water hammer giving the solid particles in the flow a passive role which means only increasing the density of the mixture (Kartvelishvili 1979; Makharadze 1986; Wallace 1972; Kyrychenko, Romanyukov, Taturevich 2009). This approach allows using simplified mathematical apparatus, based on the homogeneous model (Wallace 1972; Charny 1975) for calculations, which significantly reduces the accuracy of the results, because water and solid particles have different inertial properties. Obviously, such methods with acceptable for engineering calculations accuracy allow calculating the ground hydrotransport, which can be designed with appropriate safety margin. However, according to the authors of this paper, this approach is hardly acceptable for the calculation of such unique engineering facilities like DPH, because it does not take into account the specifics of marine mining equipment exploitation in difficult conditions of great depths.

The pipeline of DPH is the backbone of the entire subsea equipment. Because of the great length, mass and overall dimensions the pipeline is characterized by dangerous static longitudinal stresses (Kyrychenko 2001). When the carrier vessel moves, the pipeline takes a curved shape experiencing dynamic loads, caused by pitching, as well as various kinds of aerohydroelastic instability of the marine environment (aeolian vibration, galloping, flutter). It is also possible loss of the divergent stability of the pipeline and occurrence of parametric resonance due to interaction with the stationary and pulsating flow of transported fluid. In addition, the processes associated with starting and stopping of pumping units in case of wrong management may be accompanied by the phenomenon of water hammer. These factors will inevitably give rise to additional dynamic stresses that may impair the integrity of the system.

Thus, one of the limiting factors of deep-water hydrohoists development is the lack of effective controlling systems, preventing of the above-mentioned harmful effects. The development of such systems, in turn, is restrained by the lack of research results of unsteady modes and dynamic processes in the deep-water hydrohoists, as well as the mathematical description of the transitional processes and the lack of sufficiently accurate and physically grounded method for calculating the parameters of DPH and its software implementation. The controlling system must be capable of fast "tuning" of the current parameters in conditions of multivariate disturbing influences, which means to possess sufficient efficiency and performance.

Based on the above mentioned features, method of calculation of DPH must meet the following requirements:

– high accuracy due to the lack of safety margin;
– high integrity and efficiency, which means the possibility of studying the entire spectrum of nonstationary and transient processes in frames of a single mathematical apparatus, based on the differential equations of the same type;

Earlier the correct calculation of dynamic processes in hydrotransportation systems, pumping heterogeneous mixtures, was not possible mainly due to lack of adequate, physically grounded mathematical model, which would most fully take into account the specifics of deep-water hydraulics and all range of dynamic effects. Another reason is absence of the law of the speed of sound change in two-phase slurry (Wud 1934).

However, the authors of this paper have managed to obtain such mathematical model of two-phase fluid motion (Goman, Kyrychenko, Kyrychenko, 2008), the characteristic relations for it (Kyrychenko, Shvorak, Kyrychenko, Romanyukov, Taturevich 2011), and the speed of sound change laws (Goman, Kyrychenko, Kyrychenko 2008; Kyrychenko 2009). From now on, the possibility of developing an integrated method of calculating the dynamics of heterogeneous flows has been opened. This method will automatically provide the possibility of calculating the parameters of the full range of dynamic processes in the DPH from slow concentration waves that accompany the processes associated with the launch of the system to fast transients in different emergency situations.

2 FORMULATING THE PROBLEM

The aim of this paper is to develop a method for calculating the dynamics of two-phase flows in the pipelines of deep-water pumping systems for the study of nonstationary and transient processes.

Let us briefly discuss the main components of the methodology. In (Goman, Kyrychenko, Kyrychenko 2008) the flow of liquid and solid particles is reviewed. In the one-dimensional approximation, the equations describing the motion of two-phase flow obtained in (Goman, Kyrychenko, Kyrychenko 2008) look like:

$$(1-C_1)\frac{\partial p}{\partial t} - \rho_0 a_0^2 \frac{\partial C_1}{\partial t} + \rho_0 a_0^2 (1-C_1)\frac{\partial V_0}{\partial x} = 0, \quad (1)$$

$$C_1\frac{\partial p}{\partial t} + \rho_1 a_1^2 \frac{\partial C_1}{\partial t} + \rho_1 a_1^2 C_1 \frac{\partial V_1}{\partial x} = 0, \quad (2)$$

$$\left(1+\frac{C_1 k_1}{2}\right)\frac{\partial V_0}{\partial t} - \frac{C_1 k_1}{2}\frac{\partial V_1}{\partial t} + \frac{(1-C_1)}{\rho_0}\frac{\partial p}{\partial x} = \phi_0, \quad (3)$$

$$\left(\frac{\rho_1}{\rho_0}+\frac{k_1}{2}\right)\frac{\partial V_1}{\partial t} - \left(1+\frac{k_1}{2}\right)\frac{\partial V_0}{\partial t} + \frac{1}{\rho_0}\frac{\partial p}{\partial x} = \phi_1, \quad (4)$$

where

$$\phi_0 = -(1-C_1)g\sin\alpha - \frac{\lambda}{2D}\frac{\rho_{mix}}{\rho_0}|V_{mix}|V_{mix} - $$
$$-\frac{3}{8}\left[\frac{C_1 C_{xs}}{R_1}|V_0 - V_1|(V_0 - V_1)\right],$$

$$\phi_1 = -\frac{\rho_1}{\rho_0}g\sin\alpha + \frac{3}{8}\frac{C_{xs}}{R_1}|V_0 - V_1|(V_0 - V_1),$$

$$\frac{1}{a_1^2} = \frac{\rho_1}{K_1} + \frac{\rho_1}{F}\left(\frac{\partial F}{\partial p}\right),$$

$$\frac{1}{a_0^2} = \frac{1}{a_l^2} + \frac{\rho_0}{F}\left(\frac{\partial F}{\partial p}\right), \quad a_l^2 = \frac{K_l}{\rho_0},$$

$$K_1 = \frac{E_1}{3(1-2\nu_1)},$$

$$\rho_{mix} = \rho_0^* + \rho_1^* = (1-C_1)\rho_0 + C_1\rho_1,$$

$$V_{mix} = \frac{1}{\rho_{mix}}\left(\rho_0^* V_0 + \rho_1^* V_1\right).$$

K_1, E_1, ν_1 – bulk modulus of elasticity, Young modulus and Poisson ratio of solid particles correspondingly; K_l – compression modulus of liquid;

116

a_l – sound velocity in pure unbounded liquid; R_1 – equivalent radius of solid particles; k_1 – a coefficient, which describes how virtual masses are affected by nonsphericity and concentration of solid particles; g – gravitational acceleration; α – pipeline canting angle; D – pipeline diameter; λ – Darcy coefficient; t – time; C_{xs} – solid particles resistance coefficient; C_i – phase bulk volume; p – pressure; ρ_i – phase real density; ρ_i^* – phase reduced density; V_i – phase velocity; x – longitudinal coordinate; sub-indices mean: "0" –water; "1" – solid particles; "m" – mixture.

It should be noted, that derivative of concentration C_1 enters only into continuity equations (1) and (2). Hence if we expresses the derivative $\dfrac{\partial C_1}{\partial t}$ from equation (2) and substitute it into equation (1), we get general continuity equation:

$$\rho_0 a_0^2 (1-C_1)\frac{\partial V_0}{\partial x} + \rho_0 a_0^2 C_1 \frac{\partial V_1}{\partial x} +$$

$$+\left[(1-C_1)+\frac{\rho_0 a_0^2 C_1}{\rho_1 a_1^2}\right]\frac{\partial p}{\partial t} = 0. \qquad (5)$$

Now the set of equations (1)-(4) can be divided into two subsets: first subset includes equations (3)-(5) and contains only V_0, V_1 and p derivatives, while derivative of concentration C_1 is absent. Second subset includes equation (2), contains C_1 time derivative and is connected with the first subset via derivatives of p and V_1. At the same time, first subset is connected with the second one only via concentration C_1 (but not via its derivative). Concentration enters into first subset both as a coefficient and implicitly through ϕ_0 and ϕ_1 variables.

Perturbation velocity in mixture and characteristic relations on the mach front can be derived from the first subset (3)-(5). Equation (2) is in fact an ordinary differential equation for calculating changes of concentration C_1 with time in every fixed point of pipeline (x) assuming that V_0, V_1 and p are already defined as functions of x in every time layer t.

If the transported liquid contains plenty of solid particles (pulp movement), wave movement in the pipeline has some peculiarities, caused by compressibility of solid particles, relative slip between solid and liquid phases (which is present in general

case), different inertia of solid substance and transporting liquid etc.

We have to note, that the most complete expression for sound velocity in two-phase mixture D_0 is below:

$$D_0 = \frac{1}{\sqrt{\rho_y\left(\dfrac{(1-C_1)}{K_0}+\dfrac{C_1}{K_1}+\dfrac{1}{F}\dfrac{\partial F}{\partial p}\right)}}, \qquad (6)$$

where

$$\rho_y = \mu \cdot \rho_0, \quad \mu = \frac{A}{B},$$

$$A = \frac{\rho_1}{\rho_0}\left(1+\frac{C_1 k_1}{2}\right)+\frac{k_1}{2}(1-C_1),$$

$$B = \frac{\rho_1}{\rho_0}(1-C_1)^2+(2-C_1)C_1+\frac{k_1}{2}.$$

Figure 1 shows the dependence of sound velocity on real bulk concentration of solid particles in slurry (hereinafter called slurry buld concentration) for various wave numbers of pipeline and spherical solid particles densities with diameter of 0.005 m. Comparing the depicted curves one can see that behavior of curves has little dependence on density of solid substance.

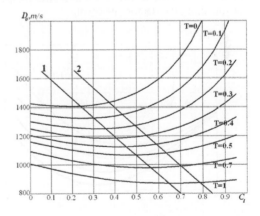

Figure 1. The dependence of perturbation velocity on concentration of solid substance in slurry at various pipeline parameters. ($C_2 = 0$; $\rho_1 = 1600$ kg / m³; $K_1 = 4.5 \cdot 10^{10}$).

Analyzing mentioned relations several conclusions can be drawn. Sound velocity in slurry in general case depends on parameters of both slurry and pipeline. The competing influence of pulp and pipeline parameters defines three specific regions on the plot.

First region is located to the left from line 1 (Figure 1) $D_0=-2124.7C_1+1630$ and corresponds to descending behavior of curves with increasing concentration of solid substance. This happens because increase of solid particles effective volume compressibility is outrunned by growth of pulp density.

The second region is located between the first line and the second line, defined by $D_0=-2101.8C_1+1870$. Second region corresponds to quasi-constant speed velocity at fixed pipeline wave number. In the given range of solid substance concentrations, which is limited by lines 1 and 2, the slurry density is proportional to its volume compressibility. In this region sound velocity depends on pipeline wave number only. Such behavior significantly simplifies calculations, necessary for engineering methodology design. The first approximation of speed velocity is the following:

$$D_0=-575T+1415,$$

where $T = \dfrac{K_0 D}{E\delta}$ – pipeline wave number, δ – pipe wall thickness.

Third region is located to the right from line 2 and corresponds to ascending behavior of curves due to overrunning growth of effective volume compressibility of slurry comparing to increase of its density.

From Figure 2 one can conclude that decreasing of solid substance density leads to growth of speed of sound, if other factors are equal.

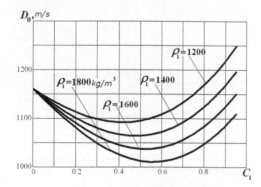

Figure 2. The dependence of sound velocity on solid substance concentration for various values of solid substance density ($C_2 = 0$; $T = 0.5077$; $K_1 = 4.5\cdot10^{10}$).

The obtained results lead us to reconsidering of a stereotypic statement, which claims that presence of solid substance in transporting liquid results in growth of sound velocity (Kyrychenko 2009).

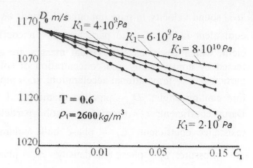

Figure 3. The dependence of sound velocity on solid substance concentration for various values of solid particles volume compressibility ($C_2 = 0$; $T = 0.5077$; $\rho_1 = 1600$ kg/m³).

Speed of sound is also affected by volume compressibility of solid particles K_1 (Figure 3): higher volume compressibility leads to growth of sound velocity, if other factors are equal.

3 CHARACTERISTIC RELATIONS

Paper (Kyrychenko, Shvorak, Kyrychenko, Romanyukov, Taturevich 2011) shows, that for system (1)-(4) characteristic relations are fulfilled on the set of three characteristics:

$$dp + \mu\rho_0 D_0\left[(1-C_1)dV_0 + C_1 dV_1\right]-$$

$$-\frac{\mu\rho_0 D_0}{A}\psi dt = 0. \tag{7}$$

$$-dp + \mu\rho_0 D_0\left[(1-C_1)dV_0 + C_1 dV_1\right]-$$

$$-\frac{\mu\rho_0 D_0}{A}\psi dt = 0. \tag{8}$$

for following acoustic characteristics

$$\left(\frac{dx}{dt}\right)_1 = D_1 = D_0, \tag{9}$$

$$\left(\frac{dx}{dt}\right)_2 = D_2 = -D_0. \tag{10}$$

and $\left[\left(1-C_1\right)\left(1+\dfrac{k_1}{2}\right)+1+\dfrac{C_1 k_1}{2}\right]dV_0 -$

$$-\left[\left(1-C_1\right)\left(\frac{\rho_1}{\rho_0}+\frac{k_1}{2}\right)+\frac{C_1 k_1}{2}\right]dV_1 - \Omega_1 dt = 0, \tag{11}$$

for characteristics like

$$D = 0, \tag{12}$$

where

$D = x'(t)$ – mach front propagation velocity,

$$\psi = \varphi_l g \sin\alpha - \frac{\lambda \rho_{mix} |V_{mix}| V_{mix}}{2 D \rho_0} \varphi_p +$$

$$+\frac{3}{8}\frac{C_{xs}}{R_1}|V_0 - V_1|(V_0 - V_1)\varphi_1,$$

$$\varphi_l = -(1 - C_1)\varphi_p - C_1 \frac{\rho_1}{\rho_0}\left(1 + \frac{k_1}{2}\right),$$

$$\varphi_p = (1 - C_1)\frac{\rho_1}{\rho_0} + C_1 + \frac{k_1}{2},$$

$$\varphi_1 = C_1(1 - C_1)\left(1 - \frac{\rho_1}{\rho_0}\right),$$

$$\Omega_1 = (1 - C_1)\left(\frac{\rho_1}{\rho_0} - 1\right)g\sin\alpha -$$

$$-\frac{\lambda \rho_{mix}|V_{mix}|V_{mix}}{2 D_p \rho_0} - \frac{3}{8}\frac{C_{xs}}{R_1}|V_0 - V_1|(V_0 - V_1).$$

It has to be emphasized, that characteristic relations (7), (8) on acoustic characteristics represent relations between total differentials of p, V_0 and V_1 functions along these characteristics, but does not include differential of concentration C_1.

Characteristic condition (11) is fulfilled along $x = const$ lines, so differentials dV_0 and dV_1, which enter the expression, stand for increment of corresponding functions with time at every fixed section of pipeline.

It should be noted, that characteristic relation (11) does not contain differentials of concentration C_1.

Thereby in general case of liquid mixed with solid dispersed phase there are three families of characteristics, and along each of them a specific relation between the total differentials of unknown functions dp, dV_0 and dV_1 is fulfilled.

Concentration differential dC_1 does not enter into these characteristic relations. Concentration C_1 is to be obtained by solving differential equation (2), which is in fact an ordinary differential equation with respect to $\frac{\partial C_1}{\partial t}$. It allows us to use numerical integration using finite-difference schemes.

The obtained characteristic relations can be used as a basis for numerical calculation of non-stationary characteristic of hydromixture using integrated methodology, which represents a combination of characteristics technique for hydrodynamic parameters p, V_0 and V_1 and finite-difference technique for calculating the concentration C_1.

4 METHODOLOGY

Using the above results, let us construct a comprehensive methodology for calculating the dynamics of two-phase flows.

1. Specifying the initial data. In order to calculate the transients in the hydraulic system the following data is used:

– a scheme of hydraulic system (lengths of separate sections and sizes of pipelines; marks of the height of their docking sections; tilting angle of each section; location of major units and valvings of the system (pumps, check valves, gate valves, etc.), system performance, concentration of solid phase and its grain composition, marks of the receiving end output sections of the system);

– data of hydraulic calculation of the stationary mode (flow/pressure characteristics of pumps, their operating points; hydraulic slopes of pipes and pressure distribution throughout the system in stationary mode; operational velocities of components of the slurry; the coefficients of hydraulic and the coefficients of local losses in valvings, etc.

These hydraulic data serves as initial data for calculating non-stationary processes and the phenomena of water hammer in the hydraulic system, resulting from the abrupt change in mode of operation of one or more units or elements of the valve system (de-energizing of the pump or alarm failure, sudden complete or partial overlap of the valves; routine or emergency operation of check valves.

2. Calculation of unsteady hydraulic parameters begins with the partition of the entire pipeline to a finite number of computational elements of a length Δl_i, the ends of which A_i are reference points for determining the hydrodynamic parameters, and each unit of the hydraulic system is considered as a separate "zero" element, which has zero length, but has its own "input" A_k and "output" $A_k +1$ (Figure 4).

All the initial parameters of the slurry are determined at all points of A_i at time t_0, at which the non-stationary process, which is to be calculated, arises. It is assumed that for each "zero" element (pumps, locking devices and other valvings) hydraulic law of this element (finite or differential equation defining the relationship between differential pressure upstream and downstream of this element and a flow rate of the mixture) is known.

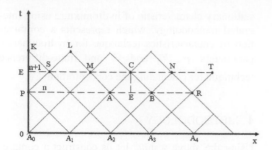

Figure 4. Scheme for using the combined method of characteristics for definition of the unsteady parameters of the slurry.

3. Calculation of unsteady hydraulic parameters goes as follows.

3.1. Determination of the initial distribution of pressure P_0, velocities V_0, V_1 and concentration of the solid particles C_1 in all the nodes A_i^0 $\left(x_i^0, t = 0\right)$

3.2. Coordinates of the points A_i^1 of the new time layer, and new points of observation (x_i^1, t_i^1) are determined from the simultaneous solution of algebraic equations of the characteristics derived from equations (9), (10) by replacing the differentials by finite-difference relations.

$$x - x_A = (D_0)_A (t - t_A), \qquad (13)$$

$$x - x_B = -(D_0)_B (t - t_B). \qquad (14)$$

The coordinates x_i^1 for all zero elements remain the same. As for the distributed elements, the coordinates of the points of observation x_i^1 vary.

3.3. Calculation of pressure P and velocity of the carrier liquid V_0 and solids V_1 at the new time layer at all internal nodes A_i^1 is carried out by solving algebraic equations

$$p_C - p_A + (a_{00})_A (V_{0C} - V_{0A}) +$$

$$+ (a_{01})_A (V_{1C} - V_{1A}) = (b_0)_A (t_C - t_A), \qquad (15)$$

$$-(p_C - p_B) + (a_{00})_B (V_{0C} - V_{0B}) +$$

$$(a_{01})_B (V_{1C} - V_{1B}) = (b_0)_B (t_C - t_B), \qquad (16)$$

$$(a_{10})_E (V_{0C} - V_{0E}) +$$

$$+ (a_{11})_E (V_{1C} - V_{1E}) = (b_1)_E (t_C - t_E), \qquad (17)$$

obtained by replacing the differentials by finite-

difference relations in the equations of characteristics (7) (8) and (11). Here, the coefficients a_{ij} and b_i are determined by comparing equations (15)-(17) with equations (7), (8) and (11), respectively.

The concentration of solid phase is calculated using equation (2)

$$(C_1)_C = \left[(C_1)_E 1 - \left(\frac{1}{\rho_1 a_1^2} \left(\frac{\partial p}{\partial t} \right)_C + \right. \right.$$

$$+ \left(\frac{\partial V_1}{\partial x} \right)_C \right) (t_C - t_E) \right], \qquad (18)$$

where C_{1E} – expressed by interpolating the neighboring nodes A and B, and the derivative $\left(\frac{\partial V_1}{\partial x} \right)_C$ is determined using the values of V at the nodes adjacent to node C, using information from the previous calculation step. The velocity of disturbances propagation (the speed of the shock wave) D is calculated via the formula (6).

3.4. Calculation of hydrodynamic parameters in the boundary nodes (input and output section) as well as "zero" elements is based on the same equations (7), (8), (11) and (18) with appropriate boundary conditions or hydraulic law of each individual element.

3.5. The calculation of each new time layer is carried out by repeating the procedure described in item 3.3-3.4.

5 METHODOLOGY APPROBATION

Numerical experiment. Based on the stated methodology program complex HydroWorks 2p, intended for calculation of dynamics of two-phase flows, is developed. The complex is compatible to CAD-platform SolidWorks 2010/2011 and supports operating systems Windows Vista (x32, x64) and Windows 7 (x32, x64). User is offered two versions of installation package: add-in for SolidWorks and stand-alone application, allowing to work without SolidWorks installed. Application consists of the following units:

– The calculation dll unit, implementing the methodology. The library has an open API interface and can be integrated into other CAD/CAE-systems.

– Dll unit integrated into SolidWorks environment.

– Executable Windows application (.exe).

– Visualizer (dll). Forms reports, displays diagrams and tables.

Among the main functional capabilities, it is possible to select:

– construction of complex parametric pipeline systems;

– complete integration to SolidWorks;

– saving results in external formats (excel, word, txt).

Using the developed software, many numerical experiments have been conducted and the distribution of pressure, velocity and concentration for dynamic tasks in different statements obtained. As an example of the developed methodology usage, we present only the most typical results for the determination of the amplitudes of pressure waves in the pipeline of DPH. One embodiment of such hydro-hoist DPH equipped with three electric submersible pumps H1-H3 (Figure 5).

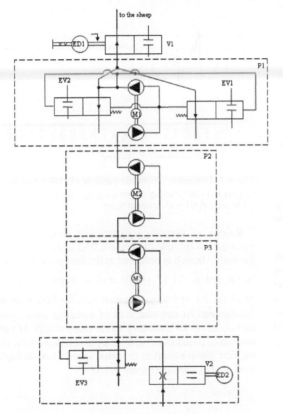

Figure 5. Hydraulic circuit of DPH.

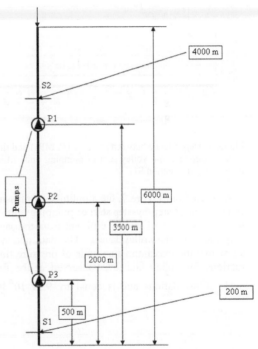

Figure 6. Arrangement of pumping units of DPH.

Slurry submission is conducted from depth of 6000 m. Pumps N1-N3 are installed sequentially in distances of 3500 m, 2000 m and 500 m from a sea-bottom accordingly (Figure 6) at the closed valves of emergency slurry escape (KAC1-KAC3) and open valves (KШ1, KШ2). Regulation of valves KSH1 and KШ2 is carried out by means of electric drives MSH1 and MШ2. Electric drives of pumping aggregates M1-M3 have possibility of regulation by means of the frequency shifter. It is admitted as well direct switching-on of electric drives in a vessel's network.

Valves of emergency slurry escape KAC1-KAC3

are installed parallel to pipeline.

Let us consider the routine sequential start of pumping units lifting up the water. This launch includes a step-by-step switching-on of the pumps, one after another, with some specified interval of time when filling out a pipeline with a homogeneous liquid.

For the first numerical experiment, step-by-step start-up of pumping units from H3 to H1 has been selected with the 5 seconds interval between launches. Alternative of step-by-step start-up of pumping aggregates is volley start-up when the power is simultaneously applied to all pumps. Research was led taking into account connection of engines of all pumps to the frequency shifters, thus acceleration of shafts of the pumps till the rated speed in both cases happens in 5 seconds.

As control sections for research of dynamic parameters of DPH sections S1 (200 m) and S2 (4000) have been selected.

On Figure 7 diagrams of dependence between pressure of a slurry and time in section S1 are shown at volley and step-by-step start-up of pumping units under the conditions described above.

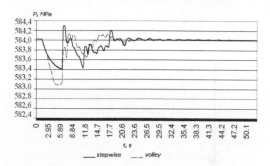

Figure 7. Dependence between pressure (P, MPa) and time (t, s) at stepwise and volley start of pumping units lifting up water in the section S1.

As seen from Figure 7, the amplitude of pressure fluctuations at step-by-step start of pumping units in the cross section S1 is very different from the same amplitude for the volley launch. The data obtained show that the maximum amplitude of pressure fluctuations for volley launch is observed at the first peak of oscillations and is equal to $9.17 \cdot 10^4$ Pa.

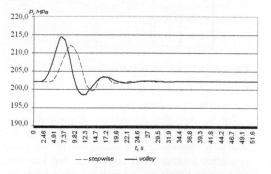

Figure 8. Dependence between pressure (P, MPa) and time (t, s) at stepwise and volley start-up of pumping aggregates lifting water in section S2.

For sequential start such maximum is much smaller ($6.08 \cdot 10^4$ Pa) and is observed in the first and the third peaks of pressure fluctuations due to the non-simultaneous occurrence of pressure waves

in the pipeline and their subsequent superposition. The difference between the maximum deviations from the hydrostatic pressure in the cross section S1 for the volley and step-by-step launches is $3.09 \cdot 10^4$ Pa.

Figure 8 shows the dependence plots between the pressure of the slurry and time in the cross section S2 for the case of stepwise and volley run of the pumps.

Figure 9. Dependence between pressure (P, MPa) and time (t, s) at stepwise start-up of pumping aggregates lifting water in section S1 with/without delay

Results of numerical experiment show that the maximum amplitude of pressure fluctuations of slurry for volley launch is observed at the first peak of oscillation and is $21.17 \cdot 10^5$ Pa and at stepwise start – $9.83 \cdot 10^5$ Pa, and also corresponds to the first peak of oscillations. At stepwise start of pumping units, system comes to its stationary mode more mildly, but at such start, it makes sense to delay acceleration time of the first pump's shaft to the rated speed to avoid high load on it at start-up (Figure 9, 10).

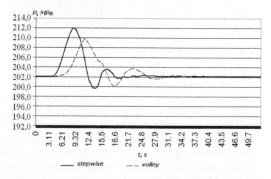

Figure 10. Dependence between pressure (P, MPa) and time (t, s) at stepwise start-up of pumping aggregates lifting water in section S2 with/without delay.

As seen from the numerical experiments, increasing the acceleration time of the pump has reduced the pressure at the peaks of up to $5.58 \cdot 10^4$ Pa and $7.55 \cdot 10^5$ Pa in sections S1 and S2 respectively.

Of particular interest is the emergency shutdown of the system, when one or more of the pumps breaks down and abruptly closes section of the pipeline by its impeller. Typical feature of this transitional process is that at the time of the stop the pipeline is filled with liquid and solid particles and water hammer occurs in two-phase mixture (Figure 11, 12, 13).

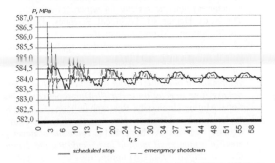

Figure 11. Dependence between pressure (P, MPa) and time (t, s) at an emergency shutdown of pumping units in section S1.

Figure 12 and 13 shows the transients at a scheduled stop of the system starting from the pump unit H1 and ending with unit H3, and vice versa with the interval between power off pumps in 20 seconds in cross sections S1 and S3, respectively.

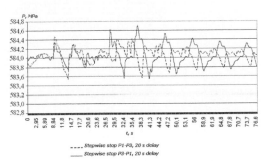

Figure 12. Dependence between pressure (P, MPa) and time (t, s) at a routine stop of pumping units in cross section S1.

In case of an emergency stop of the system, peak values of pressure several times exceed the corresponding values for routine stop. This can cause damage to both pumping units, and the pipeline with poorly predictable consequences. In case of a routine stop, the Figures 12 and 13 show that pressure oscillation amplitude at "bottom-up" algorithm of the pumps' stop is always much lower than the cor-

responding amplitude at "top down" case, which allows selecting the correct scheme of pumps' shutdown.

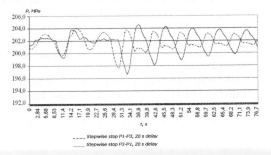

Figure 13. Dependence between pressure (P, MPa) and time (t, s) at a routine stop of pumping units in cross section S2.

Analyzing the obtained results for the three schemes of running the system, can be claimed that the maximum amplitude of pressure oscillation was observed in the case of volley launch of pumping units and reached $21.17 \cdot 10^5$ Pa. This is very unwanted in terms of possible damage to pumping equipment and pipelines in general, and also increases wear and reduces durability of the system elements. The lowest increase in pressure corresponds to step-by-step start-up of the system delaying the launch of the first pump reaching $7.55 \cdot 10^5$ Pa. It is not dangerous and does not create problems in terms of negative effects of water hammer. Such scheme ensures maximum efficiency of start/stop of the system and the risk of damage is minimal due to the avoidance of direct water hammer.

6 CONCLUSIONS

Based on the presented material the following scientific and practical results can be formulated. For the first time a comprehensive methodology of calculating the dynamics of two-phase flows has been developed, allowing quickly and with high accuracy study the full range of stationary and transient processes in the deepwater pumping installations within a single mathematical formalism.

Based on the foregoing, following can be concluded. The investigation of the issue showed that to date there is no method of calculation of nonstationary and transient processes in elements of DPH, taking into account the specifics of their operation in difficult conditions of great depths and providing sufficient accuracy for this class of problems.

Based on the developed methodology, software

package HydroWorks 2p is compiled, which allows to solve various problems associated with the two-phase flow and to determine the parameters of non-stationary and transitive modes in the deep-water mining installations. The developed complex is able to calculate the entire spectrum of transients from the launch of the system (pumping only water) up to the processes associated with the regulation and shutdown (operating with slurry).

Using this software package, different transients in deep-water pumping installations have been studied and the basic parameters of the systems depending on the time in different sections of the pipeline are obtained. It is established that the pressure oscillations caused by these dynamic stresses may reach critical values and significantly affect the efficiency of the installation up to the violation of its integrity.

Next stage of work is the use of the developed application HydroWorks 2p for different calculations of deepwater mining installations and making appropriate improvements and extensions to the existing software package.

REFERENCES

Fox, D. 1981. *Hydraulic analysis of unsteady flow in pipelines*. Moscow: Energoizdat.

Nigmatulin, R. 1987. *Dynamics of multiphase flows*. Moscow: Nauka.

Kartvelishvili, N. 1979. *Dynamics of pressure pipelines*. Moscow: Energiya.

Makharadze, L. 1986. *Nonstationary processes in the pressure hydrotransport systems and protection against water hammer*. Tbilisi: Metsniereba.

Wallace, H. 1972. *One-dimensional two-phase flows*. Moscow: Mir.

Kyrychenko, V., Romanyukov, A., Taturevich, A. 2009. *Study of parameters of water hammer under transient modes in deep-water hydrohoists*. Dnipropetrovsk: National Mining University. Research bulletin of NMU, 1.

Charny, I. 1975. *Unsteady motion of a real fluid in pipes*. Moscow: Nedra.

Kyrychenko, Y. 2001. *Scientific substantiation of the parameters of pipeline systems for minerals hydrohoisting*. Thesis of Doctor of Technical Sciences. Dnipropetrovs'k: National Mining University.

Wud, A. 1934. Sound waves and their application. Moscow, Leningrad: Gostekhizdat.

Goman, O., Kyrychenko, Y., Kyrychenko, V. 2008. *Development of multi-functional dynamic model of a multiphase flow in relation to airlift*. Dnipropetrovs'k: National Mining University. Research bulletin of NMU, 8.

Kyrychenko, Y., Shvorak, V., Kyrychenko, V., Romanyukov, A., Taturevich, A. 2011. *The issue of development of a numerical method for calculating the dynamics of multiphase flows*. Dnipropetrovs'k: National Mining University. Research bulletin of NMU, 2.

Goman, O., Kyrychenko, V., Kyrychenko, Y. 2008. *Determining the propagation velocity of pressure waves in the elements of deep-water hydrohoist*. Dnipropetrovs'k: National Mining University. Research bulletin of NMU, 9.

Kyrychenko, V. 2009. *Substantiation of the rational parameters of deep-water airlift taking into account the transitional processes*. Thesis of Candidate of Technical Sciences. Dnipropetrovs'k: National Mining University.

Technical and Geoinformational Systems in Mining – Pivnyak, Bondarenko & Kovalevs'ka (eds)
© *2011 Taylor & Francis Group, London, ISBN 978-0-415-68877-2*

Electric stimulation of chemical reactions in coal

V. Soboliev, N. Bilan & A. Filippov
National Mining University, Dnipropetrovs'k, Ukraine

A. Baskevich
State higher educational institution "Ukrainian state chemical and technological university"

ABSTRACT: Experimental research has shown that a part of the solid phase of black coal transmuted into gas at action of the weak electric field. It is supposed, that destruction of the carbonaceous and hydrocarbonaceous chains leads to formation of the mobile components. The quantum-mechanical decision of a problem concerning chemical bonds stability of coal chain structure is given.

1 INTRODUCTION

Nowadays coal is exploited largely as energy fuel and technological raw materials for various productions. Coal as a natural nanosystem is becoming an object of the fundamental research. A number of the problems related to a task of coal origin still remains a riddle, and existing theories, versions, hypotheses are not convincing and, as a rule, are sated by paradoxes. Mechanisms, which form a variety of physical and chemical properties of coal, are also contradictory in the same way as the reasons of properties variation within a coal layer. Decision of these tasks is extremely important, especially for development of new technologies of controllable coal processing and making the most use of its power and chemical potential.

Development of new power-saving and eco-friendly technologies of coal treatment initiates necessity for creation of new ways of chemical bonds destabilization in coal components. The behaviour of the coal nanostructure especially at simultaneous action more than two physical parameters (pressure, temperature, electric field, magnetic field, and fluids) characteristic for natural processes of mineral formation has not been studied till now. Each of the numbered parameters could break balanced state in system, initiate development of various chemical processes and have considerable influence on chemical and physicotechnical properties of coal.

Up-to-date results in the field of physical mechanics of coal, which investigate its properties, reasons and mechanisms of unstable state origin in nanostructure, are obtained and applied for elaboration of technical, social and ecological projects in the coal-mining industry (Soboliev 2003; Frolkov & Frolkov 2005; Alekseev 2010). Therefore, actuality of investigations of origin kinetics and new formed hydrocarbonaceous and carbonaceous phases growth, role of the surface in chemical reactions, dynamics of properties changing of coal organic mass in the process of coalification, quantum-mechanical regularities during formation of coal organic mass component composition under action of different physical fields is evident.

Investigations of the weak electric field influence on phase transition in coal have the greatest scientific interest because in nature tectonic activization is accompanied not only by complicated rocks deformation, but also by increasing of electrical and magnetic fields intensity.

2 FORMULATING THE PROBLEM

In (Soboliev 2003) there was suggested for the first time the mechanism of coal organic mass transition into gas as a result of action of pressure with shearing. Experiments confirmed it later (Frolkov & Frolkov 2005). The effect of mechanical energy transmutation into chemical one has been chosen as the mechanism of bonds destruction (Butyagin 1971). In recent times by investigations of structural and phase transformation in coal it is substantiated, that the part of coal mass transfers into gas in temperature range 300-400 K under action of the weak electric current (Soboliev, Chernay & Chernyak 2006). Thus, it is supposed, that transitions "coal → ga" under mechanical and electrophysical action are identical and could be described the unique physical mechanism.

The purpose is creation of physical and mathematical model, which would fulfil requirements of chemical bonds stability in coal components in case of electrophysical action.

Problems of carbonaceous-hydrogen chains stability were considered from the point of view of quantum-mechanical systems stability on the example of a typical carbonaceous-hydrogen chain

$$
\begin{array}{cccc}
\text{H} & \text{H} & \text{H} & \text{H} \\
| & | & | & | \\
\text{H}-\text{C} - \text{C} - \text{C} - \text{C}-\text{H} \\
| & | & | & | \\
\text{H} & \text{H} & \text{H} & \text{H}
\end{array}
$$

3 CALCULATION

For examination of a problem on stability of carbon-hydrogen molecules consisting of linearly arranged atoms of carbon, it is necessary to solve problems about movement of an electron in the field of two Coulomb fluctuating isotropic harmonic oscillators, and about movement in the field of N-Coulomb centres arranged linearly and about charges interaction with a chain of atoms.

Common energy of the given carbon-hydrogen chain consists:

$$ E = E_0 + W_1 + W_2 , $$

where E_0 – interaction energy of particles in the decision of a two-centric problem, W_1 and W_2 – disturbances caused by oscillating movement of Coulomb centres and atoms of a chain.

For the solution a Schrodinger equation in ellipsoidal coordinates

$$
\left\{ \frac{4}{R^2\left(\lambda^2 - \mu^2\right)} \left[\frac{\partial}{\partial\lambda}\left(\lambda^2 - 1\right)\frac{\partial}{\partial\lambda} + \frac{\partial}{\partial\lambda}\left(1 - \mu^2\right)\frac{\partial}{\partial\lambda} + \right. \right.
$$

$$
\left. \left. + \frac{4}{R^2\left(\lambda^2 - 1\right)\left(1 - \mu^2\right)}\frac{\partial^2}{\partial\varphi^2} \right]\psi + \right.
$$

$$
+ 2[E + U(\lambda,\mu)]\psi = 0 \qquad (1)
$$

where

it is necessary that variables λ, μ were parted and the requirement satisfied:

$$ U(\lambda,\mu,\varphi) = \frac{\Phi_1(\lambda) + \Phi_2(\mu)}{\lambda^2 - \mu^2} . $$

Potentials, which allow the equation (1) to be parted:
– a Coulomb potential

$$ U_{coul}(\lambda,\mu,\varphi) = \frac{2}{R}\left[\frac{Z_1}{\lambda + \mu} + \frac{Z_2}{\lambda - \mu} \right] ; $$

– harmoniously fluctuating Coulomb potentials

$$ U_{fluct}(\lambda,\mu,\varphi) = \frac{R^2\omega^2}{8}\left(\lambda^2 + \mu^2\right), \text{ where } \omega \text{ is fre-} $$

quency of the basic oscillations of carbon atoms.

Calculation of energy of diatomic molecule C-C ground states was carried out under the formula

$$ E_{k,\Lambda,n} = \frac{\left\langle \psi_{k,\Lambda,n}\left|H_0\right|\psi_{k,\Lambda,n}^*\right\rangle}{\left\langle \psi_{k,\Lambda,n}\left|\psi_{k,\Lambda,n}^*\right\rangle} , $$

where H_0 – Hamiltonian of a two-centric problem

$$
H_0 = \frac{4}{R^2\left(\lambda^2 - \mu^2\right)}\left[\frac{\partial}{\partial\lambda}\left(\lambda^2 - 1\right)\frac{\partial}{\partial\lambda} + \frac{\partial}{\partial\lambda}\left(1 - \mu^2\right)\frac{\partial}{\partial\lambda} + \right.
$$

$$
\left. + \frac{4}{R^2\left(\lambda^2 - 1\right)\left(1 - \mu^2\right)}\frac{\partial^2}{\partial\varphi^2} \right] ,
$$

where $\psi_{k,\Lambda,n}$ – wave function of a two-centric problem, k,Λ,n – main quantum numbers.

Energy of diatomic molecule C-C:

$$ E = E_{\frac{1}{2},0,0} + E_{e-e} + \frac{Z_1 Z_2}{R} , $$

$$
E_{\frac{1}{2},0,0} = \frac{4\left[\frac{1}{2}\left(a - Z^+\right) + e^{4a} \cdot E_i(-4a) \cdot \left(a^2 - a \cdot Z^+ - \frac{1}{4}a\right) \right]}{R^2\left[\frac{1}{2a} - \frac{4}{3}a \cdot e^{4a} \cdot E_i(-4a) \right]} ;
$$

$$
E_{e-e} = \left\langle \psi_{det}\left|\frac{1}{r_{1,2}}\right|\psi_{det}\right\rangle = \frac{4}{R\left[\frac{1}{2a} - \frac{4}{3}a \cdot e^{4a} \cdot E_i(-4a) \right]^2}\left[\left(\frac{3}{40a^2} + \frac{1}{20a^2} \right)(C + \ln 2a) + e^{8a}E_i^2(-8a)\times \right.
$$

$$
\times\left(\frac{3}{40a^2} + \frac{11}{20a} + \frac{7}{5} + \frac{8a}{15} \right) + e^{4a}E_i^2(-8a)\frac{4a^2}{15} + e^{4a}E_i^2(-4a)\left(-\frac{3}{20a^2} + \frac{1}{2a} - \frac{1}{5} \right) + \frac{1}{8a} - \frac{1}{10} \left. \right] .
$$

For a finite linear chain of atoms in length R, a charge of the first atom designates Z_a and a charge of Nth one – Z_b. In ellipsoidal coordinates (λ, μ, φ):

$$\lambda = \frac{r_a + r_b}{R}; \quad \lambda = \frac{r_a - r_b}{R},$$

where r_a – distance from an electron to the first atom; and r_b is the same one to N^{th} atom accordingly.

The operator of potential energy $U(\lambda, \mu, \varphi)$ of N – Coulomb centres system arranged along a line is expressed in the following way:

$$U(\lambda, \mu, \varphi) = \frac{2}{R}\left[\frac{Z_1}{\lambda + \mu} + \frac{Z_2}{\lambda - \mu}\right] + \sum_{i=2}^{n-1} \frac{2Z_i}{R(\lambda + \mu)} \times$$

$$\times \left[1 - \frac{1 + 2\lambda\mu + \dfrac{4i\lambda\mu}{N-1} + \dfrac{4i^2}{(N-1)^2}}{(\lambda + \mu)^2}\right]^{-\frac{1}{2}}.$$

Last equation show that the potential consists of a Coulomb potential plus disturbance:

$$W_2 = \sum_{i=2}^{N-1} \frac{2Z_i}{R(\lambda + \mu)}\left[1 - \frac{1 + 2\lambda\mu + \dfrac{4i\lambda\mu}{N-1} + \dfrac{4i^2}{(N-1)^2}}{(\lambda + \mu)^2}\right]^{\frac{1}{2}}.$$

In case of $R_1 = R_2$ last expression becomes:

$$W_1 = Z^+ \sum_{i=1}^{\infty} (-1)^i \frac{a^i R^{i-1}}{i!}\left[2I_{i+1} + \frac{2E_i(-4a)}{3}\right] +$$

$$+ Z^- \sum_{i=1}^{\infty} (-1)^i \frac{a^i R^{i-1}}{i!}\left[2I_2 + \frac{2E_i(-4a)}{3}\right] +$$

$$+ Z^+ \sum_{j=1}^{\infty} \sum_{i=4}^{\infty} (-1)^{i+j+1} \frac{a^{i+2j} R^{i+j}}{(8 + 32(i+j-4))} \times$$

$$\times \left[\frac{2I_{i-1}}{(2 + (i+j)+1)} + \frac{2E_i(-4a)}{(2(i+j)+3)}\right],$$

where $Z^{\pm} = Z_1 + Z_2$, $I_i = \int_1^{\infty} \frac{\lambda^i e^{-a(\lambda-1)}}{\lambda + 1} d\lambda$.

To submit the influence of the third Coulomb centre on separately chosen chemical bond, the third centre is presented as some disturbance, which operates on it. Then it is possible to express a Hamiltonian in a view:

$$H_0 = -\frac{h^2}{2M_1}\Delta \vec{R}_1 - \frac{h^2}{2M_2}\Delta \vec{R}_2 - \frac{h^2}{2M_3}\Delta \vec{R}_3 +$$

$$+ \frac{Z_1 Z_2}{|\vec{R}_2 - \vec{R}_1|} + \frac{Z_1 Z_3}{|\vec{R}_3 - \vec{R}_1|} + \frac{Z_3 Z_2}{|\vec{R}_3 - \vec{R}_2|},$$

where H_0 – Hamiltonian of interaction of three particles; $\varPhi$ – full wave function of a temporary Schrodinger equation.

Taking into account the estimation and using two Coulomb centres as the base, the disturbance caused by influence of the third centre is expresses in Neumann decomposition:

$$W_3 = \frac{2Z_3}{R}\sum_{p=0}^{\infty}\sum_{m=-p}^{p}(-1)^m (2p+1)\left[\frac{(p-|m|)!}{(p+|m|)!}\right] \times$$

$$P_p^{|m|}(\lambda_<)Q_p^{|m|}(\lambda_>)P_p^{|m|}(\mu_3)Q_p^{|m|}(\mu_3)e^{im(\varphi-\varphi_3)},$$

where $\lambda_3 = \frac{R_2 + R_3}{R_1}$; $\mu_3 = \frac{R_2 - R_3}{R_1}$; $\lambda_{<,>}$ – major or smaller quantities; $P_p^{|m|}(\lambda_<)$ and $Q_p^{|m|}(\lambda_>)$ – associated functions of Legendre I and II sorts.

Then a problem of three particles taking into account disturbance energy will be following:

$$E_0(R_1, R_2, R_3) = \frac{\langle \psi | H + W | \psi^* \rangle}{\langle \psi | \psi^* \rangle} + E_0(R_1) + \sum_{i=1}^{N} \frac{4aZ_i}{R_i\left[\dfrac{1}{2a} - \dfrac{4a}{3}e^{4a}E_i(-4a)\right]} \times$$

$$\times \left\{Q_0^0(\lambda_3)\left[\frac{2}{3}e^{4a}E_i(-4a) - \frac{1}{4a^2}\right] - \frac{1}{8a^2 e^{2a(\lambda_3-1)}}[P_2(\lambda_3)P_2(\mu_3) - 1] \times\right.$$

$$\times \left[e^{2a(\lambda_3+1)}E_i(-2a(\lambda_3+1)) - e^{2a(\lambda_3+1)}E_i(-2a(\lambda_3-1))\right] - Q_2(\lambda_3)P(\mu_3)\left[\frac{2}{3}e^{4a}E_i(-4a) +\right.$$

127

$$+ \left(\lambda_3^2 - 1\right)e^{2a(\lambda_3 - 1)}E_i\left(-2a(\lambda_3 - 1)\right) + e^{-2a(\lambda_3 - 1)}\left(\frac{1}{2a} - \frac{1}{4a^2} - \frac{\lambda_3}{2a}\right) + \frac{1}{4a^2}\right] +$$

$$+ Q_0(\lambda_3)P_2(\lambda_3)P_2(\mu_3)\left[\left(\lambda_3^2 - 1\right)e^{2a(\lambda_3 + 1)}E_i\left(-2a(\lambda_3 + 1)\right) - \left(\frac{1}{2a} - \frac{1}{4a^2} - \frac{\lambda_3}{2a}\right)e^{-2a(\lambda_3 - 1)}\right. \times$$

$$\times P_2(\lambda_3)P_2(\mu_3)\cdot\left[e^{4a}E_i(\lambda_3 + 1)E_i\left(-2a(\lambda_3 + 1)\right) + \frac{1}{2a}e^{-2a(\lambda_3 - 1)}\right]\right\} .$$

4 DISCUSSION AND CONCLUSION

Calculation of C-C bond energy revealed that the presence of a superfluous electron results in "anti-bonding" of the chemical bond (Figure 1, curve 2). Moreover, the chemical bond practically ceases in case of influence on it more than two electrons (Figure 1, curve 3).

Figure 2 shows that interaction of a positive electric charge with C-C chemical bond increases the distance between atoms and accordingly reduces bond energy and its stability.

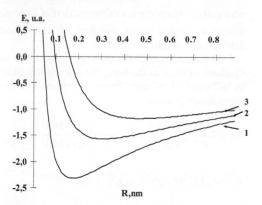

Figure 1. Influence of "superfluous" electrons on bond energy of the next atoms of a carbon molecule: 1 – energy of unperturbed bond C-C; 2 – bond energy C-C taking into account influence on it of a superfluous electron; 3 – bond energy C-C taking into account influence on it of two superfluous electrons.

Experimental research on the weak electric field influence on coal structure was carried out for substantiation of above-mentioned calculation. Destructive processes accompanying by formation of movable components (radicals, gas) take place while weak electric current passing. An electronic paramagnetic resonance (EPR) showed the high concentration of paramagnetic centres about $6.5 \cdot 10^{19}$. The analysis of X-ray diffractograms testifies unequivocally that a degree of coal amorphism is getting higher after electric current passing.

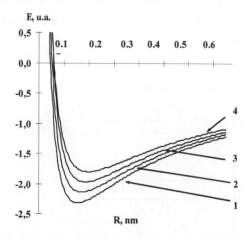

Figure 2. Changing of bond energy from charge $Z = 2$, which is at a distance H from the centre of this bond: $1 - H = 6$; $2 - H = 5$, $3 - H = 4$; $4 - H = 3$.

According to infrared spectroscopy (IRS), destruction of bridge aliphatic chains is confirmed by decline of an optical density of bands 2920 and 2860 sm^{-1} corresponding to the valence and deformation oscillations of bonds C-H in structures, which contain CH_2-groups. Destruction of oxygen-methylene bridges is accompanied by break of CH_3-methanal groups, which related with them (the band 1370 sm^{-1} decreases). Besides, growth of bands 1025 and 1080 sm^{-1} on infrared spectrum, which is typical for primary $(-CN_2ON)$ and secondary $(>CHOH)$ alcoholic groups, points to destruction.

By results of action electrical stimulation of chemical processes in coal is similar to mechanochemical activation; however in case of mechanochemical activation process of coal transition into gas is accompanied by high-speed decomposition of coal.

Quantum-mechanical estimation of influence of exterior elementary electric charges on a chemical bond stability testify to an energy drop between atoms of carbon and bond breaking in case of "superfluous" electrons increase, which is confirmed by EPR and IRS data.

REFERENCES

Soboliev, V.V. 2003. *To the Question about Nature of Outburst Coal Formation.* Collection of scientific works of NMU, 17. V.1: 374-383.

Frolkov, G.D. & Frolkov, A.G. 2005. *Mechanochemical Conception of Outburst in Coal Layers.* Coal, 2: 18-22.

Alekseev, A.D. 2010. *Physics of Coal and Mining.* Kyiv: Naukova dumka: 424.

Butyagin, P.Yu. 1971. *Kinetics and Nature of Mechanochemical Reactions.* Successes of Chemistry. V.40: 1935-1959.

Soboliev, V.V., Chernay, V.V. & Chernyak, S.A. 2006. *Role of Electric Current in Stimulation of Destructive Processes in coal.* Bulletin of Higher educational institutions. The Northern Caucasians region. Techn. Sciences. Appendix, 9: 45-51.

Substabtiation of the parameters of elements of mine vent systems while exploiting bedded deposits of horizontal occurence

V. Golinko, O. Yavors'ka & Y. Lebedev
National Mining University, Dnipropetrovs'k, Ukraine

ABSTRACT: The results of research aimed at increasing efficiency of vent systems of mines and pits are given. There is a substantiation of parameters of such elements of mine vent systems as vertical mine working (mine shafts, holes, holes cluster) built on the area of mining works to remove outgoing air stream at mine with horizontal layer occurrence on comparatively small depth. Economic and mathematical model is developed on which basis formula for determining optimal parameters of vertical mine workings is obtained. Transfer from optimal parameters of elements of mine vent system to rational ones is substantiated.

1 INTRODUCTION

Reducing the way of air current movement through mine ventilation network can be definitely the only way of economic solution of the ventilation problems while exploiting large mine field.

Analysis of the research results and recommendations for Donbass coal mines shows that using mine ventilation network with vertical mine working for removing outgoing air stream (fresh air supply) within mining allows to reduce considerably the way of air current movement through the ventilation network, to reduce depression of ventilation network, to increase the amount of incoming air, to decrease ventilation loss etc. As a rule, use of such ventilation networks without necessary scientific and economic grounding results in decrease of ventilating efficiency.

2 FORMULATING THE PROBLEM

While exploiting deposits with 70-100 m occurrence of horizontal seam (for example, Podmoskovya and Nikopol basins) it is possible to use direct-flow ventilation with moving air taken away at a flank of mine field near mine workings. Such ventilation networks can be implemented only under condition of consecutive vertical shaft sinking (air shaft or hole) within stoping. It helps take into account dynamics of mining and remove return ventilation current straight in the area of mining instead of transporting contaminated air through the whole ventilation network in the opposite direction. In such a way it is possible to reduce considerably air

flow path, to increase the scheme reliability, and to decrease ventilation loss. While advancing mining air shafts or holes are being liquidated with new ones replacing them.

Under advance mining protective pillar are left, the scheme of mine working ventilation is done in the following way. Fresh air is supplied to the face through central and air shafts, and then through airway and gate road. Outgoing air current is removed up to the surface through airway and hole. When mining operations approach to the hole the latter is sealed and outgoing air current is removed through a newly built air hole (n is the number of air shafts or holes built consecutively from the surface to a drift within the mining period with L length).

To put the ventilation schemes into operation it is important to know such parameters as quantity of return air shafts or holes, their location in the scheme and diameter or section. The matter is that the parameters influence greatly capitalized expenses while constructing ventilation schemes under conditions of high concentration of stoping.

3 BASIC UNIT

To calculate optimal values d and n, corresponding to the minimum of total expenditures, the economic and mathematical model of total expenditures is obtained (Golinko, Yavors'ka & Lebedev 2009).

$$f(n,d) = 1.9 \cdot 10^3 \cdot d \cdot l \cdot n + \frac{lkQ^3 T\alpha P}{2 \cdot S^3} +$$

$$+\frac{lkQ^3T\alpha P}{2\cdot S^3 n}+\frac{6.5lkQ^3T\alpha_h}{2\cdot d^5},\tag{1}$$

where l – length of air hole, m; d is diameter of hole, m; k – coefficient, taking into account annual cost of electricity for the system; Q – consumption, m^3/s; T is period of pillar mining, years; α – coefficient of aerodynamic resistance of the mine working; P – perimeter of mine working, m; S – cross-section area of mine working, m^2; α_h – coefficient of aerodynamic resistance of a hole.

The obtained model is dynamic one as it reflects dynamics of the process of ventilating boundary area of slope mine while removing outcoming air through consecutively built ventilation holes. It often turns that values of such a parameter as the quantity of ventilation holes or shafts determined according to the obtained expression (Golinko, Yavors'ka & Lebedev 2009) are not whole numbers that is why they need to be rounded. Herewith total given expenditures increase to some extent. If function is symmetrical, rounding to the near whole number is allowable. In this case costs transforming from the parameters expected values to the whole ones increase equally. If the function is asymmetrical, and total expenditures increase differently while deviating from expected value either side rounding to the near hole can result in ungrounded increase in total expenditures.

Consider function of total expenditures (1) for ventilation and hole construction throughout the mine working. After examining the function as for extreme point we get:

$$\frac{\partial f(d)}{\partial d}=1.9\cdot 10^3\cdot l\ n-\frac{32.5\cdot lkQ^3T\alpha_h}{d^6}=0$$

Whence

$$d=0.5\cdot\sqrt[6]{\frac{k\alpha_h Q^3 T}{n}}.\tag{2}$$

One may not take into account the second term in function (1), as it can not influence d and n parameters. Hence, taking into account (2) we have:

$$f(n)=\frac{LkQ^3T\alpha P}{2\cdot S^3}\frac{1}{n}+$$
$$+1.16\cdot 10^3\cdot n^{5/6}\cdot l\cdot\sqrt[6]{kQ^3T\alpha_h}.\tag{3}$$

If we set

$$a=\frac{LkQ^3T\alpha P}{2\cdot S^3}.\tag{4}$$

$$b=1.16\cdot 10^3\cdot l\ \sqrt[6]{kQ^3T\alpha_{\text{скв}}}.\tag{5}$$

We get

$$f(n)=\frac{a}{n}+bn^{0.81}.\tag{6}$$

Optimal amount of ventilation holes corresponding to minimum of function is determined from the expression

$$n_o=\left(1.23\cdot\frac{a}{b}\right)^{0.55}.\tag{7}$$

On the basis of (7), we get

$$a=0.81\cdot bn_o^{1.81}.\tag{8}$$

$$b=1.23\cdot\frac{a}{n_o^{1.81}}.\tag{9}$$

Under optimum quantity of ventilation holes with taking into account (8) and (9), the expression (6) is

$$f(n_o)=2.23\cdot a\frac{1}{n_o}=1.81\cdot bn_o^{0.81}.\tag{10}$$

Having divided expression (6) by (10) we get:

$$f\left(\frac{n}{n_o}\right)=0.45\cdot\frac{n_o}{n}+0.55\cdot\frac{n^{0.81}}{n_o^{0.81}}.\tag{11}$$

If we set $n/n_0=\delta$, we get:

$$f(\delta)=\frac{0.45}{\delta}+0.55\cdot\delta^{0.81}.\tag{12}$$

The expression is the function of fractional variation of total costs for ventilation and ventilation holes construction if their optimum quantity no is replaced by rational n.

Function graph (12) is given in Figure 1 on which basis at optimum value of no it is possible to write

$$f(\delta)=0.45\cdot\delta_o^{-1}+0.55\cdot\delta_o^{0.81}=1.\tag{13}$$

Let Δ is deviation of function value (1) from optimum

$$f(\delta)-f(\delta_o)=\Delta f(\delta_o).\tag{14}$$

If we put function value (12) into equation (11) taking into account (10) we get

$$0.45\cdot\delta^{-1}+0.55\cdot\delta^{0.83}=1+\Delta.\tag{15}$$

132

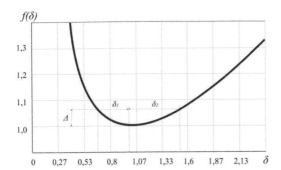

Figure 1. Change of relative expenses under deviation from optimum.

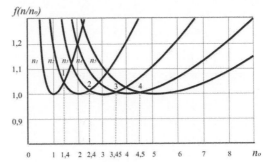

Figure 2. Graphs of function (11) for neighbouring n_o values.

Taking value Δ, and solving equation (15) we find that two quantitatively different values δ_1 and δ_2 correspond to each Δ value. It shows that function of total costs is asymmetrical. Thus transition to the near whole, while obtaining fractional value n_o, can lead to groundless cost increase. Correlation between optimum fractional and rational integral number of holes can be expressed by the inequation $n < n_o < n+1$. It means that more or less integral values n correspond to each optimum value. Rational one is the value transition to which will cause less cost increase. Graph of function (11) shows that for neighbouring goal objectives of n and $n+1$, shown in Figure 2, it can be seen that each interval from n up to $n+1$ contains such value as n_Δ. If we take any of the two neighbouring integral values n or $n+1$ instead of n_Δ it can result in similar increase of total costs. Therefore if in the process of determination n_o we will see that $n_o < n_\Delta$, it is necessary to take less value of the integral. And visa versa, if $n_o > n_\Delta$ then it is more.

Develop a formula for calculating n_Δ. It is obvious that ratio $f(n)/f(n+1) = 1$ occurs for neighbouring integral values n and $n+1$. Taking into account this ratio and basing on expressions (6) and (8) we obtain

$$\frac{f(n)}{f(n+1)} = \frac{0,81 \cdot \dfrac{n_\Delta^{1.81}}{n} + n^{0.81}}{0.81 \cdot \dfrac{n_\Delta^{1.81}}{n+1} + (n+1)^{0.81}} = 1. \qquad (16)$$

Whence

$$n_\Delta = 1.124 \cdot \left[n(n+1)^{1.81} - (n+1)n^{1.81} \right]^{0.55}. \qquad (17)$$

Table 1 gives rounding intervals determined from the expression (17).

If calculation results show that $n_o < n_\Delta$, it is necessary to take less neighbouring integral n; if $n_o > n_\Delta$, it is more.

Table 1. Rounding intervals for transition of value of no parameter to rational ones.

Rational Values	1	2	3	4	5	6	7
n_Δ	1.41	2.45	3.48	4.5	5.5	6.54	7,56
Rounding Intervals	0.7-1.41	1.41-2.45	2.45-3.48	3.48-4.5	4.5-5.5	5.5-6,54	6.54-7.56

4 CONCLUSIONS

1. Economic and mathematical model is developed on which basis formula for determining optimal parameters of vertical mine workings is obtained.

2. Transfer from optimal parameters of elements of mine vent system to rational ones is substantiated.

REFERENCES

Golinko, V.I., Yavors'ka, E.A & Lebedev, Y.Y. 2009. Estimation *of efficiency of vent networks of manganese mines*. Materials of International Research and Practice Conference "School of Underground Mining": 167-184.

Ennobling of salty coals by means of oil agglomeration

V. Beletskyi
Donetsk national technical university, Donetsk, Ukraine

T. Shendrik
Ukrainian academy of sciences, Donetsk, Ukraine

ABSTRACT: The phenomenon of coal modification at oil agglomeration has been studied, specifically, changing of its structure and physical-chemical properties of surface. It was established: agglomeration process of coal with size 0-1(3) mm is accompanying by direct adhesive contact of "coal-oil" on 75-80% from external surface of coal. The high power chemical bonds are formed together with physical bondes in the interphase zone. This modification leads to the increasing hydrophobicity of coal surface and the contrastance of mosaic liophylic-liophobic picture; internal surface of coal is hydrophobized by diffusing oil agent into pores and fissures. Infiltration phenomenon intensifies this process since light fractions of a binder penetrate into micropores of coal substance; the changes in supermolecular structure of coal organic mass (COM) have been revealed in oil agglomeration process. The experience carried out on low rank salty coals of Western Donbas.

1 INTRODUCTION

Perspective of exhausting natural reserves of petrol and gas and increase of coal consumption have conditioned an increasing interest of the world scientists to coal technologies. Special attention is given today to the study of special processes of coal preparation and ennobling, and this opens new possibilities of processing of low grade raw materials to conditional ecologically clear products (Ding & Erten 1989).

During past ten years the process of coal selective oil agglomeration is quickly developed. The interest to this problem is considerable not only from specialists in fossils ennobling, but also coal chemists, heat-power engineers, transport workers (Biletskyi, Sergeev & Papushyn 1986; Shrauti & Arnold 1995; Wheelock etc. 1994; Mishra Surendra & Klimpel Richard 1989). The oil agglomeration is considered by us as a perspective high efficient substance of preparing low quality coal to coking, burning, pyrolysis and also as polyfunctional process of coal preparation to its liquefaction. Besides, some investigators have demonstrated advantages of utilization the techniques and technologies of oil agglomeration in main hydrotransport systems of energetic and coking coal (Beletskyi, Papayani, Svitly & Vlasov 1995; Rigbi, Jones & Meiwaring 1982).

Universality and technological parameters of oil pelleting allow to use it for processing of low grade coals, in particular, oxidized, with high ash, and high salt content (salty coals).

The goals of our investigation were: to study the structural changes of coal organic mass (COM) at its oil agglomeration; to estimate the efficiency of oil pelleting using for obtaining ecologically suitable coal products.

2 EXPERIMENTAL

Primary raw materials and reagents. Energetic and coking coals from Donetsk (UKRAINE) are of different rank with ash content from 10 to 14%, have been investigated (Biletskyi, Sergeev & Papushyn 1986). Special attention was given to salty coals (SC) of Novomoskovskoe deposit, Western Donbas. The some characteristics of this sample are, %:

$$W^a = 21.4; \quad A^d = 9.9; \quad C^{daf} = 72.8; \quad H^{daf} = 5.0;$$

$V^{daf} = 42.6$. Na_2O content in ash is 11.6%. The furnace residual oil (mazut M100), oil of oiling charge (OOC) and polymer of benzene department of coke-chemical production have been used (as reagents-binders).

Parameters of oil agglomeration process: Agglomeration process has been carried out in the laboratory granulator of impeller type at solid : liquid $= 1 : 3$, $pH = 7$, $t = 18 - 20\,^\circ C$, frequency of impeller rotation 1500 min^{-1}, concentration of binder was from 3-5 to 25-30 mass. %, size of coal particles were from 0 to 200 mn.

Methods of investigation:

– Optical microscopy of agglomerates (microscope NEOPHOT-21);

– Electron parametric resonance (radiospectrometer RE-1306);

– IR-spectrometry (Specord IR-75, Perkin-Elmer);

– X-Ray diffractometry (DRON-UM-1.5);

– Photocolorimetric studying of coal surface.

3 RESULTS AND DISCUSSION

Microscopic investigations of coal-oil structures anshliffs (flocules, agglomerates, granules) allowed to select 4 principal structure types of coal aggregates (Figure 1):

I – *pellicle* – compact formations, with thin broad pellicles of oil-binder between individual grains of coal.

II – *meniscus* – structures with concave meniscus of binder between coal grains on aggregates surface.

III – *powdered* – when binder drops filled with coal grains.

IV – *bridge* – friable formations of coal grains, which are bonding with oil-binder "bridges".

With photocolorimetric methods (using methylene blue solution) (Kolbanovskaya 1959) has been established the part of coal grains surface, which are cowed with a binder for the I type structures consist of 57-79%, II – 86-95 %, III – 100%, IV – 40-44%. This case on the coal grains surface of I type oxipellicle ("white border" around coal grain) can be kept; it doesn't exist in III rd type of granules (Figure 2).

However, for salty coals that high parameters are achieved only at special treatment of raw material with oil agent.

In general, the salty coal has a hydrophilic surface and it is agglomerated by apolar agent with difficulty. To increase its agglomeration capacity, we have worked out a number of steps.

Firstly, this is oil treatment of coal (with mazut M100, for exemple) during the crushing process (Patent 1514404).

Secondly, for salty coal agglomeration, it is recommended to use aromatized agents which contain functional oxygen groups in the side links (Biletskyi 1986).

Thirdly, the agglomeration process itself should be carried out in conditions of increased turbulance of water-coal-oil mixture ($Re > 3000$) and Solid: Liquid = 1:1(3). Without special treatment of coal surface, cowing with the binder is fluctuated between 0-5% only. This fact can be interpretated as a low natural ability of SC to agglomeration.

| 1 | 2 | 3 | 4 |

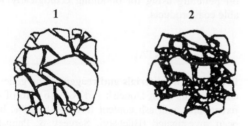  

Figure 1. Oil-coal agglomerates (×5 times).

I type

(a)

(b)

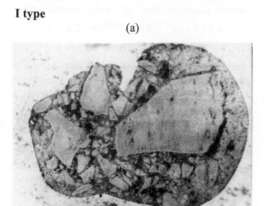 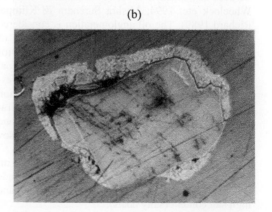

III type

(c)   (d)

Figure 2. Anshlifs of granules: I type (a, b) – × 600; III type (c, d) – × 400.

Microscopic investigations allow to confirm a penetration of a binder in pores and fissures of coal substances (Figure 3). It is obviously, that this process is accompanied by infiltration phenomena, during which light fractions of binder penetrate into micropores and more heavy ones remain on surface of coal grains. The last one promotes the formation of border solvate layer of binder on coal surface.

As the result, the cohesion of binder pellices is arised, and the stability of aggregates (agglomerates, granules) is increased.

Paramagnetic characteristics of samples were determined on serial radiospectrometer RE-1306 at wave length $\lambda = 3.2$ sm in air. Mn^{+2} in a lattice of MgO was the internal standard.

The obtained data for constituents of agglomerate, intermediate and end products are shown in Table 1. It is seen, that the first addition of oil agent (I stage of process) did not influence on nature and paramagnetic centers (PMC) concentration significantly.

Oil-agent M 100 is more complex, because its three types of PMC are characterized. They are very differed from one another (Figure 4).

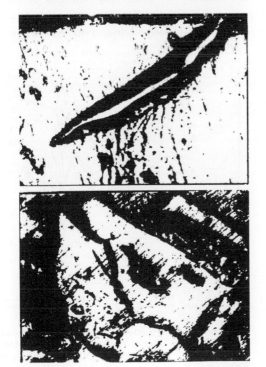

Figure 3. Anshlifs of oil-coal granules with the penetration of oil into coal substance × 600.

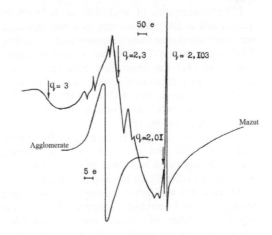

Figure 4. ERS-spectra of M100 and agglomerate, which has been obtained on the basis of salty coal (Novomoskovskoe deposit, Western Donbas).

137

Table 1.

Sample	ΔH , Gs	g -factor	N , spin / g
Initial coal	4.6	2.0035	$8.9 \cdot 10^{17}$
Coal +5% mazut:			
In air	4.6	2.0036	$8.9 \cdot 10^{17}$
In vacuums	4.9	2.0036	$9.8 \cdot 10^{17}$
Mazut:			
Signal 1	4.6	2.0031	$9.7 \cdot 10^{16}$
Signal 2	12.4	2.0169	$9.7 \cdot 10^{16}$
Signal 3	320	4	$9.7 \cdot 10^{16}$
Agglomerates:			
Agglomerate – 23	4.8	2.0035	$1.4 \cdot 10^{16}$
Agglomerate – 29	5.1	2.0035	$4.8 \cdot 10^{16}$

Table 2.

Sample	Oxygen-containing groups, mg-ekv / g			Na – contain, mg-ekv / g		Heat of combustion
	OH	$COOH$	$O = N = O$	in coal	in ash	Q , MJ / kg
Initial coal	1.60	0.20	9.3	0.30	4.08	24.3
A-23	2.44	0.11	7.1	0.10	1.77	28.2
A-29	2.40	0.10	6.5	0.10	1.88	28.6

Literature data analysis allowed to classify signal I (at $g = 2.0031$ and $\Delta H = 4.6$ e.) as π -polyconjugated systems (Berlin 1972), signal II (with $g \sim 3-4$ and $\Delta H = 320$ e.) as ferrum-containing paramagnetic structures (Ingram 1972), but signal III (with $g \sim 2.017$ and $\Delta H \sim 12$ e.), perhaps, as radicals of peroxide ($R-O-O$) or $R-S-S$ (Butuzova, Saranchuk & Shendrik 1985) types.

PM-characteristics of agglomerates A-23 (% of mazut) and A-29 (% of mazut), as it was expected, are more similar to same of parent coal, excluding PMC concentration. The last one is lower. It is testified about chemical interaction between coal and oil, or, at any rate, about sharply change of intermolecular interactions into COM. Evidently in this case, the important role belongs to water, as a strong hydrolyzing agent .

If it is so, the changes in supermolecular organization of COM and in active oxygen-containing group's composition must take place. As has been established (Table 2), in the agglomeration process of SC carboxyl and quinone groups decreasing and phenol hydroxyl increasing happened. Besides, the significant lessen of sodium concentration takes place, that is to say, desalting raw material to conditional level (Na_2O in ash < 2 %).

Evidently the annihilation of ERS signal II of M 100 (g -factor ~ 3-4; $H \sim 320$ e.) and quinone con-

centration decreasing, are connected between one and another. We admit, that formation of helate complexes take place here, the central ion is Fe^{3+} , and the ligands are electron donor quinone structures.

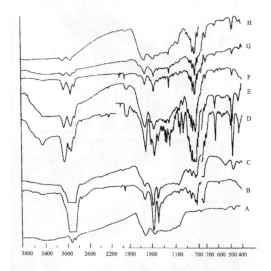

Figure 5. IR-spectra of oil-coal aggregates and their components: A – coal of mine "Inskaja", sort G; B – M100; C – agglomerate "M100-coal"; D – OOC; E – agglomerate "OOC-coal"; F – polymer; G – aggregate "polymer-coal"; H – same agglomerate after 10-days keeping on air.

The phenol groups increasing can be explained with particular quinones reduction by proton donor structures of oil or by hydrolysis of salt form (phenolates) during agglomeration process, which include the water treatment stage.

The decreasing of free carboxyl groups concentration may be explained with their particular substitution by metal or strong H-bonds formation (IR-spectra, Figure 5).

These conclusions are confirmed by IR-data and X-Ray investigations, which we are going to be published in the next article.

4 CONCLUSIONS

During oil agglomeration of coals radical changes of physic-chemical surface characteristics of coal substances take place. They are caused by interphase interaction "coal-oil". That is why it can wait sufficiently different technological properties of agglomerated coals and parent coals.

REFERENCES

Ding, Y. & Erten, M. 1990. *Selective flocculation versus oil agglomeration in removing sulfur from ultra fine coal*. Proc. and Util. High Sulfur Coals III: 3 rd. Int. Conf. Ames. Iowa: 255-264. Amsterdam.

Biletskyi, V.S., Sergeev, P.V. & Papushyn Yu.L. 1986. *Theory and practice of coal selective agregation*. Donetsk: Gran: 264.

Shrauti, S.M. & Arnold, D.W. 1995. *Recovery of waste fine coal by oil agglomeration*. Fuel. V.74 (3): 54-465.

Wheelock, T.D. etc. 1994. *The role of air in oil agglomeration of coal at a moderate shear rate*. Fuel. V.73 (7): 1103-1107.

Ed. Mishra Surendra K., Klimpel Richard R. 1989. *Fine coal processing*. N.J.: Park Ridge. Noyes Publ.: 450.

Beletskyi, V.S., Papayani, F.A., Svitly, J.G. & Vlasov, J.F. 1995. *Hydraulic Transport of Coal in Combination With Oil Granulation*. 8-th Int. Conference on Transport and Sedimentation of Solids Particles: Prague: D6.

Rigbi, G.R., Jones, C.V. & Meiwaring, D.E. 1982. *Slurry pipeline Studeson the BHP-BPA 30-tonne per hour demonstration plant*. 5-th Int. Conf. on the Hydraulic Transport of Solids in Pipes. Johannesburg: D1.

Kolbanovskaya, A.S. 1959. *Colors methods for determination of bitume adhesion with mineral materials*. Moscow: Avtotransizdat: 32.

Patent 1514404 USSR. BO3B7 100. The manner of salty coal processing V.S. Biletskyi, A.T. Elichevitch, Yu.N. Potapenko. Institute polytechnic of Donetsk. #4352939 23-03, declared 29.12.87, published 15.10.89, bul. #38.

Biletskyi, V.S. 1986. *Technological basis of rational using of oil granulation for dewatering and ennobling of hydraulical transported coals*. Ph.D. thesis. Institute polytechnic of Donetsk: 202.

Berlin, I.K. 1972. *Chemistry of polyconjugated systems*. Moscow: Mir: 272.

Ingram, E. 1972. *ERS in biology*. Moscow: Mir: 296.

Butuzova, L.F., Saranchuk, V.I. & Shendrik T.G. 1985. Water *role in themochemical destruction process of coals*. In book Geotecnological problems of fuel-energetic resources of Ukraine. Kyiv: Naukova Dumka: 108 113.

Technical and Geoinformational Systems in Mining – Pivnyak, Bondarenko & Kovalevs'ka (eds)
© 2011 Taylor & Francis Group, London, ISBN 978-0-415-68877-2

Features of the resources of the hard coal covering in thin coal-seams in Poland

A. Krowiak

Central Mining Institute, Katowice, Poland

ABSTRACT: The results of analyses of features of geological resources of the hard coal, in majority of mine located in Poland, were presented, according to adopted arbitrarily criteria. All active mines of the hard in the Silesia voivodship were provided with analysis. Apart from analysis there was only one mine stayed – "Bogdanka" in the Lublin Voivodeship. Thin coal seams are defined as the medium thickness in the range from 0.6 up to 1.6 m in the context of this analysis. The following diameters of analyses were accepted: type of coal, division into ranges of the average depth of covering of coal seams, the division into ranges of the average inclination of seams, associative criteria of the average depth of covering and the medium thickness of coal seams and associative criteria for average depth of covering, the average of the calorific value and the average of contents of sulphur. The analysis of concentrations and automatic neural networks were applied in methods of calculation.

1 INTRODUCTION

Described analysis concerns resources of the hard coal covering in active mines in Poland, located in the Silesian Voivodeship.

It is embracing almost 100% of all mines of the hard coal in Poland, apart from one mine "Bogdanka" in the Lublin Voivodeship. The Silesian voivodship is on the South of country. Coal seams are defined as the medium thickness from 0.6 up to 1.6 m.

Over 10 billion ton of the hard coal covering in thin coal- seams are in resources of analysed mines. In Poland, over 30 years ago, it was resigned from their exploitation, having thicker seams at their disposal. It comes back to idea of exploitation of thin coal-seams at present, because thick seams in some mines undergo exhaustion, which threatens these mines the closure with reason of exhaustion of resources.

It is a next argument for the return to the exploitation of thin coal-seams, that many of them, incurring little investments relatively, it is possible to exploit from mine horizon's already made available. Building new mineshafts and new levels of the output in mines is an alternative what requires very big investments.

This analysis concerns geological resources. On this stage of the works, associated with the identification of thin coal-seams, it isn't possible to settle which of them are already now available for the economically justified exploitation. It requires further research works and design. Nevertheless, accepting theoretically, that at least 10% of these resources are available, right away to efficient economical exploitation, we have over 1 billion the tone of coal at our disposal. It is An equivalent of 10 of year's production of the hard coal in Poland.

The analysis was carried out an analysis under the following criteria: type of coal, division into ranges of the average depth of covering of coal-seams, the division into ranges of the average inclination of seams, associative criteria of the average depth of covering and the medium thickness of seams and associative criteria for average depth of covering, the average of the calorific value and the average of contents of sulphur. Chosen criteria of diameters of analysis were chosen arbitrarily. The last diameter of analysis is being taken back to quality parameters of coal, taking into account its calorific value and the content of sulphur.

2 METHOD OF ANALYSIS

For aims of analysis a system of Cartesians coordinates was created in the n- dimension's space, which one kind of variables describing features of a given coal seam corresponds to every of axes in. It is possible to describe every of lines of the set of primary data in the form of the vector in the n-dimensions space, for which coordinates of the top appoint values of individual variables for. For the entire set of primary data we will receive, so, very bulk of vectors located in this space, fastened in the beginning of the system of coordinates.

A method of concentrations and automatic neural networks were used to the further data handling

(Cichosz 2000; Tadeusiewicz 1993). The application of the method of concentrations allows for assigning subsets being characterized by resemblance of variables with reason of oneself (Everitt 2001). For individual sections of analysis, from the entire set of primary data, subsets described only by analysed variables were created.

Automatic neural networks are computer programs imitating learning processes appearing in the human mind in their action. In this particular case a peculiar type of the net was exploited – Kohonena networks, and a STATISTICA program was a tool of the realization 9.0 version (Dokumentacja 2010). Primary data obtained from mines were a base of to do analyse (Materiały źródłowe 2010).

3 STRUCTURE OF RESOURCES ACCORDING TO THE CRITERION OF THE TYPE OF COAL

The criterion of types of coal was applied in this diameter of analysis. It was assumed that giving a few types of coal in characteristics of resources meant that they are in these resources quoin of different types in this analysis.

According to classification in accordance with Polish Norms the following types of coal were distinguished: Type 31 – flame coal; Type 32 – fiery – flame coal; Type 33 – fiery coal; Type 34 – fiery – cooking coal; Type 35 – orto – cooking coal; Type 36 – meta – cooking coal; Type 37 – semi – cooking coal.

In the Table 1 a structure of resources was given according to the criterion of types of coal. In figure 6 a structure of reserves according to the criterion of the type of coal for the whole tested resources was described.

In resources with the whole coals of the following types have large stakes: Type 31 – 2 946 517 th. tone i.e. 27.06% whole of stores; Type 32 – 1 683 001 th. tone i.e. 15.46% whole of stores; Type 31, 32 – 1 267 695 th. tone i.e. the 11.64% whole of stores and the Type 32, 33, 34 – 1 042 487 th. tone i.e. 9.57% whole of stores.

Table 1. Structure of sources according to the criterion of types of coal.

Id.	Types of coals	Size of resources [thousands of ton]	% of participation in resources with the whole
1	Type 31	2 946 517	27.6
2	Type 32	1 683 001	15.46
3	Type 33	70 201	0.64
4	Type 34	648 717	5.96
5	Type 35	537 166	4.93
6	Type 36	4 202	0.04
7	Type 37	37 633	0.35
8	Type 31,32	1 267 695	11.64
9	Type 32,33	445 191	4.09
10	Type 32,34	624 489	5.74
11	Type 32,35	18 820	0.17
12	Type 33,34	256 514	2.16
13	Type 33,35	13 375	0.12
14	Type 34,35	811 447	7.45
15	Type 35,36	23 729	0.22
16	Type 35,37	69 252	0.64
17	Type 36,37	5 478	0.05
18	Type 31,32,33	30 960	0.28
19	Type 31,32,34	57 766	0.53
20	Type 32,33,34	1 042 487	9.57
21	Type 32,34,35	35 494	0.33
22	Type 33,34,35	84 125	0.77
23	Type 34,35,37	5 475	0.05
24	Type 35,36,37	1 913	0.02
25	Type 31,32,33,34	70 959	0.65
27	Type 32,33,34,35	95 713	0.88
	TOTAL	10 888 319	

Source: own study.

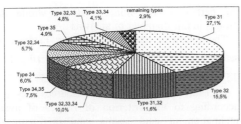

Source: own study.
Figure 1. Structure of resources according to the criterion of types of coal.

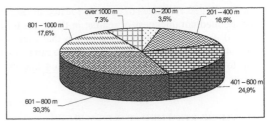

Source: own study.
Figure 2. Structure of resources according to the criterion of the average depth of covering of coal seams.

4 STRUCTURE OF RESOURCES ACCORDING TO THE CRITERION OF IDENTITY FOR RANGES OF THE AVERAGE DEPTH OF COVERING OF COAL SEAMS

In this diameter of analysis a criterion of identity was applied for ranges of the average depth of covering of restores. In analysis the following ranges of the average depth of covering were accepted: 0 – 200 m, 201 – 400 m, 401 – 600 m, 601 – 800 m, 801 – 1000 m, over 1000 m.

In the Table 2 a structure of resources was given according to the criterion of identity for ranges to the depth of covering of coal seams, and in picture 2 graphically this structure of resources was described.

In resources with the whole the largest reserves of coal are appearing in ranges of the average depth of covering: in the range from 601 up to 800 m – 3 299 842 thousands tone i.e. the 30.31% whole of stores and in the range from 401 up to 600 m – 2 708 947 thousands tone i.e. 24.88%.

Table 2. Structure of resources according to the criterion of the average depth of covering of coal seams.

Id.	Ranges of the depth of covering	Size of resources [thousands of ton]	% of participation in resources with the whole
1	0-200 m	381 258	3.50
2	201-400 m	1 791 386	16.45
3	401-600 m	2 708 947	24.88
4	601-800 m	3 299 842	30.31
5	801-1000 m	1 910 948	17.55
6	over 1000 m	795 938	7.31
	TOTAL	10 888 319	

Source: own study.

5 STRUCTURE OF RESOURCES ACCORDING TO THE CRITERION OF THE MEDIUM THICKNESS OF COAL SEAMS

In this diameter of analysis a criterion of ranges of the medium thickness of coal seams was applied. The following ranges of the medium thickness of seams were accepted: 0.6 up to 0.8 m; 0.81 to 1.0 m; 1.01 up to 1.2 m; 1.21 up to 1.4 m; 1.41 to 1.6 m and above

1.6 m. In the range resources above of the average thicknesses 1.6 m are finding underground oneself resources about the variable of thicknesses, in which the lower limit of the thickness is located in a range 0.6 up to 1.6 m.

In Table 3 a structure of sources was given according to the criterion of ranges to the medium thickness of decks, and in Figure 3 a graphical illustration of this structure was described.

Table 3. Structure of resources according to the criterion of ranges of the medium thickness of coal seams.

Id.	Ranges of the medium thickness of coal seams	Size of resources [thousands of ton]	% of participation in resources with the whole
1	0.6 to 0.8 m	821 037	7.54
2	0.81 to 1.0 m	4 120 405	37.84
3	1.01 to 1.2 m	3 492 662	32.08
4	1.21 to 1.4 m	1 577 880	14.49
5	1.41 to 1.6 m	500 760	4.60
6	over 1.6 m	375 575	3.45
	TOTAL	10 888 319	

Source: own study

The largest resources of coal are located in ranges of the medium thickness of seams: in the range 0.81 to 1.0 m – 4 120 405 thousand of ton that presents 37.84% of whole of resources and in the range 1.01 up to 1.2 m – 3 492 662 thousand of ton that presents 32.08% of whole of resources.

6 THE STRUCTURE OF RESOURCES ACCORDING TO CRITERION OF THE AVERAGE INCLINATION OF COAL SEAMS

It was applied a criterion of ranges of average inclination of coal seams as well as the resources selection into the capital groups and then its division into

the particular coal mines that are within those groups in this diameter of analysis .

The following ranges of average inclination of seams were accepted: $0°$ to $5°$; $6°$ to $10°$; $11°$ to $15°$; $16°$ to $20°$; $21°$ to $25°$; above $25°$.

In Table 4 a structure of resources was given according to the criterion of ranges of average inclination of coal seams, and in Figure 4 a graphical illustration of this structure was described.

The largest coal resources are located in range of average inclination of seams: in range $0°$ to $5°$ – 4 559 024 thousand tone i.e. the 41.87% whole of stores and in range $6°$ to $10°$ – 3 829 225 thousand tone i.e. the 35.17% whole of stores.

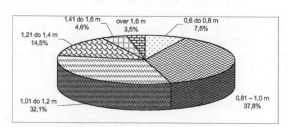

Source: own study.

Figure 3. Structure of resources according to the criterion of ranges of the medium thickness of coal seams.

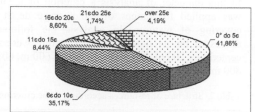

Source: own study.

Figure 4. The structure of resources according to criterion of ranges the average inclination of coal seams.

Table 4. Structure of resources according to the criterion of ranges of average inclination coal-seams.

Id.	Ranges of average inclination of coal seams	Size of resources [thousands of ton]	% of participation in resources as total
1	0° to 5°	4 559 024	41.86
2	6° to 10°	3 829 225	35.17
3	11° to 15°	918 671	8.44
4	16° to 20°	936 025	8.60
5	21° to 25°	189 195	1.74
6	above 25°	456 179	4.19
	SUM	10 888 319	

Source: own study.

7 STRUCTURE OF RESOURCES ACCORDING TO ASSOCIATIVE CRITERIA OF THE AVERAGE DEPTH OF COVERING AND THE MEDIUM THICKNESS OF COAL SEAMS

It was applied the associative criteria of the average depth of covering of coal seams and the average of thickness of these seams as well as the resources selection into the capital groups and then its division into the particular coal mines that are within those groups in this diameter of analysis. The following ranges of the average depth of covering of seams

were accepted: 0 up to 200 m; 201 up to 400 m; 401 up to 600 m; 601 up to 800 m; 801 up to 1000 m; over 1000 m.

The following ranges of the medium thickness of coal seams were accepted: 0.6 up to 0.8 m; 0.81 up to 1.0 m; 1.01 up to 1.2 m; 1.21 up to 1.4 m; 1.41 to 1.6 m and the over 1.6 m. The seam of the thickness over 1.6 m were ranked among thin coal seams about the diversified thickness, which the lower limit of the thickness is located in a range from 0.6 up to 1.6 m.

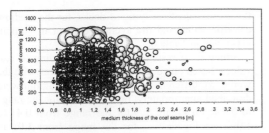

Source: own study.

Figure. 5. Structure of resources according to associative criteria of ranges of the average depth of covering and ranges of the average of thickness of seams appointed according to real values.

The structure of resources according to the associative criterion of ranges to the medium thickness and the average depths of covering of stores was given in Table 5. The illustration of structure of resources appointed according to real values was showed on Figure 5 and on Figure 6 appointed according to ranges of value.

The largest resources of coal are located in the ranges of the medium thickness of coal seams and the average depths of covering: in the range of the thickness 0.81 up to 1.0 m and the range of the depth of covering 601 to 800 m – 1 590 812 thousand ton i.e. the 14.61% whole of stores.

Table 5. The structure of resources according to associative criteria of ranges of average depth covering as well as the ranges of average of thickness seams.

Ranges of the average depth of covering of seams	Ranges of medium thicknesses of coal seams [m]						SUM
	0.6-0.8	0.81-1.0	1.01-1.2	1.21-1.4	1.41-1.6	over 1.6	
Absolute values [thousands of ton]							
0 to 200 m	66 599	126 630	100 305	58 386	9 992	4 619	1 791 386
201 to 400 m	114 000	650 579	509 224	301 317	102 503	113 763	2 712 548
401 to 600 m	227 235	1 113 775	879 427	332 538	83 671	75 902	3 314 569
601 to 800 m	228 071	1 590 812	997 034	260 852	134 432	43 368	375 575
801 to 1000 m	87 468	509 240	641 519	439 086	153 989	79 646	1 910 948
over 1000 m	37 664	129 368	365 153	185 701	16 173	58 278	792 337
SUM	821 037	4 120 404	3 492 662	1 577 880	500 760	375 576	10 888 319
Percentage share in the whole of resources							
0 to 200 m	0.61	1.16	0.92	0.54	0.09	0.04	3.36
201 to 400 m	1.05	5.98	4.68	2.77	0.94	1.04	16.46
401 to 600 m	2.09	10.22	8.08	3.05	0.77	0,70	24.91
601 to 800 m	2.65	14.61	9.16	2.40	1.23	0.40	30.45
801 to 1000 m	0.80	4.68	5.88	4.03	1.41	0.73	17.53
over 1000 m	0.35	1.19	3.35	1.71	0.15	0.54	7.29
SUM	7.55	37.84	32.07	14.50	4.19	3.45	100

Source: own study.

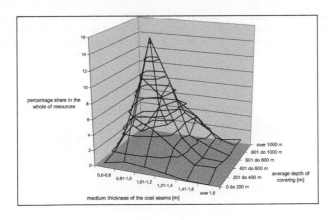

Source: own study.

Figure 6. Structure of resources according to associative criteria of ranges of the average depth of covering and ranges of the average of thickness of coal seams appointed according to ranges of value.

8 STRUCTURE OF RESOURCES ACCORDING TO ASSOCIATIVE CRITERIA OF THE AVERAGE DEPTH OF COVERING, OF THE MEDIUM CALORIFIC VALUE AND THE AVERAGE CONTENT OF SULPHUR

It was applied the associative criteria of the average depth of covering of coal seams, of the medium calorific value and the average content of sulphur in as well as the resources selection into the capital groups and then its division into the particular coal mines that are within those groups in this diameter of analysis.

The following ranges of the average depth of covering of coal seams were accepted: 0 up to 200 m; 201 up to 400 m; 401 up to 600 m; 601 up to 800 m; 801 up to 1000 m; over 1000 m. The following ranges of the medium calorific value were accepted: to 18.0 kJ / kg; 18.01 to 22.0 kJ / kg; 22.01 to 26.0 kJ / kg; 26.01 to 30.0 kJ / kg and above 30.0 kJ / kg.

The following ranges of the average content of sulphur were accepted: to 0.4%; 0.41 to 1.0%; 1.01 to 1.6%; 1.61 to 2.2% and above 2.20%.

The structure of resources for the whole of stores according to the associative criterion of ranges to the average depth of lying of decks, for the medium calorific value and the average content of sulphur was given in Table 6.

Structure of resources was set according to real values, based on adopted criteria, for stores with the whole, for different ranges contents of sulphur were described appropriately: on Figure 7 – for the range to 0.4%; on Figure 8 – for the range from 0.41 to 1.0%; on Figure 9 – for the range from 1.01 to 1.6%; on Figure 10 – for the range 1.61 to 2.2% and on Figure 11 for the content above 2.21%.

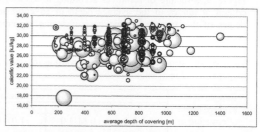

Source: own study.

Figure 7. Structure of resources according to associative criteria of the average depth of covering and the medium calorific value appointed according to real values – for the content of sulphur in coal to 0.4%.

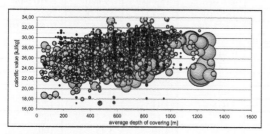

Source: own study.

Figure 8. Structure of resources according to associative criteria of the average depth of covering and the medium calorific value appointed according to real values – for the content of sulphur in coal from 0.41 to 1.0%.

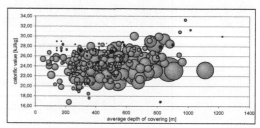

Source: own study.

Figure 9. Structure of resources according to associative criteria of the average depth of covering and the medium calorific value appointed according to real values – for the content of sulphur in coal from 1.01 to 1.6%.

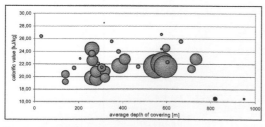

Source: own study.

Figure 10. Structure of resources according to associative criteria of the average depth of covering and the medium calorific value appointed according to real values – for the content of sulphur in coal from 1.61 to 2.2%.

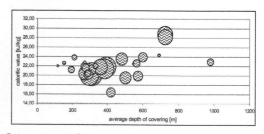

Source: own study.

Figure 11. Structure of resources according to associative criteria of the average depth of covering and the medium calorific value appointed according to real values – for the content of sulphur in coal above 2.2%.

Table 6 (a). Structure of resources according to associative criteria of ranges of the average depth of covering of coal seams, ranges of the medium calorific value and ranges of the average content of sulphur – real value.

Ranges of the average depth of coal seams	Ranges of medium calorific values [kJ / kg]					SUM
	to 18.0	18.01-22.0	22.01-26.0	26.01-30.0	over 30.0	
Absolute values [thousands of ton]						
Content of sulphur to 0.4%						
0 to 200 m	0	0	0	795	1 029	1 824
201 to 400 m	7 205	0	7 958	51 961	3 501	70 625
401 to 600 m	0	0	29 068	40 385	6 215	75 668
601 to 800 m	359	0	40 796	86 381	22 231	149 767
801 to 1000 m	0	0	27 412	35 155	34 294	96 861
over 1000 m	0	0	0	14 765	4 521	19 286
SUM	7 564	0	105 234	229 442	71 791	414 031
Content of sulphur from 0.41 to 1.0%						
0 to 200 m	21 390	0	150 094	109 751	3 003	284 238
201 to 400 m	1 746	66 949	633 341	225 984	3 335	931 355
401 to 600 m	3 729	30 901	811 865	772 411	50 173	1 669 079
601 to 800 m	0	61 804	807 316	1 436 642	219 145	2 524 907
801 to 1000 m	0	5 141	343 304	1 040 425	225 986	1 614 856
over 1000 m	0	0	252 743	453 791	28 092	734 626
SUM	26 865	164 795	2 998 663	4 039 004	529 734	7 759 061
Content of sulphur from 1,01 to 1.6%						
0 to 200 m	0	8 839	49 573	5 249	0	63 661
201 to 400 m	4 303	114 902	308 134	59 163	0	486 502
401 to 600 m	0	131 210	510 559	83 504	0	725 273
601 to 800 m	0	82 262	383 555	112 732	9 000	587 549
801 to 1000 m	640	0	148 367	42 358	2 059	193 424
over 1000 m	0	0	37 752	228	446	38 426
SUM	4 943	337 213	1 437 940	303 23	11 505	2 094 835
Content of sulphur from 1.61 to 2.2%						
0 to 200 m	0	12 425	0	1 087	0	13 512
201 to 400 m	0	0	161 694	42	0	161 736
401 to 600 m	0	127 259	80 385	586	0	208 230
601 to 800 m	0	6 823	18 064	0	0	24 887
801 to 1000 m	3 046	0	0	0	0	3 046
SUM	3 046	146 507	260 143	1 715	0	411 411
Content of sulphur above 2.2%						
0 to 200 m	0	2 688	608	0	0	3 296
201 to 400 m	0	137 611	3 557	0	0	141 168
401 to 600 m	4 625	13 386	16 304	0	0	34 297
601 to 800 m	0	0	800	26 659	0	27 459
801 to 1000 m	0	0	2761	0	0	2 761
SUM	4 625	153 667	24 030	26 659	0	208 981
TOTAL	47 043	802 182	4 826 010	4 600 054	613 030	10 888 319

Source: own study

The complete resources of coals with the content of sulphur meeting the appropriate ranges, representing the following participations : to 0.40% – 3.801% i.e. 414 031 thousands of ton; from 0.41 to 1.0% – 71.266% i.e. 7 759 061 thousands of ton; from 1.01 to 1.6% – 9.263% i.e. 2 094 835 thousands of ton; from 1.61 to 2.2% – 3.787% i.e. 411 411 thousands of ton; above 2.2% – 1.882 % i.e. 208 981 thousands of ton.

The complete resources of the coals with the calorific value meeting the appropriate ranges representing the following participations : to 18 kJ / kg – 0.439% i.e. 47 043 thousands of ton; from 18.01 to 22.0 kJ / kg – 7.373% i.e. 802 182 thousands of ton; from 22.01 to 26.0 kJ / kg – 44.306% i.e. 4 826 010 thousands of ton; from 26.01 to 30.0 kJ / kg – 42.257% i.e. 4 600 054 thousands of ton; above 30.0 kJ / kg – 5.624% i.e. 613 030 thousands of ton.

The maximum participation in the whole of resources carrying out 13.19% i.e. 1 436 642 thousand tons possess the resources fulfilling the following criteria: the content of sulphur the range from 0.41 to 1.0%; calorific value in range from 26.01 to 30.0 kJ / kg as well as depth of covered in range since 601 to 800 m.

Table 6 (b). Structure of resources according to associative criteria of ranges of the average depth of covering of coal seams, ranges of the medium calorific value and ranges of the average content of sulphur – percentage shares.

Ranges of the average depth of coal seams	Ranges of medium calorific values [kJ / kg]					SUM
	to 18.0	18.01-22.0	22.01-26.0	26.01-30.0	over 30.0	
Percentage shares in the whole of resources						
Content of sulphur to 0.4%						
0 to 200 m	0	0	0	0.0073	0.0095	0.017
201 to 400 m	0.07	0	0.073	0.48	0.0322	0.655
401 to 600 m	0	0	0.27	0.37	0.0571	0.698
601 to 800 m	0.003	0	0.375	0.793	0.20	1.371
801 to 1000 m	0	0	0.25	0.32	0.31	0.880
over 1000 m	0	0	0	0.14	0.04	0.180
SUM	0.073	0.000	0.968	2.111	0.649	3.801
Content of sulphur from 0.41 to 1.0%						
0 to 200 m	0.20	0	1.38	1.01	0.028	2.618
201 to 400 m	0.016	0.61	5.82	2.08	0.031	8.557
401 to 600 m	0.034	0.28	7.46	7.09	0.46	15.324
601 to 800 m	0	0.57	7.41	13.19	2.01	23.180
801 to 1000 m	0	0.047	3.15	9.56	2.08	14.837
over 1000 m	0	0	2.32	4.17	0.26	6.750
SUM	0.250	1.507	27.54	37.1	4.869	71.266
Content of sulphur from 1.01 to 1.6%						
0 to 200 m	0	0.081	0.46	0.048	0	0.589
201 to 400 m	0.04	1.06	2.83	0.54	0	4.470
401 to 600 m	0	1.21	4.69	0.77	0	6.670
601 to 800 m	0	0.76	3.52	1.04	0.083	5.403
801 to 1000 m	0.006	0	1.36	0.39	0.019	1.775
over 1000 m	0	0	0.35	0.0021	0.0041	0.356
SUM	0.046	3.111	13.210	2.790	0.106	19.263
Content of sulphur from 1,61 to 2,2%						
0 to 200 m	0	0.114	0	0.010	0	0.124
201 to 400 m	0	0	1.49	0.00039	0	1.490
401 to 600 m	0	1.17	0.74	0.0054	0	1.915
601 to 800 m	0	0.063	0.166	0	0	0.229
801 to 1000 m	0.028	0	0	0	0	0.028
SUM	0.028	1.347	2.396	0.016	0.000	3.787
Content of sulphur above 2.2 %						
0 to 200 m	0	0.025	0.0056	0	0	0.031
201 to 400 m	0	1,26	0.0033	0	0	1.263
401 to 600 m	0.042	0.123	0.151	0	0	0.316
601 to 800 m	0	0	0.0073	0.24	0	0.247
801 to 1000 m	0	0	0.025	0	0	0.025
SUM	0.042	1.408	0.192	0.240	0.000	1.882
TOTAL	0.439	7.373	44.306	42.257	5.624	100.00

Source: own study

CONCLUSIONS

The analysis of characteristic feature of coal seams, covering in Poland in thin seams, show that they are then the resources, in majority, very attractive to future exploitation. It is possible even to propose a thesis, that they are more attractive than, many thin coal seams exploited for decades in the Ukraine. The mining of Ukraine, with success, exploits such coal seams, then it is logic that it should also be back to their mining utilization in Poland

REFERENCES

Cichosz, P. 2000. *Systemy uczące się*. Warszawa: WNT.

Dokumentacja programu. 2010. STATISTICA v. 9.0.

Everitt, B.S., Landau, S. & Leese, M. 2001. – *Cluster analysis*. Londyn, Arnold, New York, Oxford: University Press.

Tadeusiewicz, R. 1993. *Sieci neuronowe*. Warszawa: AOW RW.

Materiały źródłowe Kompanii Węglowej S.A., Katowickiego Holdingu Węglowego S.A., Jastrzębskiej Spółki Węglowej S.A. oraz Południowego Koncernu Węglowego. 2010.

Bolt-pneumatic support for development workings with big cross-section

V. Buzilo, O. Koshka & A. Yavors'kyy
National Mining University, Dnipropetrovs'k, Ukraine

ABSTRACT: The article is concerned with problems of construction of underground workings with big cross-section. The paper is dedicated to questions of integrity face assurance and non-admission of its caving during construction of workings in soft rock, because otherwise great difficulties concerned with support and saving of their integrity occur. The results of theoretical, laboratorial and field observations of ways and means of support of faces of workings with big cross-section by bolt-pneumatic support are shown.

1 INTRODUCTION

While dealing workings with big cross-section, in underground passaging, in particular, big exposures of soft rock without support are not admitted. During construction of Kiev underground, in particular, the tunnelling was done the entry way of 1 m wide and 2 m high. The traditional technology of using steel beams and wooden laggings for face support is known to be high cost and labor intensive (Buzilo 2005).

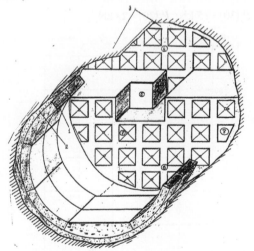

Figure 1. The technology of working with the application of bolt-pneumatic support: 1 – permanent support (lining); 2 – cement-and-sand grout slurry; 3 – rock mass; 4 – mined-out and bolted upper level; 5 – entry way being mined; 6 – lover level; 7 – bolt-pneumatic support.

The application of bolt-pneumatic support (BPS) with its soft shell of any size allows to keep face development with the entry way providing the whole area support. The suggested scheme of developing and supporting is shown in Figure 1.

The bolt-pneumatic support consists of a bolt, a soft shell and a face plate. To press properly onto the face, the soft shell must have a support. The main idea of the suggested supporting method is following: the soft shell with a face plate is held by a bolt. As the bolts the screw-in, drive, auger-type, spade, and other bolt kinds determined for soft rocks can be used here. In this context the bolt length is to be of the size with its working part located off possible caving rock zone.

The soft shell size is designed with the bolt-pneumatic support application technology and the proper holding power. Being made of metallic profile the face plate provides hard pressure of the soft shell filled with the compressed air to the rock.

2 DETERMINING OF THE BPS PARAMETERS OF FACE DEVELOPMENT WHILE WEDGE BLOCKS CAVING

While valuing the BPS holding power and the soft shell parameters we will proceed from the information about maximum caving blocks formed in the development face according to the full-scale re-searching data and results of theoretical research dealt with the determining of stress-train state zone in front of the moving face.

The analysis of the caving variants observed during building underground passages in clays shows that the wedge-shaped and prismatic blocks dumps are the most probable. Just for them we are to determine the BPS optimal parameters.

2.1 Pull-out in a bolt

While caving rock forming block is similar to wedge of dimensions D_B, L_B and a height H_B (Figure 2).

Under the boundary force balance the block is affected by the interbalanced forces system: the block weight $\vec{G}$, the reaction of unbroken rock massive $\vec{R}$, rock-to-rock friction force $\vec{F}_{TP} = f\vec{R}$, in which f – is rock-to-rock friction coefficient, and $\vec{F}$ – resultant force of shell elements.

The system balance condition is the following:

$$\vec{G} + \vec{R} + \vec{F}_{TP} + \vec{F} = 0 \qquad (1)$$

or in vertical or horizontal projections:

$$G = R\sin\alpha + fR\cos\alpha \; ; \; R\cos\alpha = F + fR\sin\alpha \, , \quad (2)$$

where $tg\,\alpha = L_B/H_B$ (see Figure 2). Resulting:

$$R = G/(\sin\alpha + f\cos\alpha) . \qquad (3)$$

$$F = G(1 - ftg\,\alpha)/(f + tg\,\alpha) . \qquad (4)$$

If we know clay rock specific γ, we may get block weight:

$$G = \gamma H_B L_B D_B /2 . \qquad (5)$$

Resulting formulas (4), (5) allow to value total loading on a bolt when forming caving block near face surface, the block form is shown on Figure 2.

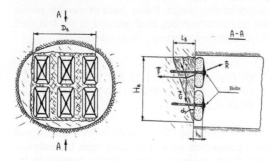

Figure 2. Design model of bolt-pneumatic support when edge blocks caving.

Joining formulas (4) and (5), we get

$$F = \gamma D_B H_B^2 \frac{1 - f L_B/H_B}{2(1 + f L_B/H_B)} . \qquad (6)$$

An average clay specific weight under natural conditions $\gamma = 19.3$ kN / m³, so

$$F[\kappa N] = 9.65 D_B H_B^2 \frac{1 - f L_B/H_B}{1 + f L_B/H_B} . \qquad (7)$$

The most unacceptable conditions for support loading appear to be when rock-to-rock friction is ignorable, what allows to get maximal total loading on support for given block caving dimensions.

$$F_{max}(H_B, D_B) = 9.65 D_B H_B^2, \text{kN} \qquad (8)$$

Then for a caving block of $H_B = 8$ m, $D_B = 6$ m:

$$F_{max}(8 \text{ m}, 6 \text{ m}) = 463, \text{kN} \qquad (9)$$

If caving block requires n soft shells, so an average force affecting a bolt makes:

$$P_3 = F_{max}(H_B, D_B)/n = 9.65 D_B H_B^2/n , \text{kN} \quad (10)$$

and with $H_B = 8$ m, $D_B = 6$ m

$$P_3(n) = 463/n , \text{kN} \qquad (11)$$

The plot of maximal load P_3 on a bolt against a number of shells ensuring caving block support is shown on Figure 3. For $n = 40$, in particular, $P_3(40) = 12$ kN. If a caving block surface is covered by fewer shells load on a bolt grows and makes in this example:

$$P_3(30) = 15 \text{ kN}; \quad P_3(20) = 23 \text{ kN}$$

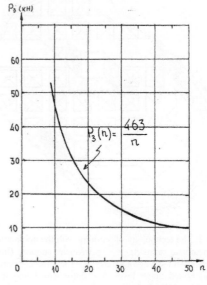

Figure 3. The plot of support load against shells number for a block dimensioning $H_B = 8$ m, $D_B = 6$ m ($\gamma = 19.3$ кN / m³).

Formulas (8-11) allow to determine extreme load on bolt depending on caving block size and soft shells number, and, thus to make requirements for supporting power bolts used. Computations made on the ground caving data and the before obtained results have the right to claim that specific supporting power of bolts used is to be not less than 20 kN.

The obtained results help to solve the opposing problem: according to given bolt supporting power $P_{3,max}$ and limit block size we are to determine a proper number of soft shells.

$$n = int\left[\frac{\gamma D_B H_B^2}{2P_{3,max}}\left(\frac{1 - f L_B / H_B}{1 + f L_B / H_B}\right)\right]. \qquad (12)$$

2.2 Determining a bolt length

In determining a bolt length we ground on that it may have non-working part equals inrush depth in the considering cross-section $\ell_x = H_x \, tg\,\varphi = L_B H_x / H_B$, where H_x – distance from block base to observing bolt, ℓ_P – working part ensuring real bolt fastening with pinning strength at break $P_{3,max}$ and a value of bolt cross-section parameter S_T:

$$\ell_P = P_{3,max} / (\sigma_T S_T). \qquad (13)$$

where σ_t clay fluidity limit at a bolt surface. A full length of bolt ℓ_a makes a value:

$$\ell_a = \ell_x + \ell_p = L_B H_x / H_B + P_{3,max} / (\sigma_T S_T). \quad (14)$$

For the above considered matter with $P_3 = 20$ kN, $S_T = 0.051$ m²: $\ell_a = 2.8$ m.

2.3 Pressure in a soft shell

To determine surplus internal pressure in a soft shell cavity ensuring a proper value F of all BPS shells pull to face surface we are to consider a design model in Figure 4.

In Figure 4 a ℓ_s и b_0 – soft shell initial length and width (in the plan), r_c – butt end corner radius, forming when come pressing of a face plate and a shell under pressure P and depending on compression value h_s.

Suppose d_a and h_a – are diameter and height of a collar, ensuring a hole in the shell centre.

On condition of inextensible of soft-shell fabrics we may determine length ℓ_s and width b_s of area contacting soft shell with a rock:

$$\ell_s = \ell_0 - h_s \,; b_s = b_0 - h_s. \qquad (15)$$

Butt end corner radius r_c is proportional to a compression value h_s and can be defined by formula $r_c = k \cdot h_s / 2$, where k – proportion coefficient.

In Figure 4 b one can see

$$d_p = h_s - h_a + d_a. \qquad (16)$$

Now we can deduce soft shell-rock contact area S_k:

$$S_k = b_s \ell_s - (1 - \pi/4) r_c^2 - \pi d_p^2 / 4.$$

or provided $k \approx 1$, we get:

$$S_k = (\ell_0 - h_s)(b_0 - h_s) - \\ - \left((1 - \pi/4) k^2 h_s^2 + \pi (h_s - h_a + d_a)^2 / 4\right). \qquad (17)$$

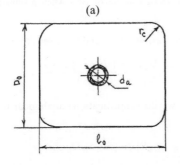

(a)

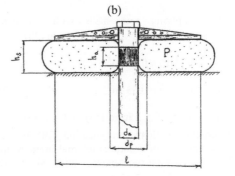

(b)

Figure 4. Design scheme of a soft shell of bolt-pneumatic support.

If n – is a total sum, number of the soft shells covering inrush area, so surplus pressure P in any shell cavity can be calculated by the formula:

$$P = \frac{F}{nS_k} = \frac{\gamma D_B H_B^2}{2nS_k}\left(\frac{1 - f L_B / H_B}{1 + f L_B / H_B}\right). \qquad (18)$$

Formula (17) for soft shell contact area can be simplified provided that for really applied soft shell $h_a \approx h_s$ and $k \approx 1$. Resulting:

$$S_k = (\ell_0 - 2r_c)(b_0 - 2r_c) -$$

$$- ((1 - \pi/4)d_p^2 + \pi d_a^2/4). \qquad (19)$$

Technologically bearable support values of h_s make: 1) $h_s = 0.07$ m; 2) $h_s = 0.15$ m. Then accordingly we have: 1) $d_a \approx h_s/2$; 2) $d_a \approx h_s$ and equation (19) for both cases:

$$S_k\,(\,h_s = 0.07\text{ m}) \approx b_0\ell_0 - h_s(b_0 + \ell_0) + 7\pi h_s^2/16 ,$$

$$S_k\,(\,h_s = 0.15\text{ m}) \approx b_0\ell_0 - h_s(b_0 + \ell_0). \qquad (20)$$

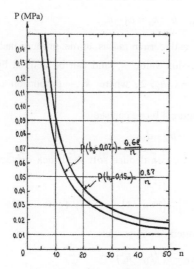

Figure 5. The dependence of surplus pressure in a soft shell cavity on soft shells number covering wedge block.

When applying soft shells with parameters $\ell_0 = 0.94$ m; $b_0 = 0.85$ m resulting from (20)

$$S_k\,(\,h_s = 0.07\text{ m}) \approx 0.68\text{ m}^2,$$

$$S_k\,(\,h_s = 0.15\text{ m}) \approx 0.53\text{ m}^2. \qquad (21)$$

Using values F_{max} from (15) and S_k from (21) and substituting them into equation (18), we will find dependence of a surplus internal pressure in a soft shell cavity on their number under above considered caving block parameters:

$$P\,(\,h_s = 0.07\text{ m}) = 0.68/n ,\text{ MPa}$$

$$P\,(\,h_s = 0.15\text{ m}) = 0.53/n ,\text{ MPa} \qquad (22)$$

The dependence diagrams provided by formulas (22), are given on Figure 5. In particular, with $h_s = 0.15$ m и $n = 40$ we have $P = 0.022$ MPa; and with $h_s = 0.07$ m и $n = 40$, $P = 0.07$ MPa.

3 DETERMINING OF SUPPORT PARAMETERS WHEN PRISMATIC BLOCKS CAVING

3.1 Pull-out in a bolt

Another type of block caving has a form approximate to the rectangular prism under the angle ($\pi/2 - \alpha$) to the face line. As a rule, such blocks are formed in the face top near roof of working (Figure 6). The block balance condition is of the kind:

$$F = G(tg\,\alpha - f)/(1 + f\,tg\,\alpha), \qquad (23)$$

where the block weight G is calculated by the formula:

$$G = \gamma H_B L_B D_B \cos\alpha . \qquad (24)$$

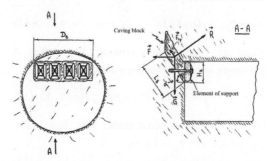

Figure. 6. Design model of BPS when prismatic blocks caving.

Joining formulas (23) and (24), resulting:

$$F = \frac{\gamma H_B L_B D_B \cos\alpha(\sin\alpha - f\cos\alpha)}{(\cos\alpha + f\sin\alpha)} . \qquad (25)$$

There we do simplifications analogical to (8) resulting:

$$F_{max}(H_B, L_B, D_B) = \gamma H_B L_B D_B \sin^2\alpha$$

or when $\varphi \approx \pi/3$

$$F_{max}(H_B, L_B, D_B) = 2.6 H_B L_B D_B ,\text{ kN} \qquad (26)$$

For the block dimensioning $D_B = 1.5$ m,

$H_B = 6$ m, $L_B = 6$ m, we have:

$$F_{max}\,(6\text{ m, }6\text{ m, }1.5\text{ m}) \le 140\text{ , kN}$$

3.2 *Pressure in the soft shell*

Surplus pressure in the soft shell cavity is calculated by the formula analogical (18):

$$P = \frac{F}{nS_k} = \frac{\gamma L_B D_B H_B \cos\alpha(\sin\alpha - f\cos\alpha)}{nS_k(\cos\alpha + f\sin\alpha)}. \quad (27)$$

that for the above-mentioned case ($h_s = 0.15$ m) becomes $P = 0.26/n$, MPa, with:

$$n \le int\left(\frac{H_B D_B}{b_0 \ell_0}\right). \quad (28)$$

The dependence diagram (27) is shown in Figure 7. In particular, for $n = 10$, we have $P = 0.026$ MPa, and for $n = 6$ – $P = 0.04$ MPa.

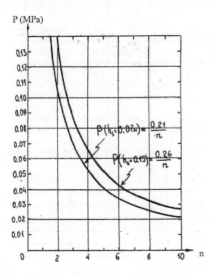

Figure 7. The plot of surplus pressure in the soft shells cavity against the shell number covering prismatic block of dimensions, when $D_B = 1.5$ m, $H_B = 8$ m, $L_B = 6$ m.

The designs made according to the described technique for bolt-pneumatic support for development in clays showed that BPS shall have the following values: shells number $n = 24$; pressure in a soft shell $P = 0.04$ MPa; pull in a bolt $P_a = 30$ кN, a bolt length $\ell_a = 2.8$ m.

4 THE RESULTS OF BPS TESTING UNDER THE FIELD OBSERVATIONS CONDITIONS

The BPS passed its industrial verification during building of the "Pecherskaya" station, zone #221. The wedge-shaped, drive and screw-in bolts were tested. The pull-out value was determined by the appliance PA-3.

For the wedge-shaped bolts installation the blast-holes of 36 mm in diameter and 1.5 m length were drilled. The drilling of one blast-hole with its further clearing took 3 minutes. The average pull-out value of the wedge-shaped bolt according to the testing results made 18 kN.

The drive bolts were made of drilling bars with the cross-section hexahedron-shaped and were mounted with the pneumatic hammer and the hydraulic prop HP-3. The drilling bar was hammered on the depth of 1 m during 1.2 minutes. The average pull-out value of the drive-bolt is – 13 кN.

The tested drive bolt parameters are given on Table 1.

The symbols used in the table 1 are the following (look at Figure 8): $S_n = \pi nt(d + 2h)$ – the surface area of the cylinder traversing the bolt thread; $S_0 = \pi h(d + h)$ – the area of one turn surface exposed face; $S_L = \pi d^2(L - nt)/4$ – the bolt surface area between the face and the threaded initial part; $S_c = \pi(d + 2h)^2/4$ – the bolt pipe cross-section area; $S = S_n + S_0 + S_L$ – the bolt surface total area.

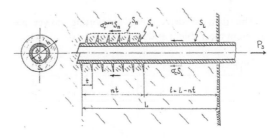

Figure 8. The scheme of drive bolt installing in the massive.

Figure 9 shows experimentally made diagrams of functional dependence of average resistance force F_c against the bolt penetration into the massive on the penetration depth (when driving bolts every 5-10 turns the force, level and depth of entry way were marked).

Table 1. The drive bolts parameters.

No of bolt	Diameter d (m)	Pace of a thread t (mm)	Height a thread h (mm)	Number of turns n	Penetration depth L (m)	$S_n \cdot 10^3$ (m²)	$S_o \cdot 10^3$ (m²)	$S_L \cdot 10^3$ (m²)	$S \cdot 10^3$ (m²)
1	0.065	0.02	0.08	9	1.00	4.58	1.885	2.725	50.360
2	0.033	0.02	0.08	6	1.52	6.20	1.030	1.265	8.495
3	0.049	0.02	0.08	9	1.20	36.8	1.433	1.923	40.150
4	0.033	0.03	0.08	9	1.00	41.6	1.030	0.700	43.330
5	0.033	0.02	0.08	9	1.15	27.7	1.030	0.825	29.560

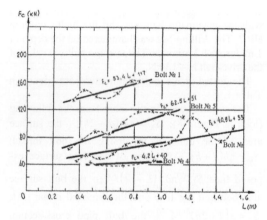

Figure 9. Functional dependence of average resistance force on penetration depth: −x−x− the results of measurements; ——— the lines of regression.

The data of drive bolts supporting power are given in the Table 2 and in the Figure 10.

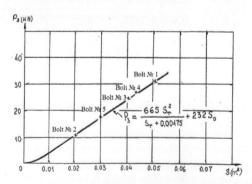

Figure 10. The plot of the drive bolt supporting power against the lateral surface area.

To compare various bolts by their supporting power in Table 3 one can find values of their spe-

cific supporting power, i.e. ratio of extreme load P_3 to the total surface resisting shift S.

Table 2. The bolts pull-out values.

No of bolt	$S_n \cdot 10^3$ (m²)	$S_o \cdot 10^3$ (m²)	$S_L \cdot 10^3$ (m²)	$S \cdot 10^3$ (m²)	P_3 (кN)
1	4.58	1.885	2.725	50.360	31
2	6.2	1.030	1.265	8.495	11
3	36.8	1.433	1.923	40.150	25
4	41.6	1.030	0.200	43.330	27
5	27.7	1.030	0.825	29.560	18

Table 3. The bolts compared of their supporting power.

No of bolt	$S \cdot 10^3$, m²	P_3 / S, MN / м²
1	50.360	0.616
2	8.495	1.295
3	40.150	0.623
4	43.330	0.623
5	29.560	0.609
bar	69.000	0.232

5 THE INDUSTRIAL VERIFICATION OF BOLT-PNEUMATIC SUPPORT (BPS) OF FACE

The BPS verification was done in the distilling tunnels of mines #221 and #225 of "Kyivmetrostroy" in accordance with the approved program and testing methodology.

The BPS verification was conducted in the workings of 5.5-8.5 m in diameter. The bolts of friction, screw-in, face plates and the soft shell were tested aiming at setting BPS parameters, defining applica-

154

tion efficiency and the bolt and soft shell labour productivity.

The bolt was mounted in the massive with its protruding part coated with soft shell and face plate both were fixed with the washer and nut. The bolt length made not less than 2.8 m while maximal "depth" of inrush could reach 2.3 m.

Being rectangular-shaped, dimensioning 850-940 mm with 20 mm width, the soft shell consisted of cap rock made of kapron fabrics and hermetic camera.

The grummet of 80 mm wide and the hole of 40-50 mm in diameter was set in the centre of the soft shell. The noise-piece with the stopper for compressed air fit-and-discharge was set in the lower lateral side.

The air for soft shell was given from mine manifold through the pressure regulator set on working pressure of 0.02-0.03 MPa. With the working pressure of 0.03 MPa in the soft shell the support section ensures the spread strength of 2.5 tons enough for holding and fastening the caving block dimensioning $6 \times 6 \times 2.3$ m. The face plate provided fast and reliable fastening to the bolt. The achieved time for the face plate and soft shell setting, fastening the bolt and filling in with compressed air made 1.5-2 minutes. The tests proved the BPS working capability and expediency of its application when working with big cross-section in soft rocks.

The BPS design Ukrainian and Russian patents defended (Petrenko 1992; Petrenko 1994; Buzilo 1994).

6 CONCLUSIONS

1. For the considered geological conditions there were determined the BPS parameters – the bolt's supporting power $P \geq 20$ кN, the bolt's length $\ell_a = 2.8 - 2.9$ m and the maximal surplus pressure

in the soft shell cavity up to $P_{max} = 0.04$ MPa.

2. The dependence allowing to determine the extreme load in the BPS bolt by the dimension of the block rocks detached from the massive and the soft shells number can be obtained.

3. The dependence of surplus pressure in the pneumatic shell of support on the bolt's fastening strength and its contact area with the surface support or face has been found.

4. Using full-scale research data there has been defined the dependence of bolt's pull-out value on its depth penetration into the massive with clay.

5. There was determined the holding power of wedge-shaped bolt, hammer bolt and some drive bolts. The drive bolt of 33 mm in diameter and thread height of 12 mm appears to have the biggest supporting power ($P_3 = 30$ кN) and it is to be recommended for the application along with BPS under industrial conditions.

REFERENCES

Buzilo, V., Rahutin, V. & Serdyuk, V. 2005. *Potential for use of pneumatic constructions in underground mining*. International Mining Forum. New Technologies in Underground Mining. Safety and Sustainable Development. Poland: Taylor & Francis Group: 83-86.

Patent of Ukraine #93010056. 1992. *The method of temporary support of permanent mine workings within the zone of active rock pressure* / Petrenko, V.I., Rahutin, V.S., Kalinichenko, G.F., Buzilo, V.I. and others. (Ukraine). - Application date 02.12.1992.

Patent of Ukraine #94020546. 1994. *Method of face mining and support while mining in soft rocks* / Petrenko, V.I., Rahutin, V.S., Kalinichenko, G.F., Buzilo, V.I. and others. (Ukraine). - Application date 02.12.1994.

Patent #3849 (RF). 1994. *Method of temporary support of soft rocks and its performing mechanism* / Buzilo, V.I., Rahutin, V.S., Kalinichenko, G.F., Likhman, S.N. and others. (Ukraine). – Publication date. 27.12.94. Bul. #6-1.

Technical and Geoinformational Systems in Mining – Pivnyak, Bondarenko & Kovalevs'ka (eds)
© 2011 Taylor & Francis Group, London, ISBN 978-0-415-68877-2

About the influence of intense fracturing on the stability of horizontal workings of Eastern Donbass mines

P. Dolzhikov & N. Paleychuk
Donbass State Technical university, Alchevsk, Ukraine

ABSTRACT: The article presents the results of mine researches of stability of workings, as well as analytical studies of the parameters intensely fractured zones. Produced by typing parts of development workings out of the zones affected by coal-face works and by tectonic disturbance in terms of stability and fracture parameters. According to researches the classification of the various zones of workings.

1 INTRODUCTION

One of the main components that determine the economic and technological efficiency of underground coal mining is the state of development workings. On the state of the underground workings is influenced a lot of geological, technological and operational factors. To one of the most influential factors fracturing of rock massif is included. Analysis of domestic and foreign research in the field of geomechanics and mine buildings (Malinin 1970; Parchevskiy & Simanovich 1966; Erofeev 1977) shows that are currently under intense fracturing, basically, means the number of fissures per unit length or area of rock massif (Shashenko & Pustovoytenko 2004). However, this definition refers to an already formed system of fissures, primarily as a result of tectonic processes, i.e. to the natural and the tectonic fracturing types. In turn, according to (Dolzhikov, Paleychuk & Kobzar 2010), under intense refers to a fracturing, which is characterized by an to increase in the number and parameters of fissures over a fixed period of time in a certain direction of space. This definition is the technological fracture corresponds more.

In this regard, the influence of intensive technological fracturing on the state of development workings is actual.

2 FORMULATING THE PROBLEM

In the framework developed in the current classification of rock outcrops on the stability (Melnikov 1988), is put a sign of "bias rocks U ", which is a consequence of the geomechanical processes in rock massif around the workings. With the last one can only state what type of stability are those of certain rocks. This option is most reasonably be used during the building, before the installation of permanent roof supports. Since the period of exploitation there is an interaction roof supports with the rock massif, define about the sustainability of workings on the basis only the displacement of rocks is not correct for several reasons. In the first, in the hard rocks (sandstones, siltstones) of the investigated region at bias of rocks 0.5 m exploitation of working can be performed in accordance with the appointment, without prejudice to the processes, convenience and safety. However, there are cases when in a small bias of rocks (up to 0.3 m) was emergency state of roof supports, and the working of required repairs. Secondly, it is unclear how to classify the parts of workings with different values the bias of rocks and the state of roof supports, on the basis of what evidence to produce a comparison of different parts with the same values the bias, but in a different exploitation state. Thirdly, the need not only refers to the rocks of various categories of stability, but also to produce prediction in space and time of an emergency condition of the latter.

Thus, for qualitative assessment of the operational status of the development workings is necessary to classify their various parts on the intensity of fracturing in combination with other factors.

3 RESEARCH OF THE DEVELOPMENT OF FRACTURING IN WORKINGS

The object of investigation were selected horizontal development workings of seams h_8 and h_{10} of mines "Komsomolskaya" and "Partizanskaya" SE "Antratcite", as well as mine name V.V. Vakhrusheva SE "Rovenkiantratcite". Host rocks of coal seams in the region represented by siltstones, power m which in the researched mines were in the range 4.7-9.5 m, and the temporary compressive strength σ_c perpendicular to the bedding planes was 53.7-64.2 MPa, by sand shale with $m = 2.8$-16.2 m, $\sigma_c = 67.5$-71 MPa, by mudstone with $m = 7$-24.6 m, $\sigma_c = 73$-96.8 MPa and by sandstones with $m = 9$-38 m, $\sigma_c = 135.7$-178 MPa. Angles to the dip of rocks in the studied workings were $\alpha = 2 \div 19°$ in the depth range $H = 694 \div 1115$ m.

Due to the fact that the length of some of the workings up to 3 km in assessing their operational status of their total length was divided into plots of 40 m. The stability measure our ω_S, defined as the ratio of the actual minimum to project of area cross-section in part of working, as well as with index ω_N calculated as the ratio of the number of able-bodied frame of metal roof supports to their total number at a length of 40 m (Dolzhikov, Paleychuk & Kobzar 2010). The study found that between sustainability indicators ω_N and ω_S depend on, so for convenience of calculation in what follows, we use the index ω_N, as it is a quantitative characteristic of stability and allows you to visually compare the operational status of various parts of workings, while the figure ω_S – qualitative characteristic reflecting the change in cross-sectional area of workings after making the displacement of rocks. An analysis of the extent and nature of the deformation of rock massif and permanent roof supports in the mines SE "Antratcite" and SE "Rovenkiantratcite" meaningfully measure ω_N identified four types of the most characteristic zones: the value of indicator less than 0.5 – Zone I-type, with $\omega_N = 0.5 \div 0.65$ – Zone II-type, with a value $\omega_N = 0.651 \div 0.8$ – Zone III-type, and parts of workings with $\omega_N > 0.8$, respectively, Zone IV-type.

To determine the change of the sustainability indicator ω_N depending on the density of fractures of the rocks massif over a fixed period of time (1 month), during the year were instrumental investigations. With the help of photo planimetry method determines the initial value of fissures density λ, then determined the increment of the number of fissures on 1 m of working over time at different parts. The greatest interest is the intensity of fracturing, at which a qualitative transition part of working from the zones of the previous generation to the next zone types.

Results of the research of fracturing in horizontal development workings, lifetime over 5 years, outside the zone of influence of coal-face works and tectonic disturbance the for conditions of the seam h_{10} are shown in Figure 1.

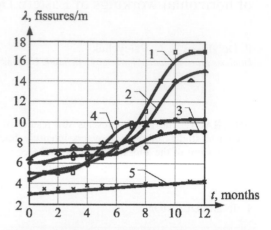

Figure 1. Graph of fissured in time zones of various types of development workings of seam h_{10}.

In Figure 1, line 5 corresponds to the zone IV-type at the end of the research. Curve 3 corresponds to the zone type II (the value of sustainability indicator $\omega_N = 0.6$) transition which occurred at the 8-th month of researches; up to this indicator ω_N, was 0.75 (Zone III type). On part of workings, represented by curve 4 in the current year indicator ω_N changed from 0.825 (May 2009) to 0.575 (April 2010). The transition from zone IV to zone III type occurred in the second month of research, and from zone III to zone II type – on the fifth. In the zones represented by curves 1 and 2 the value of the index ω_N were respectively 0.825 and 0.7 at baseline and 0.375, 0.45, after a year. On the part of working, which is characterized by curve 1, the indicator ω_N changed from 0.825 to 0.75 in the fourth month of researches, from 0.725 to 0.575 in the sixth, with 0.55 to 0.375 in the eighth. On part of working, the development of fissures roughness, which reflected the curve 2, the initial value of sustainability index was 0.7; the seventh month dropped to 0.64, while the tenth was reduced to 0.45.

Curves 1-5 are approximated by least squares corresponding functions of the form:

$$\lambda(t)_1 = 4.093 + 1.505t - 0.646t^2 + \\ + 0.115t^3 - 0.004t^4 ; \quad R^2 = 0.85 ; \tag{1}$$

$$\lambda(t)_2 = 6.214 + 1.212t - 0.458t^2 + \\ + 0.065t^3 - 0.002t^4 ; \quad R^2 = 0.78 ; \tag{2}$$

$$\lambda(t)_3 = 5.786 + 1.028t - 0.345t^2 +$$

$$+ 0.047t^3 - 0.002t^4; \ R^2 = 0.73; \quad (3)$$

$$\lambda(t)_4 = 5.205 - 1.033t + 0.583t^2 -$$

$$- 0.064t^3 + 0.002t^4; \ R^2 = 0.79; \quad (4)$$

$$\lambda(t)_5 = 3.261 + 0.011t; \ R^2 = 0.83, \quad (5)$$

where $\lambda(t)_1$, $\lambda(t)_2$, $\lambda(t)_3$, $\lambda(t)_4$, $\lambda(t)_5$ – density of fissures in time for the parts of workings represented by curves 1-5; (fissures / m) / mo.; t – commit time, the corresponding value of fissures density array, mo.

For dependencies (1)-(5) the accuracy of constant coefficients is 0,001. The choice of this value due to the necessity of appropriate accuracy of the approximation polynomial dependencies for subsequent determination of the intensity of fracturing.

Variation of density of cracks in time curves 1-4 corresponding approximated by polynomial dependences. Then, knowing the time at which there was a qualitative shift parts of workings of the bands previous to the zone of subsequent types, determine the intensity of fractures in these areas at which this transition occurred.

Because at the part, characterized by a direct 5 with time indicator of stability is not changed, for further research, this area are not interesting.

Intensity change of the x by function $y - f(x)$, provided that this function is continuous and differentiable at each point is defined as the first-order derivative:

$$v = f'(x) = \frac{dy}{dx}. \quad (6)$$

Thus, the expression for determining the intensity of fracturing in areas represented by curves 1-4, can be written accordingly:

$$v_1 = 1.505 - 1.292t + 0.345t^2 - 0.017t^3; \quad (7)$$

$$v_2 = 1.212 - 0.916t + 0.195t^2 - 0.009t^3; \quad (8)$$

$$v_3 = 1.028 - 0.69t + 0.141t^2 - 0.008t^3; \quad (9)$$

$$v_4 = -1.033 + 1.166t - 0.192t^2 + 0.008t^3. \quad (10)$$

The time at which the transition was detected parts of workings of the zones II-nd in the band I-th type for curves 1 and 2 is 8 and 10 months respectively. The intensity of fracturing in the transition zone of I-type is:

$$v_1 = 1.505 - 1.292 \cdot 8 + 0.345 \cdot 8^2 -$$

$$- 0.017 \cdot 8^3 = 4.5, \text{(fissures / m) / mo;} \quad (11)$$

$$v_2 = 1.212 - 0.916 \cdot 10 + 0.195 \cdot 10^2 -$$

$$- 0.012 \cdot 10^3 = 2.51 \text{ (fissures / m) / mo.} \quad (12)$$

Similarly, we calculated the value of the intensity of fracturing in the transition parts in Zone II-nd and III-rd types. The results are shown in Table 1.

Table 1. Calculated values of the intensity of fracturing in zones of various types for the conditions of seam h_{10}.

Parameters	Types of zones							
	I-th type		II-th type				III-th type	
Number of curve	1	2	1	2	3	4	1	4
Commit time, the transition zone of this type, t, mo.	8	10	6	7	8	5	4	2
The density of fissures, λ, fissures / m	11	14	8	8	10	8	6	5
The intensity of fracturing, v, (fissures / m) / mo.	4.5	2.51	2.5	1.2	0.95	1.0	0.8	0.6

Analysis of the dependences shows that, in the seams h_{10} most intensive development of fissures occurs in the moment of transition the part of working from the zone of II-nd in the zone of I-type, but less intense – in the transition zone of the IV-rd in the zone III-type.

In the graphs in Figure 2 shows the results of mine researches of fracturing in time for the conditions of seam h_8.

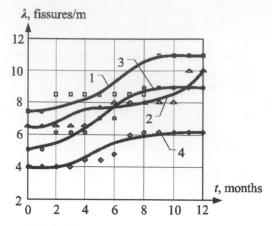

λ, fissures/m

t, months

Figure 2. Graph of fissured in time zones of various types of development workings of seam h_8.

A curve 1 and 2 corresponds to the zone of I-th type at the end of the research. On zone, characterized by curve 1, the indicator ω_N changed from 0.825 to 0.7 at the fifth month research, from 0.7 to 0.5 in the seventh, with 0.5 to 0.45 in the tenth. In local areas of working, the development of fissured which affects the curve 2, the initial value of sustainability index was 0.875, the fifth month dropped to 0.675, was 0.55 at the eighth and the eleventh had fallen to 0.375. Curve 3 corresponds to the zone II-type, (value of the index of stability $\omega_N = 0.65$) transition which occurred at the 6-th month of stud-

ies, up to this figure ω_N site generation, representation, curve 3, was 0.775 (Zone III-rd type). In section, represented by curve 4 in the current year figure ω_N changed from 0.9 to 0.75. The transition from zone IV-th zone III-first type occurred in the third month of studies, and from the zone of the III-rd in the zone II-type – on the fifth.

Curves 1-4 well approximated by the corresponding polynomial dependence:

$$\lambda(t)_1 = 7.193 + 1.370t - 0.501t^2 +$$
$$+ 0.068t^3 - 0.0025t^4 ; \quad R^2 = 0.82 ; \qquad (13)$$

$$\lambda(t)_2 = 6.609 - 0.592t + 0.312t^2 -$$
$$- 0.041t^3 + 0.002t^4 ; \quad R^2 = 0.88 ; \qquad (14)$$

$$\lambda(t)_3 = 5.086 + 0.129t + 0.078t^2 - 0.005t^3 ;$$
$$R^2 = 0.91 ; \qquad (15)$$

$$\lambda(t)_4 = 4.102 - 0.336t + 0.130t^2 - 0.007t^3 ;$$
$$R^2 = 0.78 . \qquad (16)$$

By the method of determining the intensity of fracturing for the seam h_{10} similar calculations were performed for the seam h_8. The results of calculations are summarized in Table 2.

Table 2. Calculated values of the intensity of fracturing in zones of various types for the conditions of seam h_8.

Parameters	Types of zones							
	I-th type		II-th type				III-th type		
Number of curve	1	2	1	2	3	4	1	2	4
Commit time, the transition zone of this type, t, mo.	10	11	7	8	6	5	5	5	4
The density of fissures, λ, fissures / m	11	7	8.75	8	7	4.5	8	7	4
The intensity of fracturing, v, (fissures / m) / mo.	1.75	2.0	0.9	0.62	0.5	0.4	0.2	0.39	0.35

As follows from the results in the seam h_8 the most intensive development of fissured was observed in the transition parts of the workings of the zones II-nd in the band I-type, but less intense – in the transition sections of the workings of zones IV-th zone in the III-th type .

Comparative analysis of results shows that the nature of fracture in time for the consideration seams is different. This fact is due to the substantial difference of deformation and strength properties of

host rocks. However, the most intensive development of fractures observed in the transition parts of workings in the zone I-th type of seams conditions h_{10} and h_8.

Based on these calculations, as well as analysis of field researches, zones I-th type can be described as intensely fractured, zone II-type – as a zone of active development of fissures, III-first type – the zone activation of fissures, and zone IV-type – as a zone of potential development of fracture.

4 DEVELOPMENT OF CLASSIFICATION PARTS OF MINE WORKINGS

The basis of the developed classification is laid features of the intensity of fracturing v in zones of different types of seams h_8 and h_{10}. For the most complete description of the operational status of development workings in the classification were also used parameters such as time compression strength σ_c perpendicular to the bedding planes of rocks, the average power of rocks m, averaged over the length of the parts of the offset value of displacement of rocks U, density of fissures massif λ, as well as the values indicators ω_N and ω_S, which were obtained and analyzed as a result of the mine field researches. Along with this, the most important parameter is likely inrushes, which determines the degree of development of accident and the need to be repaired of working.

Table 3. Classification zones of development workings.

Parameters	Marking of zones			
	I-th type	II-th type	III-th type	IV-th type
Sustainability indicator, ω_N	$\omega_N < 0.5$	$0.5 \leq \omega_N \leq 0.65$	$0.65 \leq \omega_N \leq 0.8$	$\omega_N > 0.8$
Sustainability indicator, ω_S	< 0.6 / < 0.65	$0.6 \div 0.72$ / $0.64 \div 0.74$	$0.73 \div 0.85$ / $0.75 \div 0.84$	> 0.85 / > 0.84
The temporary roof rock strength, σ_c; MPa	$50 \div 90$ / $80 \div 160$	$50 \div 80$ / $110 \div 153$	$62.5 \div 85$ / $92 \div 168$	$72 \div 97$ / $110 \div 180$
Power roof rocks, m; m	$5 \div 13$ / $5 \div 20$	$7 \div 14$ / $7.7 \div 25$	$8.5 \div 20$ / $10 \div 28$	$13.6 \div 25$ / $13 \div 38$
Density of fracturing, λ; fissures / m	$11 \div 20$ / $7 \div 13$	$7.5 \div 10$ / $4.5 \div 8.75$	$5 \div 8$ / $4 \div 8$	$3 \div 7$ / $1 \div 5$
The mean value of the displacement of rocks, U; m	> 0.65 / > 0.5	$0.46 \div 0.65$ / $0.34 \div 0.5$	$0.16 \div 0.45$ / $0.15 \div 0.33$	≤ 0.15 / ≤ 0.14
The intensity of fracturing, v, initiating the transition zone of this type; (fissures / m) / mo.	>2.5 / >1.0	$0.95 \div 2.5$ / $0.4 \div 1.0$	$0.4 \div 0.94$ / $0.2 \div 0.39$	--
Probability of inrushes, $P(A)$; %	87	30	2.8	0.5

By generalizing the materials mining and analysis was the classification of the zones, which is shown in Table 3. In the numerator are given values parameters for the seam h_{10}, and the denominator – for the formation h_8.

The main practical advantage of the proposed classification is the ability to predict changes in the workings of sustainability indicators, as well as the probability inrushes in bed h_8 and h_{10} Eastern Donbass mines based on the values density and intensity of the fractured edge of a rock massif that will allow efficient use of interventions to prevent inrushes and increase the stability of development workings.

5 CONCLUSIONS

1. For a characterization of the fracture of rocks, in addition to parameters such as disclosure and the density of fissures, it is advisable to use a parameter of intensity, which corresponds to the rate of fracturing is one of the criteria for determining the probability inrushes at the local areas of workings.

2. The researches established the values of density and intensity of fracturing for mining and geological conditions of seams h_{10} and h_8 deep mines of the Eastern Donbass, in which there is a change of stability parts of workings and there is a qualitative shift from one zone to another zone type.

3. The developed classification parts of workings

contains the quantitative (indicator ω_N) and qualitative (indicator ω_S) stability characteristics which depend primarily on the intensity of fracturing.

4. The advantage of the developed classification is the ability to predict changes in the indices of stability depending on the intensity of fracturing and other factors.

REFERENCES

Malinin, S.I. 1970. *Geologichyeskiye osnovy prognoza povedyeniya porod v gornykh vyrabotkakh po razvedochnym dannym.* Seriya: Gorno-geologichyeskaya. Moscow: Nyedra, 192.

Parchevskiy, L.Ya. & Simanovich, A.M. 1966. *Issledovaniye vliyaniya porodnykh polos na sostoyaniye podgotovityelnykh vyrabotok.* Kommunarsk: Donbasskiy gorno-metallurgichyeskiy institut: 168.

Erofeev, B.N. 1977. *Prognozirovaniye ustoychivosti gornykh vyrabotok.* Alma-Ata: Nauka: 81.

Shashenko, A. N. & Pustovoytenko, V.P. 2004. *Mekhanika gornykh porod.* Uchyebnik dlya VUZov. Kyiv: Novyi druk: 400.

Dolzhikov, P.N, Paleychuk, N.N. & Kobzar, Yu. I. 2010. *Isslyedovaniye osobennostey usloviy ekspluatatcyi arochnykh ramnykh kryepey v zonakh intensivnoy tryeshchinovatosti.* Zbornik nauchnykh trudov. Dnipropetrovs'k: Natcionalnyi Gornyi Universityet: 4, T. 1., 280.

Melnikov, N.I. 1988. *Provedyeniye i kreplyeniye gornykh vyrabotok.* Moscow: Nyedra: 336.

Technical and Geoinformational Systems in Mining – Pivnyak, Bondarenko & Kovalevs'ka (eds)
© 2011 Taylor & Francis Group, London, ISBN 978-0-415-68877-2

The nature and prediction of regional zones for development of dynamic phenomena in the mines of the Donets Coal Basin

A. Antsiferov & V Kanin
UkrNIMI NAS of Ukraine, Donetsk, Ukraine

M. Dovbnich & I. Viktosenko
National Mining University, Dnipropetrovs'k, Ukraine

ABSTRACT: The main notions of method and results of regional zones prediction for development of dynamic phenomena in mines of the Donets Coal Basin are considered. Researches on the basis of geological environment mechanical stresses estimation caused by the Earth equilibrium state disturbances on gravimetric data are executed. A detailed comparison of computed stress fields with gas-dynamic phenomena occurred during mining of seams in A.F. Zasyadko mine was made. The authors are convinced that in investigation of dynamic phenomena in geologic environment, independently of their scale – earthquakes, rock bursts, gas-dynamic phenomena and others – the most important element is study of all whole factors, starting from planetary and ending by local ones, which result in disturbance of equilibrium state of the planet and cause occurrence of mechanical stresses in the outer shells of the Earth.

1 INTRODUCTION

The Donets Coal Basin is the main fuel-energy region of Ukraine. For its development inestimable human resources have been used. For that reason human losses, sometimes accompanying coal production, are very appreciable. Accident prevention in mining is the main industrial-functional problem of coal producers that requires the most careful attention and science knowledge.

In coal mines of the Donets Coal Basin, as, it must be said, in any other coal regions where coal is produced underground, at all times there has long been risk of dynamic phenomena in mine workings. In the first instance the risk is due to dynamic phenomena - sudden outbursts.

Sudden coal, rock and gas outbursts are in the form of avalanche-like collapse of the breast part of coal (rock) mass as a result of which considerable material resources are spent and people die. Generally sudden outbursts occur after some preparation period duration of which is determined by several factors:

– *geological one* connected with a series of geological processes at all stages of coal deposit formation;

– *human-induced one* connected with changes in geomechanical condition of rock mass surrounding mine working as a result of advance mining of the neighboring seams;

– *technological one* connected with changes in gas-dynamic mode of the breast part of the seam in the process of its mining under the influence of different methods and techniques of rock mass impact;

– *current one* connected with energy state of the breast part of rock mass at a time.

And if three latter factors are determined mainly by the human-induced impact on rock mass being mined, geological factor is of purely natural origin. One of the most important characteristics that determine the role of geological factor is tectonic stresses. Specifically, in conditions of the Donets Coal Basin these are stresses arising in deformation of sedimentation mass during movement of the basement blocks. By its nature the natural component of the mechanism of dynamic phenomena in mine workings in many respects is similar to earthquake generating mechanism. Actually dynamic phenomena can be considered as a step-like process of coal-rock mass discontinuity during mining that disturbs natural stress state, which in its turn is determined both by weight of overlying rocks and by tectonic processes that took place and are taking place within rock mass. In these circumstances the type of dynamic phenomenon is determined by the properties of rock mass itself.

2 FORMULATING THE PROBLEM AND METHOD

Stress-deformed state of subsurface is one of the key factors in the nature of occurrence of multiscale dynamic phenomena. In general case in any point of rock mass stress state is determined by the weight of overlying rocks and tectonic factors, and in case of mining it is also determined by redistribution of stresses around mine working. The features of spatial distribution of tectonic stresses are more complex than those of lithostatic ones. Using the terms of exploration geophysics we can say that lithostatic stresses are normal and tectonic stresses are anomalous. It is important to distinguish tectonic stresses arising under the effect of planetary factors and stresses due to secondary deformation processes in subsurface, e. g. crustal block movements, folding and faulting. More over, it is important to realize that it is the case of stress field change with time. We can speak like that about recent stresses and paleo-stresses acted in geological past that have been relaxed partially or completely so far. The defining role in maintaining considerable level of recent tectonic stresses is plaid by neo-tectonic activity within any given territory. In this connection one of the topical problems in research into geologic causes for occurrence of dynamic phenomena in mine workings is prediction of geodynamically active zones.

Basically, idea to research into zones for development of dynamic phenomena within the Donets Coal Basin attracting information on neo-tectonic activity is not new. In the beginning of 1960-s of XX century G.A. Konkov (Konkov 1962) put forward a concept of grouping coal and gas outbursts into linear zones. His research was based on comprehensive analysis of coal and gas outburst distributions and idea of their connection with recent tectonic movements. In his works G.A. Konkov identified a number of zones striking northeast which, on his judgment, corresponded to the regions of the strongest movements. Later G.A. Konkov's ideas have found their development in the works of V.S. Vereda et al. (Vereda et al. 1968), who also identified regional gas-dynamic zones striking northeast, supposing that regional outburst-prone zones reflect increased tectonic stresses in coal-bearing sediments of the Donets Coal Basin connected with increased recent ground surface deformations. V.A. Privalov (Panova et al. 2009) in his research showed tectonic nature of zones of bursting liability in the Donets Coal Basin, specifically their connection with shear dislocations.

The authors of this work also hold to an opinion on the confinedness of dynamic phenomena in mine workings to geodynamically active zones of tec-tonosphere and suppose that research into stress fields is the most important stage in identifying geodynamically active zones in prediction of the zones for development of geodynamic phenomena in mine workings.

The objective of this work is to consider main points of the method and the results of prediction of the regional zones for development of dynamic phenomena in the mines of the Donets Coal Basin.

The basis for the proposed approach is the model of the rotating Earth equilibrium state proposed by K.F. Tyapkin and called *geoisostasy*, which is well covered in geological references (Tyapkin 1980 & Tyapkin 1985). Today computation algorithms for tectonosphere stress fields due to disturbance of the Earth equilibrium state are developed based on the analysis of geoid anomalies (Dovbnich 2008).

In previous work (Dovbnich & Demyanets 2009) it was shown that geodynamically active zones of tectonosphere manifest themselves in stress anomalies attributable to disturbance of the Earth equilibrium state. First and for most such zones, on the assumption of sufficient value of stresses acting therein, manifest themselves as seismically active (Demyanets & Dovbnich 2010). Elements that are tectonic basis for seismo-generating structures find their reflection in the stresses under consideration.

3 RESULTS

In conditions of the Donets Coal Basin computations for estimation of subsurface stress state attributable to disturbance of the Earth equilibrium state were made based on ground gravity survey data on a 1:200,000 scale and a digital terrain model. For the most part of the territory of the region, based on the author's method (Dovbnich & Demyanets 2009; Demyanets & Dovbnich 2010), geoid anomalies were reconstructed by ground gravity data – Faye gravity anomalies on 4x4 km grid (Figure 1). The obtained geoid anomalies served as the basis for computation of stress fields at the territory under investigation.

In order to determine position of the estimated stress fields in tectonics and geodynamics of the investigated region and also the influence of multiscale and multi-depth processes on the disturbance of the equilibrium state we divided stress fields into local and regional components. Comprehensive analysis of the estimated stress fields, tectonics of the Priazovsky block of Ukrainian Shield (US) and junction zone of US and Donets Coal Basin allows us to suggest that the regional stress field component reflects mainly block structure of crystalline basement which is in multi-stress state. Taking into account that dynamic phenomena in mines are confined to sedimentation mass of the Donets Coal Basin it becomes evident that we should look for their connection with the local stress field component attributable to the disturbance of equilibrium state, which, on the authors' opinion, reflects mainly deformation processes in sedimentation mass. At the same time, confinedness of stress field local anomalies to gradient zones of regional anomalies is being clearly identified that speaks for their genesis in the course of development of fault-block crystalline basement.

Theoretic prerequisites for connection of dynamic phenomena, zones of migration and accumulation of hydrocarbons in coal-rock mass with the features of subsurface stress state allows us to use stress fields, attributable to disturbance of the Earth's equilibrium state, as additional predicted criterion in solving the problem of prediction of regional zones for development of dynamic phenomena in mine workings. The working assumption for such constructions can be the following statement. *Regional zones for development of dynamic processes shall be determined by the degree of deformation processes occurring in sedimentation mass, which in their turn find their reflection in the local component of stress field attributable to disturbance of the Earth's equilibrium state.*

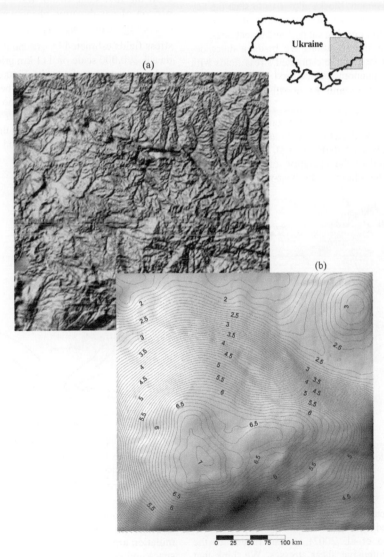

Figure 1. Light-shadow map of Faye gravity anomalies (a) and the result of reduction of geoid anomalies (b), m.

Stress field integral characteristic that reflects the whole of deformation phenomena can be energy of elastic deformations, which computation is not difficult if stress tensor is known. According to the above working assumption this characteristic can be considered as indicator of stored by coal-rock mass elastic energy related to its deformation.

As a case study of our constructions we consider predicted map of the regional zones for development of dynamic phenomena constructed on the basis of the local component of the energy of elastic deformations (Figure 2).

Comparison of the identified zones with the regional stress field component and localized boundaries of the basement blocks allows us to state the key role in their genesis of the movements of crystalline basement, both in geological past and recent ones.

During our investigations we made a more detailed comparison of the estimated stress fields with gas-dynamic phenomena occurred when extracting seams m_3, l_4, l_1 and k_8 at A.F. Zasydko Mine (Figure 3).

The Mine is methane super-hazard, prone to sudden coal, rock and gas outbursts and prone to coal dust explosibility (Antsiferov et al. 2009). Immense tragedy related to the operation of Ukrainian coal mines over the whole of the post-war history were three explosions that took place on 18.11.07, 01.12.07 and 02.12.07 at A.F. Zasydko Mine and took lives of 106 miners, tens of mine workers were hospitalized.

In tectonic terms the mine field is located in the south part of the Kalmius-Toretskaya depression, on the elevated flank of the Vetkovskaya flexure. Here Carboniferous deposits have gentle north-east dip with angles 7-25°. The mine field on the west is limited by the Vetkovsky and Panteleimonovsky Faults with amplitudes 35-55 m. At present coal seams m_3, l_4, l_1 and k_8 are being developed at the depth 1200-1400 m.

At the first stage of our research we made comparison of gas-dynamic phenomena occurred when extracting seams m_3, l_4, l_1 and k_8 with local stress fields estimated by ground gravity survey data on a 1:200,000 scale on 1x1 km grid (Figure 4).

As it is evident from the Figure 4, gas-dynamic phenomena within the Mine can be divided into two groups: 1 – majority of the phenomena is confined to the anomaly of intensive shear stresses; 2 – less of the phenomena is confined to the impact zone of the Vetkovsky Overlap Fault and is related to the anomaly of compressive stresses.

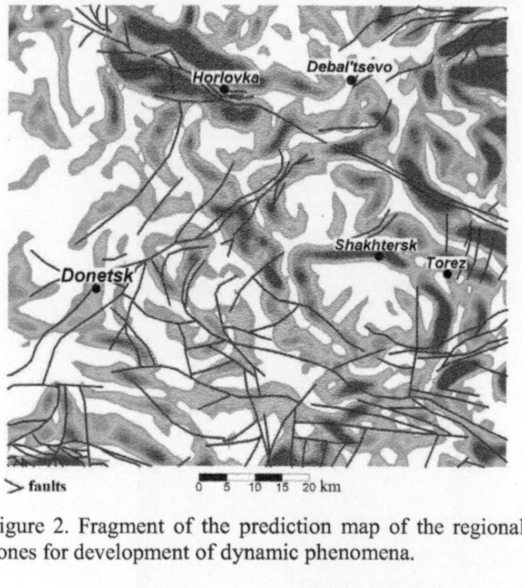

> faults   0 5 10 15 20 km

Figure 2. Fragment of the prediction map of the regional zones for development of dynamic phenomena.

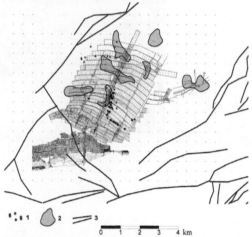

✱ 1   2   3   0 1 2 3 4 km

Figure 3. Comparison scheme of lay-out of in-seam l_1 workings with the main faults: 1 – gas-dynamic phenomena, 2 – predicted zones of methane accumulation (Goncharenko et al. 2007), 3 – faults.

The most part of the zones of methane accumulation predicted by a set of independent methods (Goncharenko et al. 2007) is also related to the anomaly of intensive shear stresses. We think that formation of regions of methane transition into free state, development of the ways of its natural migration and occurrence of the zones of methane accumulation are closely related to the increase in cavity space and coal-rock mass permeability under the effect of mechanical stresses of tectonic nature.

It should be noted that in case of the effect of shear stresses fracture opening and reservoir formation occur, gas drainage of coal-rock mass will be much lower than that of fracture opening under the effect of tension stresses. Thus we can state that within the limits of A.F. Zasyadko Mine, in addition to the human-induced component that has decisive influence on the development of gas-dynamic phenomena, of considerable importance is natural stress state of coal-rock mass responsible for both confinedness of dynamic phenomena to geodynamically active zones and development in coal-rock mass of conditions favorable for their occurrence, in particular, formation of the zones of methane accumulation (Viktosenko et al. 2011).

At the second stage of our research, for more complete analysis of deformation processes in sedimentation mass, we employed trend-analysis for the surface of coal seam m_3 within the limits of mine field as a result of which a map of local folding that complicates close monoclinal bedding of this seam (Figure 5) was obtained. This folding is the difference of the seam surface and its approximating surface which is 3rd order polynomial.

(a)   (b)

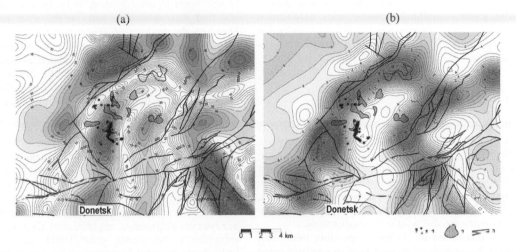

Figure 4. Comparison scheme of gas-dynamic phenomena at A.F. Zasydko Mine with the local stresses (kPa): shear stresses (a), compression-tension stresses (b); 1 – gas-dynamic phenomena, 2 – predicted zones of methane accumulation (Goncharenko et al. 2007), 3 – faults.

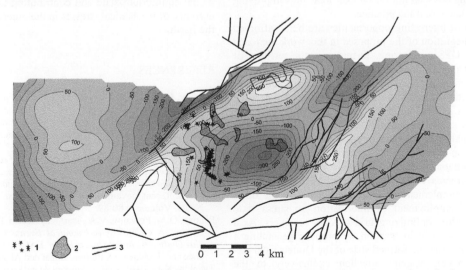

Figure 5. Comparison map of local folding of the seam m_3 with gas-dynamic phenomena: 1 – gas-dynamic phenomena, 2 – predicted zones of methane accumulation (Goncharenko et al. 2007), 3 – faults.

Comparing a map of local folds with dynamic phenomena and predicted zones of methane accumulation we see that majority of them is confined to the gradient zone of local folds, the nature of which is closely connected with the processes that find their reflection in the anomalies of intensity of the local shear stresses. These regularities dramatically confirm the fact of confinedness of gas-dynamic phenomena to the zones of seam kinks (Zabigailo et al. 1974).

Regularities identified within the limits of A.F. Zasyadko Mine field confirm previously made suggestions on the connection of certain components of stress field, attributable to disturbance of the Earth's equilibrium state, with deformation processes in sedimentation mass and reflection therein of the zone of development of dynamic phenomena.

5 CONCLUSIONS

It is important to realize that this characteristic is unique, but not exclusive, that determines geologic factor of the occurrence of dynamic phenomena in mine workings. Only comprehensive consideration of stress fields, tectonics, and features of coal seam hypsometry, depth of occurrence of coal seams, metamorphism intensity and other factors will allow improving reliability of such constructions.

Prediction on the fields of operating mines and those being built of stressed areas, potentially prone to gas-dynamic phenomena, will allow making early control improving mine safety and substantially reducing costs on non-hazardous areas thus improving efficiency of mine practice.

Most interesting opportunities are being afforded in integration of the proposed in this work approach with GPS monitoring. In this case the territory of the Donets Coal Basin shall be covered by stationary operating grid of GPS profiles on which their elevations are determined in one and the same points. Thus uplift velocity-anomalous relaxation areas shall be identified under which existence of anomalous subsurface stress is assumed. By so doing, on the one hand, stress fields connected with disturbance of equilibrium state and maps of recent movements could be independent indicators mutually complementing each other. And on the other hand, employment of information on stress state will allow optimizing grid of GPS receivers. As a result of such integration we propose generating a map of stress-deformed state of the Donets Coal Basin paying special attention exclusively to the stressed areas potentially prone to gas-dynamic phenomena. Geochemical investigations at the hazardous areas shall be conducted with the aim of determining isotopic composition of combustible gases escaping into mine workings and generating maps of distribution of subsurface gases.

Also promising in such anomalous zones is geomechanical modeling of deformation processes in sedimentation mass in order to estimate stress-deformed state followed by geologic interpretation. Today seismic survey is the only geophysical method that allows, on one hand, making detailed structural constructions of formation studied, where total deformations experienced by geologic environment during its development (from accumulation of sediments to manifestation of recent neotectonics) find their reflection, and, on the other hand, based on the analysis of elastic wave propagation velocity and density of geologic environment to give correct enough information related to subsurface elastic properties (Kozlov, 2006). As consequence, we have information required to estimate subsurface stress-deformed state due to the deformation processes therein. In recent years researches repeatedly noticed in their works a possibility to study stress-deformed state on the basis of subsurface structural velocity models by seismic data in solving problems of petroleum and coal geology (Kozlov, 2006, Dovbnich et al. 2009 & Dovbnich et al. 2008).

The authors argue that in research of subsurface geodynamic phenomena, regardless of their scale – earthquakes, rock bursts, gas-dynamic phenomena and others, the most important element is study of the whole of factors, beginning from planetary and ending with local ones that lead to disturbance of global equilibrium state and contributing to the occurrence of mechanical stresses in the outer shells of the Earth.

REFERENCES

Konkov, G.A. 1962. *On the connection of the newest and recent tectonic movements with methane-bearing and outburst-prone zones in conditions of the Donets Coal Basin.* Reports of the Academy of Sciences of the USSR, 3: 670-673.

Vereda, V.S. & Yurchenko, B.K. 1968. *On the correlation of the Donets Coal Basin gas-dynamic zones, coal fracturing and thermal behavior with recent tectonic movements.* Recent movements of the Earth's crust. Moscow: Nedra. Volume 4: 80-89.

Panova, O.A., Pryvalov, V.A., Izart, A., Alsaab, D. & Antsiferov, A.V. 2009. *Geodynamical Events (Coal-and-Gas Outbursts) in the Donets Basin.* EAGE 71th Conference and Technical Exhibition, Expanded Abstract: 222.

Tyapkin, K.F. 1980. *New rotational hypothesis of structure formation and geoisostasy.* Geophysical Journal, 5: 40-46.

Tyapkin, K.F. 1985. *New model of geoisostasy and tectonogenesis.* Geological Journal, 6: 1-10.

Dovbnich, M.M. 2008. *Disturbance of geoisostasy and stress state of tectonosphere.* Geophysical Journal, 4: 123-132.

Dovbnich, M.M. & Demyanets, S.N. 2009. *Tectonosphere stress fields attributable to geoisostasy and geodynamics of the Azov Sea-Black Sea region.* Geophysical Journal, 2: 107-116.

Demyanets, S.N., Dovbnich, M.M. 2010. *Satellite and Ground Gravimetry - The Innovative Approaches in Studying the Earthquake Nature and Prognosis.* EAGE 72th Conference and Technical Exhibition, Expanded Abstract: 582.

Antsiferov, A.V., Golubev, A.A., Kanin,V.A., Tirkel, M.G., Zadara, G.Z., Uziyuk, V.I., Antsiferov, V.A. & Suyarko, V.G. 2009. *Gas content and methane resources of Ukrainian coal basins.* Volume 1. Donetsk: Veber: 456.

Goncharenko, V.A., Svistun, V.K., Gerasimenko, T.V. & Malinovsky, A.K. 2007. *Prospects for integrated geologic-geophysical prediction of methane accumulation zones at coal deposits of the Donets Coal Basin.* Naukovy visnik of the National Mining University, 4:

73-77.

Viktosenko, I.A., Dovbnich, M.M. & Kanin, V.A. 2011. *Regional Zoning of Dynamic Phenomena in Mines – The Innovative Approaches in Gravimetry.* EAGE 73th Conference and Technical Exhibition, Expanded Abstract: 281.

Zabigailo, V.E., Shirokov, A.Z., Bely, I.S., Kudelsky, V.V., Mossur, E.A. & Rudometov, B.P. 1974. *Geological factors of bursting liability of rocks in the Donets Coal Basin.* Kyiv: Naukova dumka: 270.

Kozlov, E.A. 2006. *Subsurface models in exploration seismology.* Tver: GERS: 480.

Dovbnich, M.M., Soldatenko, V.P. & Bobylev, A.A. 2009. *Estimation of stress-deformed state on the basis of structural velocity models: new opportunities in solving petroleum geology problems.* Seismic exploration technologies, 2: 12-18.

Dovbnich, M.M., Soldatenko, V.P. & Bobylev, A.A. 2008. *Estimation of stress-deformed state of coal-rock mass on the basis of structural velocity models.* Geotechnical mechanics: Interdepartmental collection of scientific papers of the Institute of Geotechnical Mechanics of the National Academy of Sciences of Ukraine. Dnipropetrovs'k: Issue, 80: 97-101.

Language training for mining engineers: teaching, learning, assessment

S. Kostrytska & O. Shvets

National Mining University, Dnipropetrovs'k, Ukraine

ABSTRACT: Effective ways of teaching English for Specific Purposes for mining students are described. Entry and target levels of English language proficiency for engineers are given. Innovative approaches to teaching mining engineers with the emphasis on case study method and project work are highlighted. Requirements for ESP materials and tasks are given. It is shown that evaluation and assessment help students become active participants of the study process. The importance of Language Portfolio in stimulating motivation and encouraging lifelong learning is pointed out.

1 INTRODUCTION

More and more international opportunities for a professional engineering career are open in the flexible modern world, which is characterized by the intensive development of science and technology and considerable expansion of business and cultural ties with scientists and businesspeople. The need to use foreign languages especially English as an international one has risen. To meet the needs of language learners, it is necessary to apply more effective methods to teaching and learning English for Specific Purposes courses, which practical aim is to prepare students to communicate effectively in their academic and professional environments by developing their general and professionally-oriented communicative language competences (ESP Curriculum 2005).

ESP learners and teachers now understand the importance of language learning for effective communication. To better help students learn to communicate in a language, teaching methods, materials and learning activities are created to provide students with opportunities to interact.

Communicative language teaching is based on the fact that people learn languages by interacting. Therefore, ESP Curriculum guidelines are focused on communicative teaching.

Making mining students effective learners, ESP teachers implementing new strategies for using assessment as an instructional device recognize the ability of students to take control of their own success and accept responsibility for their own learning. These empowering feelings will inspire and motivate students toward greater achievement.

2 ENGLISH LANGUAGE PROFICIENCY LEVELS

Entering the European space Ukraine is expected to follow the European standards in language proficiency. The exit level of proficiency for Bachelors is B2 (Independent User), which should be assessed at the end of the course using the B2 level descriptors given in the Common European Framework of Reference for Languages (CEFR) (CEF 2001). B2 level descriptors that determine students' learning outcomes are already translated into subject-related skills (Descriptors of Professional Language Proficiency) (ESP Curriculum 2005). Communicative language competences, in particular, rely on students' ability to learn which mobilises existential competence, declarative knowledge and skills, and draws on various types of competence (CEF 2001). With its regard, students develop their study skills and their acceptance of responsibility for their own learning.

Based on professional and academic skills the course helps our students match national qualification levels of achievement, i.e. Education and Qualification Standards which describe the targets students should be able to reach. The content of the ESP course is determined by those communicative needs which are required for the learner's purposes.

In order to achieve the learning outcomes at this level derived from the Common European Framework of Reference for Languages (CEF 2001), rather than developing a course around an analysis of the language, we start instead with an analysis of the learner's needs.

For the identification of our students' fairly specific needs, we proceed by first identifying the target situation and then carrying out a detailed analysis of the linguistic features of that situation (Hutchinson & Waters 1986).

At the Department of Foreign Languages of the National Mining University we use OUP Quick Placement Test when placing students into an appropriate level/group before the course starts. It is a flexible test of English language proficiency, quick and easy to administer.

The entry level of mining students ranges from A1 to B1. Table 1 shows that the majority of students have A1 and A2 levels of language proficiency.

Field of spe-cialization	Common reference levels		
	A1	A2	B1
Engineering (Mining)	42.5%	41.8%	13.4%

Table 1. The entry level in language proficiency of mining students.

The data indicate that mining students need better and more effective ways of learning foreign languages and their teachers need to employ more effective teaching methods. The most significant change that has been taking place in the university language classroom is the shift from traditional teacher-fronted models of learning to learner-centred ones. In the latter model the learner is central to the learning process and cooperates rather than competes with the other learners.

Academic mobility Bachelor's qualification level presupposes that by the end of the course mining students will be able to read with a large degree of independence, adapting strategy to a range of study and specialism-related texts, communicate within academic or professional environment with a degree of fluency and spontaneity, follow the essentials of talks, lectures, reports, presentations, discussions on study and specialism-related topics, convey detailed study and specialism-related information in writing, drawing on a range of sources etc.

According to the objectives for B2 level (ESP Curriculum 2005), the study skills which will be developed throughout the ESP course are grouped into the following categories: information location, academic speaking, academic writing, organization and self-awareness, and assessment.

The focus on developing students' study skills reflects the values of the Bologna Process with its emphasis on individual responsibility for learning (Bologna1999).

The checklist of study skills from the CEF (Byrne 1984) was given to students:
1. I can identify my own needs and goals in language learning.
2. I can plan my own learning.
3. I can reflect on my own learning.
4. I can be aware of my own strengths and weaknesses as a learner.
5. I can organise my own strategies in order to activate skills and maximise effectiveness.
6. I can search for appropriate learning materials and use them for independent learning.
7. I can co-operate effectively in pair and group work.
8. I can monitor and evaluate my own progress that helps me to watch the changes in my progress and learn to correct myself.
9. I can take responsibility for my own progress and work independently.

Being focused on very useful practical objectives, students can improve their study skills both in ESP instructional settings and during self-study in which a learner works alone or with other learners without the control of a teacher. Learners can make choices about what to learn, what strategies to use and the amount of time to spend on a learning task. They can take charge of their own learning progress. Study skills in the study situations in which they are likely to be needed are given as a list of abilities, techniques, and strategies which are used when reading, writing or listening for study purposes (Jordan 1997).

The students need to master a number of study skills and strategies. As the data from research findings show, the majority of ESP learners need English to have access to information via the Internet, to use e-communication, to extract information from the specialist literature and to exchange scientific and technical information (ESP 2004).

They learn how to use the library resources as an aid to learning, for example, as a resource for independent research projects; how to read with the greatest efficiency and the least wasted time and energy.

3 INNOVATIVE APPROACHES TO TEACHING MINING ENGINEERS

Effective ESP course can focus on improving students' linguistic skills. We identify productive (oral and written) and receptive (auditory and reading) performance. All those four primary modes of performance are seen as integrated language skills.

English for Specific Purposes (ESP) Curriculum for Universities highlights the necessity of an integrated approach to the teaching of the primary language skills. The process of integrating language skills to reflect and match real-life use involves linking them together in such a way that what has been learned and practiced through one skill is reinforced and extended through further language activities which bring one or more of the other skills into use (Byrne 1984). Though listed as separate core objectives for Bachelor's language proficiency level B2, they are seen as integration of skills incorporating professional communicative competence developed with the performance of academic and job-related tasks.

In practice, a module that deals with the reading skills will also deal with related listening, speaking and writing skills. Taking a whole language approach which seeks to focus on language in its entirety rather than breaking it down into separate components, it might include pre-reading discussions, listening to a series of informative statements or a lecture, a focus on a certain reading strategy (e.g., scanning), writing a paraphrase of a section of a reading passage, i.e. extensive use of the four skills in an academic setting.

Instead of telling students about how language works, teachers give students opportunities to use language. Student-centered experiential techniques would include simulations, role plays, research projects, hands-on projects, i.e. a variety of highly motivating task-based and communicative activities so that skills are learnt effectively in an integrated manner.

Besides using an integrated approach for developing macro-skills (reading, listening, spoken interaction, spoken production, and writing), the following ones are also applied:

– skills-based approach where students acquire such skills as generic job-related skills (writing e-mails, CV, letters; giving presentations; socializing etc.), reflection, self-study, self-assessment and self-evaluation;

– communicative approach to teaching/learning a language in order to realise the practical aim of the curriculum, i.e. to facilitate students to use the language in various academic, social and professional contexts.

– learner-centred approach where students have a more active and responsible role, and in which they often need to work together to complete a task;

– task-based approach: role plays, simulations, case studies, projects and oral presentations are to be involved.

In this way students in engineering acquire knowledge and skills needed both for their successful studies at the universities and for future target situation, long term goals. It allows for the complete integration of language skills.

Case study which is considered to be an ideal method of inducting students into their professional world improves students' motivation to learn a language, develops responsibility, problem-solving skills as well as all the language skills. Cases reflect typical real life situations which mining students will have to encounter with in their professional activity. Teaching through the case studies supported with the language related exercises and follow-up activities enhances both linguistic and cross-cultural awareness of the learners.

There are some advantages of the case study method that make it effective and, consequently, commonly used in teaching English for specific purposes:

– active student involvement;

– high degree of interaction among students;

– use of authentic materials (manuals, specifications, instructions, business letters, technical journals, etc.) improves the reading comprehension and intensifies the relationship to the students' job;

– each student is an individual and they work in their own manner.

Project work is also increasingly used nowadays to promote meaningful student engagement with language and content learning. Projects can be successfully integrated into a skills-based thematic unit, or introduced into a special sequence of activities in a more traditional classroom.

The general objectives of the project are to:

– provide intrinsically motivating activities;

– allow learners to take responsibility for their own language education;

– enhance the learner's presentation skills;

– provide opportunities to work in small groups.

Project work centres around the completion of a task. It usually requires an extended amount of independent work. Much of this work takes place outside the classroom. Being involved in the project work students go through the following stages. Everything starts with a classroom planning, initial discussion of the idea, definition of the project objective, creating general outline. After the discussion of the content, scope of the project and the needs basic research around the topic is done. While carrying out the project to complete the tasks planned students collect information, design a questionnaire which will be used to investigate the opinions of a specific target group. Following this, the learners must go beyond the boundaries of the classroom and administer their questionnaire to the target group, conduct interviews, organize the material. Then comes reviewing and monitoring. The teams have discussions and feedback sessions. They assess and evaluate the work done.

Project work promotes collaborative learning. Classrooms are organized so that students work together in small co-operative teams. Taking into account the low level of language proficiency of some of the students, it is less threatening to them. Besides, such an approach enhances students' learning in the sense of increasing the amount of their participation.

They work together to produce a product, but the value of such work lies not just in the final product but in the process of working on it. The final product (an oral presentation, a poster presentation, a bulletin board display, a written paper, a report) is shared with others, giving the project a real purpose.

Project work is student-centred, though the teacher plays a major role in offering support and guidance throughout the process. When organized in the way described project work places responsibility for learning on the students themselves. They take control of what and how they learn.

The language is used for authentic communicative purposes. Real-world subject matter and topics of professional interest to students become central to projects. Much language use occurs in a communicative context. By encouraging students to move out of the university classroom and into the world of work, project work helps to bridge the gap between the language study and language use.

Participants of a project are engaged in interaction, production, reception or mediation, or a combination of two or more of these. They interact with an interviewee and complete a form, read a report and discuss it with peers in order to arrive at a decision on a course of action, etc. Such work leads to the authentic integration of skills and processing of information from various sources, mirroring real-life tasks which students will encounter in their future jobs.

Learners are actively involved in using communication strategies, such as clarification, confirmation, comprehension checks, requests, repairing, reacting, and turn-taking.

Curriculum and syllabus design should involve a never-ending process of making adjustments aimed at enhancing the projects' pedagogical usefulness to learners.

The effectiveness of the learning process depends a lot on students' contribution to it and their responsibility for the outcome. The ability to take charge of one's learning is defined as autonomy which means that the learner has the responsibility for all the decisions concerning their learning. Autonomy and responsibility require active involvement, and they are very much interrelated. Autonomous learners understand the purpose of their learning programme, explicitly accept responsibility for their learning, share in the setting of learning goals, take initiatives in planning and executing learning activities, and regularly review their learning and evaluate

its effectiveness (Holec 1981).

Autonomous learning is a natural, common-sense approach to ESP learners who have different needs. Thinking about learning how to learn, we have been concerned with encouraging ESP learners:

- to work out their objectives;
- to think through a range of strategies they need;
- to plan their learning;
- to give a time scale to it;
- to select materials to meet their objectives;
- to undertake self-assessment, etc.

The main reason for advocating learner autonomy is that students need to be able to continue their ESP learning without ESP teachers after finishing the course. Movement towards autonomous learning requires awareness raising and learner training (Jordan 1997). The properly trained autonomous learners can understand the aims and purposes of learning, they can make them their own and work on them. They see the importance of being concerned about what they are trying to do. The learners of this kind are aware of the teacher's objectives. Besides, they are able to formulate their own objectives which can either coincide with the teacher's or be additional to what the ESP teacher is doing.

Effective autonomous learners can select and implement appropriate learning strategies. They can monitor their own use of learning strategies, are able to identify strategies that are not working for them, that are not appropriate, and use other strategies.

Engineering students, for instance, choose learning strategies that are more analytic than those selected by humanities students (Oxford 2002).

4 MATERIALS AND THEIR ROLE IN LANGUAGE TEACHING

Broadly, materials are defined as "any systematic description of the techniques and exercises" to be used for learning/teaching the language (Brown 1995).

According to Huthchinson and Waters (1986), there are three possible ways of turning your course design into actual teaching materials: materials evaluation, materials development, materials adaptation. Materials can be adopted, developed, or adapted, or some combination of the three. The choice of overall strategy will depend on the programme's overall orientation.

The learning materials could be a good motivating factor. The coursebook for mining engineers 'English for Study and Work' (2010) is designed to meet the needs of ESP students majoring in mining and to help students to achieve target B2 language proficiency level as required for Bachelor' Degree.

Each unit of the coursebook module starts with the objectives and expected outcomes. Part II 'Resources for Self-study' is aimed at independent and autonomous learning of general and professionally-oriented English and developing communicative language competences. It is a pack of resources to be used by students individually during their self-study. As self-study is an integral part of the ESP course and takes 30-60 per cent of overall students' load, the main aim of this part of the coursebook is to develop students' study skills, enhance their job-related skills developed in the class, as well as cognitive skills and learning strategies, including self-organization. Each section starts with the expected outcomes, i.e. what students should be able to do by the end of the section and finishes with Self-assessment section with the help of which students can check their progress by using the key answers to the end-of-module test. The key answers are accompanied with the explanations and can be seen as an additional input. The tasks and activities relate to the course content and students' interests and experiences and are aimed to develop a whole range of competences – language skills and language knowledge, communicative, socio-linguistic, pragmatic, socio-cultural competences and study skills.

Students will learn and change if the tasks are meaningful, relevant, motivating, challenging, have a clear purpose and clear instructions, meet students' needs, make them think and share their opinions and own experiences, and allow to develop their confidence and fluency.

Besides using the teacher-generated materials or already existing in textbooks, students working in groups can produce those for the whole class to use. Each group can do a different task. The task of the teacher is to help groups with advice and guidance in the preparation of the exercises. Student-based materials foster group cohesion and a spirit of group solidarity. Relevant ESP teacher/learner materials should provide a systematic means for independent self-study.

Effective materials offer a clear, systematic, coherent and flexible enough unit structure to allow for creativity and variety. Variety is essential to practice a number of micro-skills; to introduce a range of activity types; and to vary the type of interaction taking place during the class.

The tasks that reflect the students' specialist world should be meaningful, relevant, motivating, challenging, have a clear purpose and clear instructions, meet students' needs, make them think and share their opinions and own experiences, and allow to develop their confidence and fluency.

5 ESP COURSE ASSESSMENT AND EVALUATION

In order to assess and improve student learning, ESP teachers need to select the right assessment tools. Traditional tests are frequently used as summative evaluation to grade students. They are an effective way to define the goals of the course. Students concentrate on learning whatever they think will be on the test. Tests are not often used to provide feedback to both students and teachers on whether learning goals are being met. Formative mid-course/module feedback at the classroom level, repeated at regular intervals, helps both learners and teachers clarify their goals and assess progress towards those goals. There is still time to make changes based on the feedback.

Assessment is an integral part of learning. Communicative competence can be tested both on the receptive level and on the production level. Nowadays, modern tests include tasks that assess the four primary language skills in an integrated way. It is considered to be one of the most valid ways to test language proficiency.

Ability to monitor the students' own learning progress can be effectively developed through the use of Language Portfolio for Professional Communication which is a learning as well as assessment tool. Students are involved in deciding what to include in the portfolio. They may revise material in the portfolio after feedback from the teacher or peers. They reflect on the work in the portfolio, thus becoming aware of personal development. Language portfolio not only helps students develop their capacity for reflection, it enhances "learning to learn" and promotes the development of critical thinking skills (Curriculum, 2005). Thanks to a purposeful collection of work there is evidence of mastery of language. Records prepared by students of their learning experiences and describing what activities they have done and the progress they have made help them become more involved in and responsible for their own learning. They help them develop their language learning strategies.

Monitoring their own learning (being involved in self-assessment) is a very important characteristic. Effective autonomous learners are consciously involved with assessment and recognize its importance. They might make a more detailed analysis of successes and failures than other learners. Autonomy is the case when the learner is much more in control of his or her learning (Dickinson 1997). One aspect of this control is the area of assessment. The students are invited to assess themselves. They may submit their self-assessed grade with the assignment. In case the grades of the student and the teacher are different, the negotiating criteria may be used.

The success in learning depends a lot on how motivated the students are during the study process and how many hours they are willing to spend daily on the learning materials. In order to stimulate student motivation and involvement to become better language learners and encourage them to become language learners for life, the teacher uses self-assessment grids included into the Language Portfolio.

Below is an extract of a sample of a B2 (Vantage) self-assessment checklist of language skills for professional needs for spoken interaction:

– I can take an active part in conversations and discussions on most general topics in my academic or professional context.

– I can make my ideas and opinions on academic and specialism-related topics clearly understood by providing relevant explanations, arguments and comments convincingly in both formal and informal discussion.

– I can pass on detailed study and specialism-related information reliably, synthesizing and reporting it from a number of sources.

– I can understand detailed professional instructions well and respond adequately (CEF 2001).

Students assess their language achievement and set personal learning goals. It helps students make the language learning process more transparent, reflect on and assess their progress to identify problems, emphasize achievements and motivate.

Regular goal setting and self-assessment are central to the LP. A language passport requires learners to assess their own proficiency using the scales and descriptors derived from the Common European Framework; and a dossier, in which the owner collects evidence of his or her developing proficiency in a foreign language. This emphasis on self-assessment coincides with the Council of Europe's concern to promote autonomous lifelong learning. The introduction of self-assessment can lead to an open dialogue with students and give them a better understanding of students' problems.

The language portfolio enhances 'learning to learn' and promotes the development of critical thinking skills as it is the basis for efficient and autonomous lifelong learning of languages after school (ESP Curriculum 2005).

In the ESP course students learn how to understand assessment requirements. It helps them to be involved into designing the criteria for assessing themselves. Their decisions then are used as the criteria for self-assessment. Students also evaluate the process of learning and reflect on what is happening. Each aspect of the module and course design is evaluated: the aims, objectives and outcomes of the module/course, the content of the module/course, the materials, and the methods.

To incorporate the integrated-skills approach in assessing communicative competence, it is necessary to keep in mind that items to test communicative competence should be closely related to real-life situations, for instance academic course topics or campus situation topics.

Both achievement and proficiency tests are designed to measure the degree of mastery in the production skills (speaking and writing) and the level of proficiency in the comprehension skills (listening and reading).

The tasks often require test takers to combine more than one skill. For example, the integrated tasks ask test takers to read, listen, and then speak in response to a question. Another example can be listening and then speaking in response to a question. It can also be reading, listening, and then writing in response to a question. The test takers may be required to type/write a response to material they have heard and read.

To measure communicative language skills, the listening-reading comprehension test is used much more often than any other combination: listening-speaking, reading-speaking, listening-writing, reading-writing (Finnochiaro, M. and S. Sako 1983). In the listening-reading comprehension tests, the aim is to measure not only the student's ability to comprehend the meaning of the orally presented material, but also the ability to understand the meaning of the responses presented in written form. In this respect, the listening-reading comprehension test item differs from the item used for pure listening comprehension.

Evaluation is an important and integral part of any course design. Evaluation of the course helps the course designers, teachers and students decide which parts of the course have been working successfully and which have not. Evaluation is also a judgment of the success of the course as a whole (Moon 2001). The purposes of evaluation are:

– to recognise what is effective and what is ineffective;

– to improve what is ineffective;

– to improve, modify or to replan the course in order to make it effectively meet the students' needs.

The purpose of the pre-evaluation or initial evaluation is to evaluate whether the aims, objectives and learning outcomes are matched with students' needs. Before the course begins, students write the entry test to modify the course to determine where they are and to suit their needs. Their responses give the course designers and teachers the information that makes the course more effective in satisfying the students' needs. The entry test, student interviews, administrative consultation, and collection of materials and books, etc. are used at this stage.

At the initial stage students' input is vital. Therefore they are asked to comment on the following: I expect ..., I look forward to ..., I hope for ... .

In the ESP course the needs analysis is of paramount importance because every individual student has their own 'specific purpose' (Jordan 1997). It implies that ESP practitioners and course designers are to take students' needs and goals more seriously when planning the course and tailor everything to their particular requirements. The information obtained from the students will help define and adjust the module/course content according to the students' expectations. Ongoing needs assessment and formative course evaluation overlap, since they help to gauge students' needs while the course is in progress so that it can be modified, as appropriate, to promote learning (Moon 2001). Evaluation during the course or formative evaluation gives information about the degree to which aims and learning outcomes are being met in order to change ineffective aspects while teaching a separate module or the course. Graves (2000) defines the formative evaluation as an ongoing, periodic evaluation of the individual units that assists the course team in modifying future units based on the feedback that teachers receive from students, so that the remainder of the course is tailored to their needs and expectations. Formative evaluation focuses on the process of teaching and learning (Rea-Dickins, P. and K. Germaine 1992). Its procedures look at a number of features including meetings, self-reporting by students and staff, some observation of classes, informal conversations and group discussions.

It is natural for conscientious teachers to ask themselves whether a lesson (a module or a course) was successful. Formative self-evaluation is needed as the basis for change and development. The following self-evaluation form can be used by teachers, which may help them to reflect on the lessons, modules and courses they have just run.

1. What were the objectives of the lesson/module/course? Did I achieve them?

2. What did I like most? What is the evidence for it?

3. What do I need to improve?

4. What should I do next to improve?

The purpose of the end-of-course evaluation or summative evaluation is to give information about the degree to which course learning outcomes have been met, to assess the achievement of the course and to provide the information for the redesign of the course. The summative evaluation takes place immediately after the course has ended and later when students are into their undergraduate studies. The summative evaluation provides the students with an overall view of progression of the course from beginning to end as perceived (Graves 2000).

The evaluation process provides:

1. Written reports, systematic classroom observations, and comments from the teaching staff.

2. Questionnaires, completed at the end of the module and the course. Students are asked to comment on the skills they felt they had learnt, materials and methods used in the course.

3. Student observation and comments, collected informally and formally during the course.

4. End of course exams.

5. Feedback from teaching staff, collected during the course, at the end of each module and at the end of the course. An analysis of student performance in the exam is also made.

The information the course designers obtain from both formative and summative evaluation will make the course more effective in meeting students' needs.

Alongside the teachers, the students are equally participated in the evaluation process that requires them to reflect on their learning. They can bring some changes and innovations to make the module/course more effective in the future. The task of the ESP teacher is to encourage students to be more active and involved in all the stages of the course evaluation in order to see what is important to students and what is of little consequence. In their feedback they may write what they are doing and what they would like to be doing. The information the students provide at the end of the module or the course will be used to determine the effectiveness of them and to decide what should be changed the following year. They make comments on the following:

1. Things I greatly liked.

2. Questions I would like to ask.

3. Suggestions I would make to improve.

4. What was challenging.

Students can record and reflect on their language learning in the language portfolio. It helps them reflect their objectives, ways of learning and success in language learning as well as plan their learning and learn autonomously. The pedagogic function of the language portfolio is to make the language learning process more transparent to students, to help them develop their capacity for reflection and self-assessment, and thus to enable them gradually to assume more responsibility for their own learning. The language portfolio enhances 'learning to learn' and promotes the development of critical thinking skills as it is the basis for efficient and autonomous life-long learning of languages after school (ESP Curriculum 2005). Students will be able to continue working efficiently and usefully even when away from their teacher and the classroom. That is why throughout the course they evaluate the process of learning and reflect on what is happening.

Eventually each aspect of the module and course design is evaluated:

– the aims, objectives and outcomes of the module/course (whether they are clearly stated, appropriate and achievable);

– the content of the module/course (whether it meets students' needs, complete and focused enough, at the right level);

– the materials (whether they are engaging and relevant);

– the methods (whether the students are comfortable with their roles and the teacher's roles).

6 CONCLUSIONS

University mining students need/have a desire to be able to communicate successfully in the English-speaking academic settings/other academic instruction environments in English and develop career-related communication skills. The task of the teacher is to raise student awareness about how they are learning and help them to find more effective ways of working, so that they can continue working efficiently and usefully even when away from their teacher and the classroom.

Organization and self-awareness are included in the study skills objectives of the ESP Core Curriculum to help students achieve B2 as the standard of achievement for the Bachelor's degree, the level recognizable within Europe in line with the levels identified in Common European Framework of Reference for Languages. Study skills are incorporated into the learning objectives of the ESP Curriculum to develop professional communicative competences of students thus becoming an integral part in teaching and learning a language. Study skills are also professional transferable skills which can be used by students in learning other subjects as well as in many other different contexts including their careers and personal lives. In order to continue working efficiently and usefully after graduating from university, when there will be no teaching, students need to develop their study skills in order to be able to learn autonomously.

Methodology and content of the materials need to be sensitive to students' previous learning experience. The use of project work and case study in ESP stimulates students' self-study. In ESP, project work is of

great importance. Students can access the source material in their own time, work through the material at their own pace, choosing topics to match their own interests. They can work on their own, in pairs or small groups or as a class to complete a project sharing resources, ideas, and expertise along the way.

Materials provide a stimulus to learning. Good materials do not teach: they encourage to learn. They need to contain challenging and interesting texts, enjoyable activities that stimulate the learners' thinking capacities, offering opportunities for learners to use their previous knowledge and skills, content which both learner and teacher can cope with. Materials provide models of correct models of language use. We should always keep in mind that the materials that we produce aim at students' better use of the language they have learned.

Assessment does not only determine students' progress and achievement in acquiring skills and knowledge, it also enables staff to evaluate the effectiveness of their teaching.

REFERENCES

Bologna. 1999. *The European Higher Education Area. Joint Declaration of the European Ministers of Education* [online] Available from: http://www.cepes.ro/information_services/sources/on_line/bologna.htm.

Brown, J.D. 1995. *The Elements of Language Curriculum. A Systematic Approach to Program Development,* Heinle and Heinle Publishers.

Byrne, D. Integrating Skills in Johnson K. & K. Morrow. 1984. *Communication in the Language Classroom.* London: Longman.

Common European Framework of Reference for Languages: Learning, teaching, assessment. 2001. Council of Europe, Modern Languages Division, Strasbourg.

Dickinson, L. 1997. *Talking Shop: Aspects of Autonomous Learning* in Hedge, T. and N. Whitney (ed.) Power Pedagogy and Practice. Oxford: Oxford University Press.

English for Specific Purposes (ESP) in Ukraine. A Baseline Study. 2005. Kyiv: Lenvit.

English for Specific Purposes (ESP) National Curriculum for Universities. 2005. Kyiv: Lenvit.

Engineers). 2010. S. I. Kostrytska, I.I. Zuyenok, O.D. Shvets, N.V. Poperechna. Dnipropetrovsk.

Finnochiaro English for Study and Work (A Coursebook for Mining, M. and S. Sako. 1983. Foreign Language Testing. A practical approach: Regents Publishing Company, Inc.

Graves, K. 2000. Designing Language Courses: A Guide for Teachers. Heinle & Heinle Publishers.

Holec, H. 1981. Autonomy and Foreign Language Learning. Oxford: Pergamon.

Hutchinson, T. & Waters, A. 1986. English for Specific Purposes. A Learning- centred Approach. Cambridge: Cambridge University Press.

Jordan, R. R. 1997. English for Academic Purposes. A

Guide and Resource Book for Teachers. Cambridge University Press.

Moon, J. 2001. Short Courses and Workshops. London: Kogan Page Limited.

Oxford, R.L. 2002. *Language Learning Strategies in a Nutshell: Update and ESL Suggestions* in Richards, J. C. and W. A. Renandya (ed.) Methodology in Language Teaching: an Anthology of Current Practice. Cambridge: Cambridge University Press.

Rea-Dickins, P. and K. Germaine. 1992. Evaluation. Oxford University Press.

Three-dimensional model creation of ground water seepage in mining zones (Kryvyi Rig iron ore basin)

Y. Sherstuk, T. Perkova & Y. Demchenko
National Mining University, Dnipropetrovs'k, Ukraine

ABSTRACT: Three-dimensional model of ground water seepage is developed in order to study man-caused changing of geologic environment. It carries out ground water seepage through permeable-impermeable rock mass and processes of layer dewatering and rewetting. Estimation of natural hydrodynamic flow regime is performed by results of numerical modeling.

1 INTRODUCTION

There are areas that are extraordinarily affected by high technogenic impacts in mining regions of central and east Ukraine. The hazard of technogenic influence consists of a variety of negative factors, which often results in irreversible changes of hydrogeomechanical and hydrochemical modes. There are impounding, day surface violations in consequence of activation of suffusion and karst formation, landslides widespread among typical violations (Lushchyk et al. 1997).

The literature widely covers issues of predictive solutions of flow and transport problems in violated conditions with analytical methods (Myronenko et al. 1980, Norvatov et al. 1976). Existing equations of groundwater flow were used to solve cases of steady and unsteady seepage in a relatively homogeneous environment (Bear et al. 1971). However, analytical calculation methods require a substantial simplification of natural conditions compared with mathematical modeling methods in complicated hydrogeological conditions with many variables characterizing different seepage aspects in a heterogeneous environment. It can lead to unreliable hydrogeological forecasts (Gavych 1988).

Common approach for numerical modeling of groundwater flow is moving to plan or profile models, reflecting the heterogeneity of conductivity. For example, the seepage was considered without taking into account changes in flow structure and solutions of transport problems performed along streamlines in a number of studies devoted to the rational groundwater use and protection in mining areas of central Ukraine (Report 2000 & Yevgrashkina 2003).

A large number of software packages based on finite-difference solution of equation systems for mathematical modeling of groundwater seepage were created. Currently, the most tested and reliable global standard for solving problems of groundwater flow and transport is software Visual Modflow (McDonald 1988). The product is based on the following assumptions: Dupuis assumption about constant vertical head in the permeable layers and Myatiev-Girinskiy assumption about the vertical seepage in confining layers.

2 SITE DESCRIPTIONS

The object of research is the area adjacent to the Novoselivka settlement located on the southern outskirts of Kryvyi Rih city (Figure 1).

Figure 1. Simulated area boundaries. M 1:100000. 1, 2 – tailings of UGOK Voikove and Obiednane, 3 – dumps Livoberezhni, 4 – pond Svistunovo.

The studied area is characterized by significant development of gullies and ravines. Interstratifiction of sand, limestone and loess of Neogene-Quaternary age is a structure feature of upper part of cross-section (Report 2000). Quaternary aquifer is sporadically spread within the study area; it represented mainly with loess loam, sometimes sandy, and the aquifer of Neogene sediments confined to the sands is ubiquitous. Aquifers are separated by heavy loam, reddish-brown clay of Lower Pliocene (Figure 2).

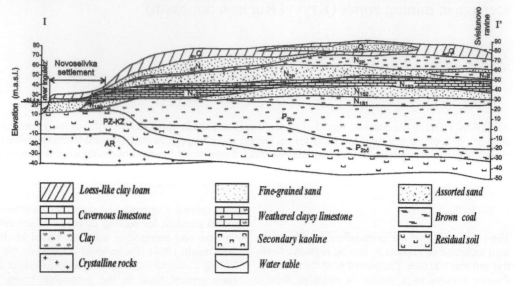

Figure. 2. Schematic geological and hydrogeological section of area along the line I – I'. Scale 1:20000.

Adverse geotechnical conditions of area and intensity of technogenic impact upon geological environment caused an intensification of natural geological processes and phenomena: the suffusion, mineralization increase of groundwater used for water supply, karst of water-bearing rocks, landslides (Report 2000).

Taking into account the strong violations of groundwater regime and high degree of hydrogeological parameter heterogeneity in Krivbass, numerical modeling is practically the only method of quantitative assessment of hydrogeological conditions.

Therefore, the purpose of work is three-dimensional model of groundwater seepage developing and identification at area in impact zone of mining industry to assess changes in hydrodynamic regime of groundwater. It will make it possible to determine the causes and mechanisms of hazardous geological processes development in the disturbed conditions.

3 NUMERICAL GROUNDWATER FLOW MODELING

The seepage model in layered strata is based on the equation of groundwater flow in the porous medium (Gavych 1988):

$$\frac{\partial}{\partial x}\left(K_{xs}m_s\frac{\partial H_s}{\partial x}\right)+\frac{\partial}{\partial y}\left(K_{ys}m_s\frac{\partial H_s}{\partial y}\right)+$$

$$+\frac{K'_{sl}}{m'_{sl}}\left(H_s-H_l\right)+Q=n_0\frac{\partial H_s}{\partial t}, \qquad (1)$$

where s and l – layers index; K_{xs}, $K_{ys} = Ox$ and Oy-axial permeability coefficient; m_s – layer s thickness; H_s, H_l – hydraulic head in layer s and l; K'_{sl} – averaged permeability coefficient between layers s and l; m'_{sl} – its average thickness, that is correspond with vertical flow path; Q – seepage rate or area flow; n_0 – active porosity of rocks in the aquifer; t – time.

The groundwater flow was simulated using special software Visual Modflow 2009.1, that is carried out three-dimensional finite difference solution of equation (1). The main feature of it is capability to compute flow and transport characteristics in common in the conditions of partial saturation, both in permeable and low-permeable layers, and also cycles of rock massifs "dewatering-rewetting".

Groundwater simulation is executed to the territory from ravine Gordovataya to the outfall of ravine Svistynovo and from river Ingyletz' to the head of ravine Shyroka, covering the recharge, transport and groundwater discharge area (Figure 1).

The model area is 37.4 km² with a grid resolution of 200 m ×200 m. Because of significant variability of hydraulic conductivity and rocks storage properties composing a geological cross-section, model domain is approximated by five effective layers: 1 – loess rocks; 2 – clay soils; 3 – coarse and fine-grained sand; 4 – weathered cavernous limestone 5 – fine-grained sand. In nature layers 3, 4 and 5 form a common rock complex of Neogene aquifer. As a result of schematization Quaternary strata is quantified by the values of conductivity varying from 0.3 to 0.5 m / day and thickness equal 2-32 m. Low-permeable clays of 2-18 m thickness and conductivity 0.01 m / day covered model area ubiquitous. Total thickness of Neogene water-bearing rock varying from 5 to 80 m, permeability coefficient ranged from 2.5 to 40 m / day in whole, and separate zone was assigned with permeability coefficient equal to 300 m / day to ravine Svistynovo site. The total amount of various permeability zones makes 31 that describe structural heterogeneity of rocks more detailed.

Hydrogeological boundaries of territory are represented by river Ingyletz' and ravine Shyroka where average annual water level (recharge boundary) was specified. The recharge was assigned separately to 5 zones allowing the significant surface erosional form development and depending on volume of surface flow it ranged from 20 mm / year to 80 mm / year.

The developed model takes into consideration vertical and horizontal components of water seepage through permeable and low-permeable rock mass in common and processes of layer dewatering and re-wetting.

4 RESULTS AND DISCUSSION

Model validity has been proved by inverse modeling of steady-state flow. Simulation veracity is proved by precision of computed natural water level to monitoring indices typical for period before tailing dump formation (Report 2000).

It has been established, that layers 1 up to 4 represent unsaturated stratum in nature except of east part of seepage area within computed layer 4 where Sarmatian limestones were insignificant watered.

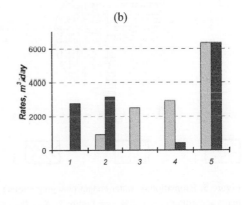

Figure. 3. Inflow I and outflow II budget components in layers 4 (a) and 5 (b): 1 – river boundary; 2 – groundwater inflow and outflow; 3 – recharge; 4 – water flow exchange between contiguous layers; 5 – total balance.

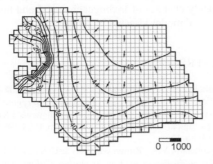

Figure 4. Water table and velocity map of Neogene aquifer in nature (according to simulating data).

Maturation of water budget in layers 3 and 4 occurs similarly to upper rock stratum and is realized by rainfall and water flow exchange between contiguous layers (Figure 3a). Groundwater discharge in river Ingyletz' and water flow relationship with temporal ravine stream considerably effects on proportions between budget components and it is a specialty of layer 5 (Figure 3b).

Groundwater flow in Neogene aquifer is characterized by variability (Figure 4), owing to significant rock heterogeneity, which composes a geological cross-section and spatial geomorphologic structure, while flow direction and computed levels correlate with observation data (Report 2000).

Verification of model stability is made by estimated deviation of balance component subject to input data changing. Data of seepage W that is corresponded to dry and high-water year, and also mudding and suffusion rock deformation K were considered (Figure 5). Because of mining influence privation, parameters K and W are the most important to estimate the model stability.

It has been established that increase or decrease inflow seepage W at 30% leads to small deflections of budget components and deflections vary within 7% (Figure 5). Water flow exchange between layers 4 and 5 is the most sensitive to input data changing. Budget components vary from 1% to 6% because of hydraulic conductivity changing K at ±30% (Figure 5). The findings prove the reasonableness and reliability of inverse modeling.

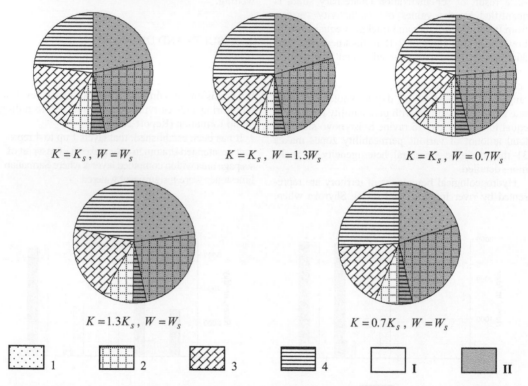

Figure 5. Estimation of water budget changing subject to input data variation: **I, II** – inflow and outflow budget components accordingly; 1, 2, 3, 4 – see Figure 3; K_s, W_s – average K and W of layer 5.

5 CONCLUSIONS

A numerical model which takes into account the heterogeneity of hydrogeological parameters and comprehensively describes the 3D seepage within the permeable and impermeable layers with realizing of dewatering-rewetting processes was developed. Hydrodynamic groundwater regime assessment for all evaluated layers in nature is made by the results of simulation. It has been established that groundwater flow is directed primarily towards the river Inguletz and ravine Shyroka. Deflection assessment of budget components is made and model ability to maintain the adequacy with significant changes in the input data is proven. Proposed nu-

merical model allows to estimate reliable the development of negative processes and to predict adequately changes of hydrogeochemical and hydrodynamical modes affected by mining.

REFERENCES

Lushchyk, A.V., Davydenko, I.P., Shvyrlo, M.I. & Yakovliev, Y.O. 1997. *Engineering situation in region of Kryvyi Ryg iron ore basin.* Information newsteller about geological environment of Ukraine in 1994-1995. Issue 14: 36-41.

Myronenko, V.A., Rumynyn, V.G. & Uchaiev, V.K. 1980. *Groundwater protection in mining regions (geological engineering survey experience).* Lenyngrad: Nedra: 320.

Myronenko, V.A., Norvatov, Y.A. & Serdukov, L.I. 1976. *Hydrogeological reseaches in mining.* Moscow: Nedra: 352.

Bear, J., Zaslavsky, D. & Irmay, S. 1968. *Physical principles of water percolation and seepage.* Paris: UNESCO: 451.

Gavych, I.K. 1988. *Hydrogeodynamics: Textbook for universities.* Moscow: Nedra: 349.

Report of geological engineering surveying for finding out reasons of landslides formation in Novoselivka settlement, Dnipropetrovs'k region. 2000. PI "DniproGIINTIZ": 43.

Yevgrashkina, G.P. 2003. *Mining enterprises impacts on hydrogeological and soil-reclamation facilities of territories.* Dnipropetrovs'k: Monolyt: 200.

McDonald, M.C. & Harbaugh, A.W. 1988. *Open-file report of MODFLOW, a modular three-dimensional finite difference ground-water flow model.* U.S. Geological Survey: 83-875.

New labor remuneration system of miners at coal mine

O. Ponomarenko
National Mining University, Dnipropetrovs'k, Ukraine

ABSTRACT: Analysis results of labor remuneration system at mines of Ukraine are shown. Advantages and disadvantages of existing miners' labor remuneration tariff system are analyzed. Differences between tariff system and grades are compared. Introduction peculiarities of labor remuneration based on ball-factor method are presented.

1 INTRODUCTION

Ukraine's entrance in WTO obliges to considerably increase competitiveness of Ukrainian coal both within the country and abroad. Despite the fact that regulatory and legal framework does not require obligatory and immediate abolishment of state support for unprofitable branches, subsidies practice, existed earlier, for unprofitable mines in order to exceed coal extraction prime cost over its selling price, it seems, will finally cease its existence (Amosha, Kabanov & Starichenko 2007).

One of the most important ways to solve this problem is to introduce effective motivation and labor remuneration system. Despite the fact that there is a countless number of works of domestic and foreign researchers dedicated to considerations of salary as a material component of labor motivation system, numerical evaluation of role, importance and, of the most significance, place of the salary in labor motivation system are insufficiently examined. Partial elimination of this gap is one of the tasks of this work.

There is a great number of approaches to labor remuneration system building described in scientific literature regarding enterprise specific character and its corporative culture.

Significant contribution into labor remuneration theory development, elaboration and improvement of its applied aspects has been done by Ukrainian and Russian scientists O. Amosha, D. Bogynya, I. Buleyev, O. Galushka, V. Boyko, O. Es'kov, A. Kolot, R. Kulykiv, O. Novikova, A. Zdravomyslov, S. Zanyuk, O. Umans'kyy, Y. Ivanova, E. Libanova, O. Pavlovs'kyy, Y. Surmina, T. Reshetllova, R. Hayet, E. Utkin, N. Chumachenko, R. Fatkhutdinov, S. Shurko and others. In works of these or other scientists there are individual questions of this problem are considered. It is worth mentioning that there is an idea among western researches that despite salary being an important factor stimulating high-productive work, one should not exaggerate its importance when developing motivation labor system at an enterprise. For example, one of the founders and directors of the world's biggest transnational corporations "Sony" – Akio Morita, studying moving force of Japanese entrepreneurship, writes "We risk in Japan, promising to people constant employment, and then we have to stimulate them all the time. However, I consider it to be a big mistake thinking that the money is the single method of rewarding people's work" (Morito 2007).

Not underestimating in any way the role of non-material component of labor motivation it can be mentioned that in economically developed countries standard of living is considerably higher than in Ukraine. So it is no wonder that, based on the social-economic researches, these countries workers' need in worthy salary does not stand on the first place among other needs.

In Ukraine the salary is considered as hygienic necessity, dissatisfaction of which makes weaker or totally nullifies motivating factors action. And if people do not get money for their job, they will not have motivation to work efficiently.

It is established by our investigations that salary is considered to be the major factor of work motivation by 95% of workers and by 66% of heads of production units in coal mines (Ponomarenko 2007). It is interesting to point out that even in such country as Poland that is in the EU and in which the living standard is higher than in Ukraine, salary is on the first place among other factors of work motivations. Professor of Catowice economic school Stefan Durka has reached this conclusion, he conducted his investigations similar to ours among Polish miners (Ponomarenko 2009).

Thus, salary structure must serve as a basis for stimulation and rewarding of co-workers of motivational work system. In connection with that, we will examine an existing structure of coal mines workers' salaries from the position of when it performs its stimulating function.

2 EXISTING LABOR REMUNERATION SYSTEM

In the article 96 of labor Code of Ukraine it is mentioned that the basis of labor organization in Ukraine, including coal industry, is tariff system that comprises tariff scales, tariff rates, schemes of remuneration rates and tariff-qualified characteristics (reference books) (Code of laws...).

Tariff labor remuneration system is used to distribute jobs depending on their complexity, and concerning workers – depending on their qualification (position) by the ratio of tariff rates amounts (remuneration rates).

Tariff scale (scheme) formation of remuneration rates is carried out based on the first labor grade worker's tariff rate that is determined as a value that increases legally-established amount of minimal salary, and qualification ratios of tariff rates amounts (remuneration rates).

The advantages of tariff remuneration system are:
– when defining amount of labor remuneration, it (the system) allows to take into account its complexity and working conditions;
– provides individualization of labor remuneration, taking into account working experience, professional skills, continuous working record at an enterprise;
– gives an opportunity to take into account factors of increased work intensity (combination of professions, management of brigade and others);
– implementation of work in conditions declining from normal (night and over-time hours, days off and holidays). Accounting of these factors when making the remuneration is performed by additional payments and premiums added to the tariff rates and salaries.

The disadvantages of tariff system are:
– differentiation in salary by tariff system is done predominantly based on formal indices that can, with more or less authenticity, affirm high labor quality of a concrete worker but insufficiently reflect real achievements and labor quality;
– low possibilities are created to encourage concrete workers and motivate them for qualitative work;
– the possibilities of stimulating motives system formation at the expense of salary are absent for mass improvement of production units quality and management work of the units heads, deputies, assistants, overmen;

– low amount of tariff rates (salaries);
– weak possibility of some criteria formation within tariff system frames that would allow to evaluate and reflect real complexity of miners work;
– insufficient level of stimulation by the heads and engineering staff to implement complex and responsible works at production units.

One of the serious drawbacks of tariff rates scales that are active at the moment is insufficient division of complexity and labor conditions peculiar to coal industry. Unfavourable working conditions in mine are accounted "in average" by the way of general raise (regarding other productions) of tariff rates and salaries, based on the undefined notion of "industry-wide conditions". As it is known from the practice, these conditions can be considerably different not only at neighboring mines but even at the mine where seams of various thickness are mined, at various depth and with wall rocks having various strength and stability. Besides, work complexity significantly differentiates depending on concrete working place. Thus, it is worth mentioning that basic principles and elements of tariff system that is implied in coal industry were laid when plan-distributive system was used and do not meet modern conditions of economics.

3 NEW APPROACH TO LABOR REMUNERATION

New approach concerning labor remuneration improvement and its stimulating function at coal mines is necessary to carry out taking into account objective tendencies that were formed in coal industry of Ukraine, borrowing advanced domestic and foreign experience from coal and other industries, regarding modern conditions of production and economics (Ponomarenko 2009).

Salary analysis of production units' workers has shown that tariff remuneration system does not meet modern market relations conditions and labor motivation.

Thus, taking modern approach, it is possible to formulate a line of requirements:
– objectivity: worker's remuneration amount has to be determined based on objective evaluation of his working results;
– prediction: a worker should know which remuneration he will receive depending on his working results;
– adequacy: the remuneration should be adequate to working contribution of each worker as a result of the whole team work, its experience and qualification level;
– timeliness: the remuneration should follow the

result achievement as soon as possible (if it does not have direct remuneration, at least as a recognition for further remuneration);

– significance: the remuneration must be important for the worker;

– justice: rules of the remuneration calculation should be understandable for each worker of organization and be fair, especially from his point of view.

As the practice shows, inobservance of this requirements leads to instability in the team and causes strong demotivating effect. Presence of fair labor remuneration structure, objectiveness of differences in salary influences the workers' attitude towards the job, their behavior, labor activity effectiveness and thus, success of the enterprise, organization or company. At present, one of the best, proved and such, as recommended itself on practice is the labor remuneration system when counting of remuneration rates is performed based on ball-factor method and matrix-mathematical models. Author of this method is an American scientist Edward Hay (Vetluzhskikh 2007). The basis of this method is the Hay's method (The Hay Guide Chart Profile Method), having taken as a basis factor system of positions assessment. Edward Hay has developed level system, where each profession receives specific number of balls, and depending on the amount of balls, the position has definite level (grade).

Grading is the classification, sorting and ordering. The grader is the positioning of positions, that is, their distribution in hierarchic structure of an enterprise corresponding to given position value for the enterprise. This labor remuneration method is one of the most universal which takes into account interests both of the employer and employee.

In 1943, Hay established consulting company "Hay Group". At present, this is the biggest international consulting company which has 88 offices in 47 countries of the world. More than 7000 clients are using its services in all over the world.

The company helps directors of enterprises to realize business strategy, increase organization effectiveness, and motivate employees for work with maximum self-commitment. This company also works in Ukraine. Its strategy has been used by such famous Ukrainian enterprises as "Azovstal'", Northern and Central ore-dressing plants, mining division of "Metinvest" company, also "Kyivstar", "Interpipe". Among the clients of "Hay Group" – banks, private companies, state and non-commercial organizations representing almost all branches of economics.

In neighboring Poland, there are also mines working based on this method. Firstly, in coal industry of Ukraine, the new labor remuneration system based on "Hay Group" program was introduced in 2010 at the enterprises of DTEK – OJSC "Pavlogradvugillya" and at "Komsomolets Donbasu" mine.

The new system is an hourly-based labor remuneration system. Each position has its own rate, and premium rewarding is performed according to Regulation about premiums.

Thus, the first part of the salary is the basic rate that contains sort of an analogue of tariff system, since tariff-grade grit and grades are hierarchic structure of positions where salaries are build based on the ascending principle. But there are some differences (Table 1).

Besides, this system is convenient for big and middle-size enterprises, since, unlike vertical career building, it allows to build career horizontally within its level. For instance, the increase of qualification and education of workers will reflect on salary level, as knowledge importance factor will increase and salary will increase, despite worker staying at his position.

The second part is the premium, amount of which will depend on productivity and effectiveness of each employee.

The basic principles on which the labor remuneration system is built on are:

– justice – remuneration (salary) depends on the position value and increases depending on working efficiency and position value growth;

– motivation (stimulation) – perspectives of remuneration growth, which the employee can influence;

– competitiveness – remuneration is competitive (attractive) for employees and potential candidates, that ensures one of the factors of mine job prestige;

– simplicity and transparency – simple and understandable for employees;

– flexibility – allows to solve organizational questions both now and in perspective.

One of the most important features lying in the basis of "Hay Group" grade system is that it is more transparent and fair. It allows to evaluate each position based on the market measures. So, when introducing grade system, it is necessary to describe positions at the first stage, evaluate these positions by specially created evaluation committee, and then to form position raters and connect employees to the grades.

All positions and professions despite influence on the enterprise work end result are divided into two groups:

– individual (managing) positions;

– model (executive) positions.

At the individual positions, the freedom to set aims and methods of their achievement is foreseen, also planning in middle and long-term perspective, solutions search is carried out, optimization of means and processes is performed.

Table 1. Differences between tariff system and grades (Slipchak).

Tariff systems	System of grades
1. Built based on professional knowledge, skills and working experience.	1. Wider line of criteria is anticipated that includes such indices as: management, communications, responsibility, work complexity, independence, error cost and others.
2. Positions are ordered based on ascending principle.	2. Grading admits intersection of two adjacent grades. As a result, a worker or a master of a lower grade, due to his professionalism, can have higher position salary than, for example, specialist in labor protection that is located near in grade of a higher order.
3. Hierarchic structure of tariff grit is based on minimal salary multiplied by coefficients (inter-grade, inter-industry, inter-positional and inter-qualification).	3. Grades structure is built only depending on the position importance that is counted in balls.
4. All positions are ordered by strict vertical line ascending (from worker to manager).	4. Positions are ordered only depending on importance for company.

Model (executive) positions embrace managing positions of initial level (units' chiefs, masters, leading specialists and so on) and working professions. These positions (professions) are characterized by jobs that are repeated, procedures presence and work instructions dominates, setting of goals "from top", restriction of possibilities of methods selection for results achievement.

Periodic evaluation at a mine takes place once a year. Evaluation of professional and personal features of worker is performed by direct supervisor. The supervisor puts from 1 to 6 balls for each evaluation criterion with substantiation to each worker. On the same sheet he marks what knowledge and skills are necessary to receive for employer in order to achieve set goals in the next evaluation period. This increases employee labor motivation.

After the periodic evaluation of personnel takes place, rating lists of employees based on the number of balls given them by supervisor are created. Rating creation is performed by the personnel evaluation department specialist by means of determination of percentage ratio between gained by the worker amount of balls and maximal possible amount of balls corresponding to established criteria, taking into account their importance. Total importance of criteria (factors) – 100%. Importance of each criterion for all workers is shown below (Table 2).

As seen from given table analysis, competency is various, depending on their importance.

Depending on the rating results, workers are divided into three ranging groups:

– group "A" – 20% from the number of workers with maximum percentage ratio;

– group "B" – 60% from the number of workers with average percentage ratio;

– group "C" – 20% from the number of workers with minimum percentage ratio.

As a result of ranging, anticipated worker's salary increase is defined. Attachment to each group provides following worker's development principles and their salary increase:

– group "A" – stably high level of professional and personal features and also effectiveness of employees during long time (one year) is a reason for significant increase of remuneration rates and are considered as candidates to inclusion in personnel reserve of OJSC "Pavlogradugol";

– group "B" – possibility of remuneration rates increase is anticipated for this group employees;

– group "C" – increase of salary is not anticipated.

When building policy of basic remuneration during rater formation, each position is developed taking into account market analysis, that is, real salary of this position in coal industry of Ukraine. After such objective evaluation raters and "forks" of salaries were formed - range of basic remuneration amounts, defined for given position from minimal to maximal value. Remuneration rates of employees are corrected

corresponding to appropriate grade. Salaries lower than average value of grade "fork" increase up to minimum grade (but not more than 30%), salaries within bounds of grade will increase by 12%, and those which exceed maximum of grade – by 3%.

Table 2. Importance of workers' competency.

Name of the competence/professional group	Workers	Employees of administrative services	Specialists and engineers	Masters	Managers (executives)
Orientation to result	35%	20%	30%	30%	30%
Orientation to quality	20%	-	-	-	-
Orientation to client	-	35%	-	-	-
Iniativeness and independence	-	-	20%	-	-
Organization and work planning	-	-	-	20%	-
Striving for professional growth	20%	-	20%	15%	20%
Professional knowledge and skills	-	25%	20%	-	-
Development of subordinates	-	-	-	15%	20%
Leadership	-	-	-	-	10%
Heading to standards and norms	-	20%	10%	-	-
Working discipline and labor safety	25%	-	-	20%	-
Working discipline and subordinates' labor safety	-	-	-	-	20%
Total importance	100%	100%	100%	100%	100%

Such new system is built exceptionally based on objective significance evaluation of either this or that activity. Firstly, it motivates workers, secondly, it is more socially-oriented because fixed (constant) part of salary is guaranteed.

Second variable part of salary is the premium that is paid depending on mine workers' activity efficiency. It depends on implementation and over implementation of the task directed to coal volume extraction on mine and its quality (ash content). Amount of this premium is calculated accordingly to regulation about labor remuneration that is constituted at each mine by Board of enterprise working collective.

4 CONCLUSIONS

1. Introduction of new labor remuneration at mines of Ukraine is one of the important trends of competency growth of coal both within the country and abroad.

2. Existing tariff system of salary does not meet modern conditions of market relations and work motivation.

3. New approach to labor remuneration at mines of OJSC "Pavlogradugol" is implemented taking into account objective trends formed in coal industry of Ukraine, borrowing advanced domestic and foreign experience in coal and other industries, in conformity with modern conditions of production and economics.

4. As of today, one of the best proved and recommended on practice is the labor remuneration system with remuneration rates calculation based on ball-factor method of Edward Hay.

5. The new labor remuneration system with grades system is built exceptionally based on objective evaluation of either this or that activity. It motivates workers and is more socially-oriented since the fixed part of salary is guaranteed.

REFERENCES

Amosha, O.I., Kabanov, A.I. & Starichenko L.L. 2007. *Problems of domestic coal industry in terms of Ukraine's entrance into World Trade Organization.* Donetsk: Institute of industry economics of NAS of Ukraine: 67.

Vetluzhskikh, E. 2007. *Labor motivation and remuneration. Instruments. Methodology. Practice.* – Moscow: Alpina Business Books: 133.

Code of laws about Ukraine's labor: MPP Sirin, 1998: 153.

Morito, A. 2007. *Sony. Made in Japan.* Trans. from English. 2nd edition. Moscow: Alpina Business Books: 290.

Ponomarenko. A.P. 2007. *Pecuniary stimulation of labor of production units workers at coal mines.* 36th intl. scientific-practical conference "Problems and perspectives of an enterprise development. Kharkiv: part. 2: 56-58.

Ponomarenko, A.P. 2009. *Peculiarity of workers labor remuneration at production units and necessity of its stimulative function improvement.* Donetsk: Industry economics of Ukraine's NAS: 1(44): 211-222.

Slipchak, S. Grades system. *Methodology of official rates / salary.* #8: 40-54.

Mechanism of ores selective flotation containing Au и Pt

O. Svetkina
National Mining University, Dnipropetrovs'k, Ukraine

ABSTRACT: Separation of the received during ore processing mixtures of copper, silver, gold, platinum, and also rare-earth elements considering great resemblance of their physical and chemical properties is a very complex operation from technological point of view even when using ion-exchange tars.

Extraction method by various chemical reagents, in particular, by phenols is a quite perspective method (Arnold, Crouse & Brown 1965). But use of flotation reagents containing donor atoms N, O, S are quite effective reagents-collectors during clayish ores flotation (Khayduk 1961). The mechanism of such flotation includes stages of coordination compounds formation due to reagents electrons transfer onto vacant orbitals of electrons.

It is known (Khayduk 1961 & Moscovits 1979) that some inorganic salts of metals and also some minerals can form so-called inorganic polymers or polymer aggregates-clusters. For example, $CuCl_2$ forms foliated lattices consisting of the chains of halogens atoms serving as the bridges and copper atoms (II):

Interaction of dichloride copper with organosilicic methylides of pyridinium and their hydroxides at various conditions is described in the work (Svetkin, Kolesnik & Myagchenko 1984).

Vibroimpact activation or fine ore grinding is a quite significant factor of the process although the concentration in size-sorted ore does not change depending on particles size.

However, after the process of vibroimpact load, the extraction of needed metals, particularly copper and silver, increases from 18% with average size being 27.3 micrometers to 93% with average particle size being 9.6 micrometers.

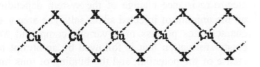

Figure1. Foliated compounds of Cu.

This is connected with the fact that during fine grinding of the minerals the ore activation occurs which leads to alteration of the parameters in crystal lattice. In connection with that, molecules of reacting matters can diffuse inside and on the crystal surface. This process is carried out when all the atoms of the crystal take part in the process.

The results of the potentiometric researches conducted by us i.e. change of the potential of azodien salt solutions titration and compound calculation based on Silen method using Bodlender equation, and also stability based on Leden method, show that adducts of variable compound form in the solution.

In connection with that the complex compounds effect of the azodien with minerals including metals of the I-IV groups on the matrix consisting of polyelectrolytes (PE) that allowed to suppose the formation of clusters with polymers.

In this case, the cluster "body" is defined by the following ratio: 1 mole of a complex compound to 80-200 elementary links of PE.

Thus, the following idea is forming: one molecule of a complex compound can favor chemical conversion along an extensive area of the polymer chain if not following any stoichiometric ratio but defining the aggregatization degree, i.e. vibroimpact activation.

Natural analogues of such model are the metal enzymes containing 4, 12 and more atoms of metal in a molecule that creates so-called "coordinating cell" – limited area within which the reactions either do not evolve or limited by it without having exit to the system medium.

Aggregation formation is thermodynamically limited; it means that under the influence of temperature and other external factors, the complex is in the balance:

$$MeR_2 \cdot 2\,Azodien \Leftrightarrow MeR_2 + 2\,Azodien .$$

Kinetically, two free moles of azodien enter into interaction with sensitized areas of the polymer chain forming cluster embracing four carbon atoms.

Profitable geometry of this cluster (chair type) consisting of 12 atoms gives it energy stability.

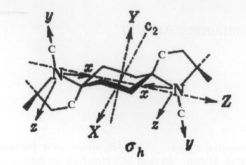

General picture of the cluster "body can be expressed by the following balance:"

Content and constants of the forming adducts stability depend on the initial components concentration that differs them from polyatomic complexes and testifies about cluster nature of the flotation mechanism. When introducing azodien's molecules, aggregatization degree of the system, due to binding of the complexes $(M_eX_2)_n$ (where $X = Cl^-$, NO_3^-, O^{2-} so on) into clusters of Azodien $(M_eX_2)_n$, increases that causes change of physicochemical properties of the solution (electro conductivity, viscosity).

$$\underbrace{\Pi \Theta \cdot [MeR_2 \cdot 2Azodien]}_{cluster} \Leftrightarrow \Pi \Theta + [MeR_2 \cdot 2Azodien] \Leftrightarrow MeR_2 + 2Azodien \Leftrightarrow \underbrace{\Pi \Theta \cdot 2Azodien}_{cluster}$$

$$\updownarrow$$

$$\underbrace{\Pi \Theta \cdot MeR_2}_{cluster}$$

Process of the clusters formation was considered as a private case of the polymer-analogue transformation – reaction of inorganic salt reaction $(M_eX_2)_n$ along energy profitable centers formed during activation from one side and by double diene links from the other side.

The interaction product is the new metal-silicon-organic (MS) cluster. There are significant differences in the formation process polyassociations copper and silver: silver is less inclined to this process. If under conditions of an experiment we succeed in receiving clusters for copper with $n = 850$ then this value is smaller for silver and is equal to $n = 650$. This process depends on Ph solution.

Thus, clustering of the surface will provide reaction, i.e. the higher reaction rate, the more intensive the MS cluster forms. This defines the possibility of new compounds formation connected with structure change. The possibilities of surface reaction restructuring become richer with complication of a solid body compound.

Going from the structure to clusters properties, it is necessary to emphasize, first of all, that they depend on the number of particles in the cluster.

It seems to be obvious that the properties change with the size should be the most abrupt for small clusters where an addition of one particle means big relative increase of the group.

Unlike copper dichloride, nitric acid silver in the solutions is less prone to formation of polyassociation.

Not possessing high generalization degree they form less spacious clusters with organic agents, in-

teraction process carries step-type character and is realized through the row of structures, the properties of which are being recorded by the potential-metric curves (Figure 2).

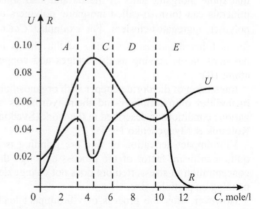

Figure 2. Dependence of the potential Зависимость (U) and electro resistance (R) of the ore $+AgNO_3$ system.

Figure 2 presents the law of the potential and electro-resistance change of the system depending on the amount of injected silver salt. The area (A) characterizes process of a various compound MS cluster formation that is followed by an abrupt increase of the potential, and the binding of ions into associates leads to the medium electrical resistance increase.

Zone (B) responds to creation of the complex with compound of 2:1 (Me / salt) with constant stability 1.19. Further increase of the pyridine's salt

194

concentration leads to its destruction (area C) and reconstruction into more stable complex of the 1:1 compound with $K_{est} = 2.39$ (area D). Differences in potential-metric measures allow to selectively extract ions of silver and copper, and also to control formation of iron oxide.

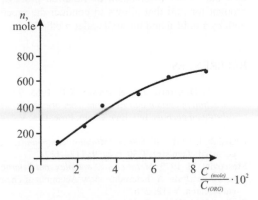

Figure 3. Dependence of MS cluster composition on concentration of organic extractant at concentration of $CuCl_2$ equal to 10^{-4} mole / l.

Figure 3 represents relation of the MS cluster depending on the concentration ratio of $CuCl_2$ and organic agent.

Composition of MS clusters is limited not by stoichiometric ratio but by the associates composition of inorganic salts. Introduction of a low-molecular electrolyte, for example, KNO_3, dislocates direction of interaction practically towards the stable complex formation with 1:1 composition that is connected with the increase of solution ion force.

Azodien having low concentration and possessing the bifunctionality is capable of binding the salts associates $(CuCl_2)_n$ forming extensive clusters. In water-organic and water media, due to hydrolysis, the associates of inorganic salts are not big, and this is indicated by moderate value of the complex composition index (Figure 4).

Concentration decrease of a pyridine's compound solution favors formation of MS clusters with high content of copper-chloride links. In area of ylides concentrations equal to $10^{-7} \div 10^{-6}$ mole / l with concentration of $CuCl_2$ equal to 10^{-2} mole / l, the condition when introduction of a donor does not cause destruction of clusters $(CuCl_2)_n$, and that is confirmed by the line 1 (Figure 4) exit on the plateau. Under such conditions it is possible to receive MS clusters with maximal content of metal.

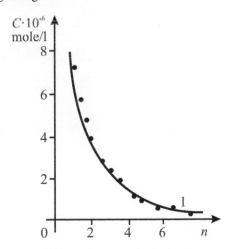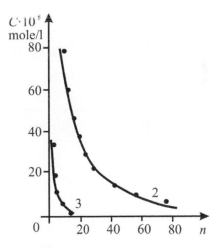

Figure 4. Dependence of the cluster composition (ratio of a metal to a molecule of Azodien$\cdot (MeX_2)_n$ for Cu) $- n -$ from concentration: 1 – in organic medium; 2 – in water; 3 – in $0.5N$ KNO_3.

Thus, received polymer clusters containing ions of metals were washed twice with $0.5N$ KNO_3. Distribution coefficient varied from 4 to 26 at extraction stages and had value of ~ 3 at the stage of washing.

Extracted ions can be individually sorted out by washing at controlled pH level; each of the elements turns out to be well cleared from other ions.

Since the structure of the polymers is not stoichiometric then it leads to their "long-range action" and energy reasonability. Depending on molecular mass of the complex compound and its amount in matrix, the content of, for example, pyrazole ranges from 1 mole per 160 to 1 mole per 800 elementary links. If to take into account that the matrix contains 1.5×10^{-4} mole / mole of carbonyl groups, 2.5×10^{-3} mole / mole of double connections at the ends of a micro-molecule and 8-14 ramifications per 1000 links then it becomes obvious that the boundaries surrounding the "body" of the cluster are extremely fuzzy, and its structure, it seems, is layered, substantiated by the shell (cover) structure (matrix).

Another side of this question is that the cluster of a $\left(M_e R_2\right)_{act} \cdot 2P_{yr}$ compound can overlap the ways of diene groups formation due to chelation by the above shown scheme without preliminary break of coordination connection.

In this case, activated ore forms the "first layer" of a shell or zone of "short-action" at the expense of active centers formation in a shape of radicals during grinding into different mills and this zone prevents oxidation processes and reactions of elimination by the radical mechanism.

Thus, indirectly, it can be proved that during vibro-load the radicals are formed. Based on the experimental data analysis, the following results can be presented: properties of complex compounds, rates of metals extraction, formation of polyolefinic sequences and study of IR-spectr. Based on these data the cluster mechanism of flotation process is brought forward that allows to conduct the process both by a radical and ion-molecular methods.

REFERENCES

Arnold, W.D, Crouse, D.J. & Brown, K.B. 1965. *Solvent extraction of cesium and rubidium from ore liquors with substituted phenols.* Ind. Eng. Chem. Proc. Des. V.4, #3: 249.

Khayduk, I. 1961. *Polymer coordination compounds.* Successes of chemistry. T.30, issue 9: 1124-1174.

Moscovits, M. 1979. *Metal cluster complex and heterogeneous catalysis.* A. heterodox view Accounts of chemical research. V. 12: 236.

Svetkin, Y.V., Kolesnik, Y.R. & Myagchenko, A.P. 1984. *Clusters in polymers.* Leningrad: Thesis of the IV international symposium dedicated to homogeneous catalysis: 168.

Determination of ventilation and degassing rational parameters at extraction areas of coal mines

O. Muha & I. Pugach
National Mining University, Dnipropetrovs'k, Ukraine

ABSTRACT: Selection of ventilation and degassing rational parameters for implementation of extraction areas stable work at coal mines is an actual task. The aim of the given work is to determine air consumption at an extraction area and degassing parameters of underworked seams in order to provide planned longwall output and safe concentrations of methane in an outgoing air current. As a result of the developed methodology of ventilation and degassing regimes calculations during their simultaneous use, analytical dependences of longwall face output on air consumption and degassing boreholes number have been received. Use of these dependences during provision of coal stable extraction on a mine will allow to manage process of gas release into mine workings of extraction areas.

1 INTRODUCTION

In recent years in Ukraine there is a tendency of providing the stopes with high-productive coal complexes in order to increase coal output volumes. Planning of stope output supposes work stability of all technological links of coal enterprises. Mining of coal seams in Donetsk basin is carried out under complex mining-geological conditions. This is connected with low thickness of mined seams, unstable roof rocks, mining depth increase and natural gas content increase of seams and host rocks. As a result of these factors influence, the possibility of mineral extraction rates decrease at some areas is possible as well as the occurrence of necessity of mine coal output temporary distribution between the stopes. Longwall output growth leads to the necessity of changing parameters of ventilation and degassing when using them simultaneously.

2 FORMULATING THE PROBLEM

The aim of the given work is to determine rational parameters of ventilation and degassing at coal mines extraction areas when changing stope output. In order to achieve the set task it is required to establish dependence of mineral extraction at the area on air consumption at this area and degassing level of methane-release sources.

3 MANAGEMENT OF GAS REGIME AT EXTRACTION AREAS OF THE MINE

Gas regime at an extraction area can be managed by changing ventilation regime (due to variation of delivered volume of gas) and degassing level (by way of boreholes number variation and distance between them).

Expected average methane release in a stope is calculated by the following formula (3.76) from (Guidance... 1994)

$$\bar{I}_s = \bar{I}_{s.f}\left(\frac{l_{s.l}}{l_{s.f}}\right)^{0.4}\left(\frac{A_p}{A_f}\right)^{0.6} k_{m.m}k_{c.m}, \text{ m}^3/\text{min}, \quad (1)$$

where $\bar{I}_{s.f}$ – average factual methane release in a stope, m³/min; $l_{s.l}$ – length of a stope for which expected methane release is calculated, m; $l_{s.f}$ – length of a stope for which average factual methane release is determined, m; A_p – planned coal production, t/day; A_f – average coal production at which factual methane release was determined, t/day; $k_{m.m}$ – coefficient considering mining method change (calculated using formulas shown in table 3.10 (Guidance... 1994); $k_{c.m}$ – coefficient considering methane content change in a mine working with depth, when mining depth changes, the $k_{c.m}$ coefficient is calculated by the equation (3.78) from (Guidance... 1994).

Expected average methane release for extracted area is calculated by the following formula

$$\bar{I}_{area} = \bar{I}_{area.f}\left(\frac{l_{s.l}}{l_{s.f}}\right)^{0.4} \times$$

$$\times \left(\frac{A_p}{A_f}\right)^{0.6} k_{m.m}k_{c.m} \text{ , m}^3 / \text{min,} \qquad (2)$$

where $\bar{I}_{area.f}$ – average factual gas release at extraction area, m³/ min.

Let us consider gas regime management at unchangeable length of a stope and at the same depth,

i.e. $\left(\dfrac{l_{s.l}}{l_{s.f}}\right)^{0.4} = 1$, $k_{c.m} = 1$.

For accepted conditions the formula (1) will have a look:

$$\bar{I}_s = \bar{I}_{s.f}\left(\frac{A_p}{A_f}\right)^{0.6} k_{m.m} . \qquad (3)$$

With increase of planned mineral extraction the necessity of gas regime parameters management occurs at an extraction area.

Planned stope output is calculated by the following equation

$$A_p = l_{s.l}V_s m_{ex}\gamma k_l \text{ , t / day,} \qquad (4)$$

where V_s – advance rate of a stope, m / day; m_{ex} – extracted thickness of a seam, m; γ – coal density, t / m³; k_l – coefficient considering coal losses, fractions.

When $A_p > A_f$ it is required to determine maximal permissible stope output by gas factor

$$A_{max} = A_p \bar{I}_i^{-1.67}\left[\frac{Q_p(c-c_0)}{194}\right]^{1.93} \text{, t / day,} \qquad (5)$$

where $\bar{I}_i$ – average absolute methane inflow of a stope $\left(\bar{I}_s\right)$ or extraction area $\left(\bar{I}_{area}\right)$, m³/ min; it is accepted depending on ventilation scheme of an extraction area according to Table 7.1 (Guidance... 1994); Q_p – maximal air consumption in a stope $\left(Q_s\right)$ or at an extraction area $\left(Q_{area}\right)$ that can be used to dissolve methane down to norms permitted by Safety Rules (NPAOP, 2010), m³/min; taken according to table 7.1. from (Guidance... 1994); c – permissible gas concentration in an outgoing current, % (volume-wise); c_0 – gas concentration in incoming to the extraction area current, %.

If $A_{max} < A_i$ then as the output volume increases – gas regime management can be carried out by:
1) air consumption increase at an extraction area;
2) degassing efficiency increase;
3) simultaneously with air consumption increase at an extraction area and degassing efficiency increase.

Air amount, coming to an extraction area, is calculated by the next formula

$$Q_{area} = \frac{100\bar{I}_{area}k_{н}}{c-c_0} \text{ , m}^3 / \text{min,} \qquad (6)$$

where $\bar{I}_{area}$ – average gas release at an extraction area, m³/ min; $k_{н}$ – coefficient of methane release nonuniformity, fractions.

Value of methane release nonuniformity coefficient is calculated by the formula

$$k_{н} = 1{,}94 I_{area}^{-0.14} . \qquad (7)$$

According to the formulas (3) и (6)

$$Q_{area} = \frac{194}{c-c_0}\left[\bar{I}_{area.f}\left(\frac{A_i}{A_f}\right)^{0.6} k_{m.m}\right]^{0.86} . \qquad (8)$$

In order to carry out calculation assessment of output increase on a stope, the output increase coefficient is introduced

$$k_p = \frac{A_p}{A_f} . \qquad (9)$$

Then, considering (7) and (9), equation (8) will have the following form:

$$Q_{area} = \frac{194 k_p^{0.52}}{c-c_0}\left(\bar{I}_{area.f}k_{m.m}\right)^{0.86} . \qquad (10)$$

Received dependence allows to determine air consumption necessary for ventilation of an extraction area with specified stope output. Thus, if annual output of the production unit increases by 30% ($k_p = 1.3$), and methane concentration in incoming and outgoing currents is equal to 0.05% and 1% correspondingly, then dependence of air amount needed for ventilation of an extraction area on absolute gas inflow of the area will look like this:

$$Q_{area} = 234\bar{I}_{area.f}^{0.86} . \qquad (11)$$

This air consumption has to be checked by permissible air current movement speed in face area of

198

a longwall and at a mine working having outgoing air current, also it has to be checked for possible local methane accumulations.

When there are ventilation schemes of extraction areas with ventilation openings filling (schemes 1–M) are used, the possibility of methane local accumulations with concentration above the norm at an abutment (at the end of extinguishment) is excluded, if the following condition is met:

Ventilation schemes of extraction areas with outgoing air current onto coal massif and $k_o \leq 1$, where k_o – coefficient, considering danger of local methane accumulations at an abutment of longwall with ventilation drift, and is calculated by the formula (6.1) (Guidance... 1994).

In ventilation schemes of extraction areas with outgoing current delivered out into worked-out area (schemes of 1-B, 1-K, 2-B, 3 – B types), the possibility of methane dangerous accumulations in a stope within worked-out area under ventilation drift (with presence of packs, ferroconcrete pedestals, stump wall, pack-chocks, chocks at their nearest edge) is excluded if the following condition is met:

$$k_o' \leq 1,$$

where k_o' – coefficient considering danger of methane local accumulations near the worked-out area under ventilation drift, and is calculated by the following formula (6.2) (Guidance...1994).

If at the abutment of longwall with ventilation drift there is a possibility of methane dangerous accumulations formation it is required to reconsider degassing scheme of adjacent seams and host rocks (Guidance...1994) in order to provide higher efficiency of degassing and after that carry out testing calculations for determination of dangerous methane accumulations.

As the longwall output increases, the required safety conditions can be provided by degassing efficiency increase, primarily by means of larger number of simultaneously working boreholes.

In order to prevent methane release volume from changing at an extraction area at increased longwall output it is needed to increase methane volume captured by degassing system.

Formula (3) taking into account equation (9) at constant longwall length, mining depth and mining method, will have the look

$$\bar{I}_{area} = \bar{I}_{area.f} k_p^{0.6}. \qquad (12)$$

At $k_p = 1.3$:

$$\bar{I}_{area} = 1{,}17 \bar{I}_{area.f}.$$

The last dependence means that as the coal production increases by 30%, computational expected methane release into a stope will grow by 17%.

In case of decrease of air consumption coming into a face, it is required to decrease minerals production or to increase degassing efficiency keeping the same factual output.

If to accept $Q_f = Q_p k_d$, then formula (8) will have the following look

$$A_p = 0.00046 A_f \left[\frac{\dfrac{Q_f}{k_d}(c-c_0)}{I_{area.f} k_{m.m} k_{\text{н}}} \right]^{1.67}, \qquad (13)$$

where k_d – coefficient considering value of air consumption decrease in a stope.

Formula (13) allows to determine possible production volume in a stope with known coefficient of air consumption decrease k_d.

If it is necessary to keep longwall output volume, i.e. $A_p = A_f$, degassing efficiency can be increased by way of degassing boreholes number growth.

According to the researches, results of which are shown in the work (Muha & Pugach 2009), dependence of methane total debit on number of working degassing boreholes has been established (Figure 1).

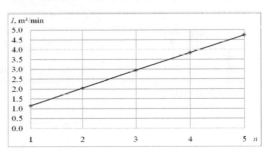

Figure 1. Dependence diagram of methane total debit on number of working degassing boreholes.

Received dependence can be described by linear equation:

$$I_{deg} = a n_b + b, \text{ m}^3 / \text{min} \qquad (14)$$

where I_{deg} – total methane amount captured by degassing boreholes, m^3 / min; a и b – empirical coefficients (for diagram shown on Figure 1, $a = 0.9$ and $b = 0.25$); n_b – number of simultaneously working degassing boreholes, pieces.

Amount of methane coming to an extraction area, considering work of degassing system, can be defined by the following equation:

$$I_{area.f} = I_{area.w} - I_{deg} , \text{m}^3 / \text{min}, \tag{15}$$

where $I_{area.w}$ – amount of methane coming to an extraction area without degassing operations used, m^3 / min.

Equation (13) considering (14) and (15) can be presented as

$$A_p = 0.00046 A_f \times$$

$$\times \left[\frac{Q_{area}(c - c_0)}{\left[I_{area.w} - (an_b + b) \right] k_{m.m} k_{\text{н}}} \right]^{1.67} . \tag{16}$$

According to (16), dependences of longwall output on air consumption delivered to an extraction area have been received, with change of number of simultaneously working degassing boreholes (Figure 2).

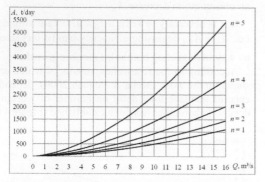

Figure 2. Dependence diagram of longwall output on air consumption at the area and number of simultaneously degassing boreholes.

Analytical dependences, shown on Figure 2 allow to define required levels of ventilation and degassing at their combined use for provision of longwall daily output. At this, it has to be taken into account that received air consumption at an extraction area must be checked by maximum allowable movement speed in a stope and by possibility of methane local accumulations formation at an abutment of longwall with ventilation working.

4 CONCLUSIONS

Based on experimental data, dependence of total methane debit on the number of simultaneously working degassing boreholes and dependence of longwall output on air consumption at the area and on number of working degassing boreholes have been established. This allows to provide planned longwall output and safe methane concentrations in an outgoing ventilation current during management of "ventilation-degassing" system at extraction areas of coal mines.

REFERENCES

Guidance for designing of coal mines ventilation. 1994. Kyiv: Osnova: 311.

NPAOP 10.0-1.01-10. Safety rules in coal mines. 2010. – Kyiv. Derzhgirpromnaglyad: 432.

Muha, O.A. & Pugach, I.I. 2009. *Degassing systems parameters calculation.* Monograph. Dnipropetrovs'k: National Mining University: 182.

Technical and Geoinformational Systems in Mining – Pivnyak, Bondarenko & Kovalevs'ka (eds)
© *2011 Taylor & Francis Group, London, ISBN 978-0-415-68877-2*

Economic indicators of BUCG on an experimental station in the OJSC "Pavlogradvugillia" conditions

V. Falshtynskyi & R. Dychkovskyi
National Mining University, Dnipropetrovs'k, Ukraine

O. Zasedatelev
OJSK "Pavlogradugol", Pavlograd, Ukraine

ABSTRACT: The study of the application feasibility of borehole underground coal gasification is given according to the conditions of mine "N.I. Stashkova", OJSC "Pavlogradvugillia". The basic parameters for effective work of the planned stations "Pidzemgaz" are taken for estimation in the article. Potential composition of employers and term of enterprise recoupment is expected.

1 INTRODUCTION

Western Donbass is the most young coal region of the Donetsk basin. The first mine – "Pershotravneva" (now blocked) – was put into exploitation in 1963, the last one "N.I. Stashkova" – in 1982. From 10 now operating mines seven of them started working in the 70's, that are in production activity hardly more than 30 years.

Subdivisions of OJSC "Pavlogradvugillia" (entering in the complement of the largest power company in country – DTEK) have a number of advantages by comparison to other coal mining enterprises of Ukraine, namely:

– more favourable structure of mine fund – an average annual load on a mine approximately in 2.5 time higher, than on coal industry of Ukraine;

– high level in mechanization of basal production processes and more perfect technologies, adapted to the difficult geological conditions;

– lower than on many other coal mining enterprises, prime cost of coal production.

In Western Donbass approximately 30% of coal goes to the addition agent to the coking charge. Depth of mining – 200-600 m, that is considerably less than in old Donbass. Geological conditions can be described as average on a degree complexities. The mines of region are provided with geological reserves on 50-70 years, although these reserves are unequivalent on the terms of their development (Bondarenko 2003).

Western Donbass is occupied by a considerable habitat in industry of Ukraine: about 240 million tons of coal have been obtained here for the last 20 years (1991-2010 years), thus during 7 years annual production activity made more than 12 million tons. While for the same period the annual mining in Ukraine declined from 135 to 80 million tons in a year that almost on 40%, there was a gradual increase of extents of production activity in Western Donbass. Such dynamics resulted in that the role of Western Donbass in the general count of the obtained coal in Ukraine grew from 7.9% in 1991 to 20% in 2010.

The mine fund of Western Donbass is higher after comparing to other coal mining regions of Ukraine. The level of mechanization and technology, small depth of development, the considerable geological reserves of coals ground to make a conclusion that this region is very perspective and has all of capabilities for development of economic indicators. In connection with mines closing in the Donetsk basin and large complexity of producing on many enterprises of industry, the role of Western Donbass will be increase. It is related to that OJSC "Pavlogradvugillia" has all of chances to save the attained production capacity, and at certain terms also to increase it (Bondarenko 2003).

In spite of economic stable state, direction of enterprise, are spared by a large value technological development of all processes in coal mining, development of innovative developments and transfer of technologies in a generation. For that was created Management on perspective development of OJSC "Pavlogradvugillia".

Together with the employees of the National mining university and other representatives of scientific schools of Ukraine, near and far abroad, exploring the area of installation of the new mining engineering and technological development of mining process are conducted. Thus the strategic charts of development are examined not only traditional mechanical technologies but also capability of application of new radical methods at which production activity takes place with the conversion of coal from solid to gaseous state. For that the capability of borehole in-situ gasification application of thin coal seams are estimated on the mines of Western Donbass (Falshtynskyi 2010).

The process of an in-situ coal seam gasification are accompanied by a high temperature, pressure, rocks deformation, formation of the depressed cones, gas and blowing escapes. Therefore determination the criteria of coal deposit taking into account influence of geological, hydrogeological, technological conditions and technogenic factors on the change of rock massif and superficial landscape was conducted.

At coal deposit, determining the following technogenic factors of BUCG process was taken into account:

• the process of BUCG artificially effects environmental conditions with formation of the local unloading, with weakening or buildup of mining pressure, by formation of rock deformations, hydrodynamic, temperature, chemical anomalies in aquifers, other physical and chemical and geological phenomena;

• in-situ coal gasification is carried out through drilling from a surface towards to a coal seam injection and production boreholes (oriented, vertical or sloping boreholes), on which blowing mixture are sent;

• the reactionary channels of gasification are created in coal seam by boring of channels on coal;

• it is planned to conduct the process of BUCG on the air blowing at working blowing pressure 0.6-0.8 MPa, or air and steam (oxygen) blowing (1-1.5 MPa);

• at coal seam gasification there are structural changes of country rocks, filtration characteristics change due to high temperatures, moving of rock massif of rocks and infilling of goaf;

• high pressure and high temperature in the zone of combustion (1100-1300 °C) change thermal - and hydrodynamic processes in country rocks. So, under act of this temperature the rocks of roof warmed up to 100-120 °C on the depth of order of 3.6-4.7 m. The ground rocks are warmed up to 100-120 on a depth 2.8-3.2 m;

• in the process of gas creation, participate moisture of the most coal seam, containing rocks, static and dynamic aquifers;

• aquifers in the underground gasgenerator zone are heated and chemical composition changes, satiated fluids, phenols, gums and other harmful contaminants;

• high pressure in underground gasgenerators at particular regime of blowing or leadthrough the boreholes caused breakthroughs gases and blowing in permeable rocks.

Considering that underground coal gasification is a difficult process with the active affecting environment, that requires the ground of acceptability of coal deposits appraisal to UCG on criteria, and also working off the characteristics of development and production activity of below ground gas generators, on stand union and laboratory options with the imitation of similarity:

– technological scheme of UCG;

– constructions of gas generators;

– geological conditions;

– characteristics of coal seam gasification process;

As a result of comprehensive assessment, for a generation experimentally the industrial station, 10 departments of the mine fields were preliminary chosen. In obedience to confessedly criteria most full for in-situ gasification fits on coal seam C_5 of the mine field "N.I. Stashkova".

A department is limited the following tectonic initiations: fault #5, fault #3, Bogdanovski fault, longitudinal fault and fault "A".

Coal seam C_5 with average power 1.05 m, simple, rarer difficult structure, relatively self-possessed, beds approximately in 300 m from the daily surface of this area.

Immediate roof is presented mainly BY an argillite, rarer by sandstone and siltstone. An argillite is massive, dense, interference with coal seam is sharp. Tensile strength on contraction changes from 66 to 460 kgf / sm^2.

An argillite in generations is unsteady, and at water encroachment very unsteady. Siltstone and sandstone behave to more immune rocks. The main roof is presented alternation of claystone, siltstone and sandstone.

Immediate roof is presented mainly an argillite and siltstone (0-0.8 m). Tensile strength on contraction – 50-524 kgf / sm^2. The argillite is unsteady.

Hydrogeological conditions relatively favourable. A department is isolated from every quarter by tectonic dislocations from a hydraulic bond with incumbent aquifers. An argillite and siltstone being waterproof rocks. The basal technical and economic indicators of building and production the station of underground coal gasification are based on definition: capital investments, quantity of worker, labour productivity, prime price of gas, profit, date of recoupment and other characteristics.

Along with the assessment of efficiency of basal

production activity of enterprise, the ecological assessment of effect consequences of this activity is made on a natural environment. As a methodical manual next sources were used (Kolokolov 2000 & Falshtynskyi 2010).

The basal initial technological characteristics of work the potential station of underground coal gasification are resulted in Table 1. Information about the potential quantity of worker and labour productivity are resulted in Tables 2 and 3. Necessary capital investments for BUCG realization on the conductors of costs are resulted in Table 4.

Table 1. Initial technological characteristics of the UCG station.

Name of coefficients	Unit of measurement	Values of coefficients
Planed power	mln. m³ fluid in a year	598.1
Commodity products	-	598.1
Average power of coal seam	m	1.05
Density of coal	t / m³	1.16
A count of gas generators in simultaneous work	gas generator	5
A number of inclined boreholes in-process		
In all	boreholes	25
Including	-	
Injection	-	10
Production	-	10
For the discharge of goaf stowing	-	5
Vertical ignition	-	5
The length of boreholes	m	9880
Expenditure of blowing	mil.m³ in a year	273.4
Calculation heat of gas combustion	MJ / m³	7.5
A number of working days in a year		
Enterprises	days	365
Worker	-	260
Duration of workweek		41
Duration of shift		8

Table 2. Quantity of worker and labour productivity.

Categories of personnel, departments	Quantity of worker, brows.	
	secret	on a bill
Industrially-production personnel		
Workers		
Gasgenerator department	51	82
Blowing department	4	7
Department of gas discharge	4	7
Heat utilize installation	35	55
Boring department	130	175
Stowing complex	30	44
Attendance and repair action of electrical equipment and facilities of automation	9	12
Repair personnel	14	19
Maintenance of off-site gridirons	25	38
Department of transport	19	26
Other processes	28	41
Complex of stowing preparation	31	46
Total workers	380	552
Directors, experts, employee	86	86
Total manufacturing staff	466	638
Non-commercial personnel	24	32
Total	490	670

Table 3. Labour productivity.

Category of worker	Labour productivity 1 brows. on a bill	
	gas (thousand m³)	in a reference fuel
Workers	2924/2445	877.2/709.1

Resulted in Table 4 the cost of production building does not take into account an expense on the construction of processing facilities of the station (and the chemical processing complex related to them, without which presently building of enterprises on UCG is impossible).

Table 4. Capital investments on the conductors of costs.

Conductors of costs	Calculation cost, mil. UAH
Building works	
Entrance ways, domestic and industrial communications,	28.76
Boring, installation and development of underground gasgenerator	86.88
Assembling works	
Power superficial complex and complex of cleaning and conversion	45.32
Installation	
Power superficial complex of cleaning and complex of conversion	102.98
Other works and costs	81.8
Total	318.74

Economic calculations of chemical complex UCG, calculations, related to the costs and economic efficiency of conversion of the so-called by-products of UCG, are also attended with considerable difficulties, explained deficiency now of the concrete technologies worked out in detail on disengagement from UCG gas, certain, economic attractive, chemical products. In addition, presently considerable vibrations take a place both in rates on separate chemical products and in the prime price of different technological processes.

Most fit for comparing to UCG there is a method of coal coking conversion. Experience of development the technology of coals coking, use of the chemical products got here can serve as a base for the generation of technology of chemical generations on the basis underground coal gasification.

Summary information of capital investments on building of the BUCG station is resulted in Table 5. Costs on a social sphere are inspected in size of 50% from the cost of production building.

A production of 1000 m³ cost is designed on the components of costs and makes 516.2 UaH / 1000 m³.

The cost of the realized products consists of two features:
– costs made and released gas consumer;
– costs of thermal energy, received due to utilization of heat.

Table 5. Summary table of capital costs on building of the UCG station.

Conductors of costs	Cost, mil. UaH
production building of the BUCG station	318.74
costs on the generation of power gas (power, technological)	87.5
costs on chemical conversion	105.9
costs on a social sphere	10.2
Total	522.34

A vacation cost of enterprise and cost of the realized commodity products is certain in Table 6. In addition, it is necessary to take into account a profit from realization of chemical products. Necessary information is presented in a Table 7.

Table 6. Supposed a vacation cost of enterprise and cost of the realized commodity products.

Name of coefficients	Unit of measurement	Values of coefficients
Prime price	UaH / 1000 m³	516.2
Income	-	120
Wholesale price	-	636.2
Tax (20% from a wholesale price)	-	127.2
Vacation cost	-	763.4
Annual extent of commodity products (fluid)	million m³	598.1
Cost of the realized gas	million UaH / year	366
Count of thermal energy, released consumers	thousand GKal / year	893.5
Cost of thermal energy	UaH / GKal	224
Cost of the realized thermal energy	million UaH / year	200.1
Total worth of the realized commodity products	million Uah / year	566.1

Table 7.

Type of chemical product	Escape of chemical products on the BUCG station (t)			
	from one gas generator (g / g)	station of BUCG (5 g / g)	cost, UaH	total worth, thousand, UaH
Coal gum	3878.7	19439.3	7268	141285
Benzene	1213	6065	6520	395438
Ammonia	2252	11264	940	105882
Grey	274	1371	1300	17823
Total				66042.8

Total, have a profit from the sale of chemical products on the BUCG station – 66042.8 thousand UaH, from them 12350.0 thousands UAH are costs (18.7%), 53692.8 thousands UAH – is an income (81.3%).

A general income on an enterprise can make 311.1 million UAH

The date of recoupment of capital investments will make thus:

$522.34/311.1 = 1.7$ year.

The economic effect of UCG consists of the followings factors:
  – economics of fuel;
  – cutbacks of capital and operating costs on production activity of fuel;
  – cutbacks of a transport spending on passing to of fuel the consumers and export of ash at a generation and use of fuel;
  – economies of operating charges for a consumer on development of fuel to incineration;

Economy of fuel. The annual generation of tank gas at the station of underground coal gasification can make 598.1 million m^3 in a year, that equivalently:

$598.1 \cdot 2000 = 170.9$ thousand t / year.

If to accept for these calculations coal by calorie content 6000 KKl / kg, the economy of coal will be equal:

$(170.9 + 26) \cdot \dfrac{70000}{60000} = 230$ thousand t / year.

Cutback of a transport spending. A prime price of hauling the UCG gas is considerably below than costs on the transport of coal. To the economy of transport charges it is necessary to take the exception of costs on the transport of rocks in a dump. At mine production 800 thousand tons of coal in a year and ash-content 34% an annual rock yield will make no less than 300 thousand tons. Hauling of it in a dump with space between them equal to 5 km.

$0.6 \cdot 15 \cdot 300 = 2700$ thousand UaN / year,

where 0.60 – paying for hauling of rocks on a spacing interval a 5 km, UaH / tons.

Accordingly, the total economy of transport charges make:

$(25 - 1) \cdot 0.23 + 2.7 = 8.22$ million UaN / year.

Economy of operating charges for a consumer. The economy of operating costs during work in power station on a gas produced at underground coal gasification will make 360 UAH / t reference fuel, and during work of boiler rooms 310 UAH / t reference fuel.

CONCLUSIONS

Underground gasification is a radical technology of mining with complex conversion of coal seam in place of bedding. Existent experience of the stations of "Pidzemgaz" and underground gas generators in many countries of the world shows a capability and efficiency of thermo-chemical technology of coal seam conversion.

Technology of BUCG with development of new technological schemes and construction of gas generator provides ecofriendlyness and economy of underground coal gasification process with development of gas generators from a terrene and in mine conditions.

UCG has the potential to provide a clean source of energy from coal seams where traditional mining methods are either impossible or uneconomical:
  – allows to realize the idea of manless mining;
  – a capability to master unconditional reserves of coal;
  – a general income for the concretely taken enterprise can make 311.1 million UAH in a year, date of recoupment of capital investments – 1.7 year;
  – there is an economy of fuel in a count 230 thousand tons reference fuel in a year;
  – transport charges are abbreviated on delivery of

fuel consumers and on hauling of rocks in dumps (8.22 million UAH in a year).

The technical and economic assessment of capital, operating, transport costs and ecological charges, at a generation and production activity of the BUCG station, confirms efficiency and expedience of application of this technology in the Western Donbass conditions.

REFERENCES

Bondarenko, V.I, Porotinikov, V.V. & Dychkovskyi, R.O. 2003. *Prospects of SC "Pavlogradvygillia" develop-* *ment.* Doneck: DonNU: 198-201.

Falshtynskyi, V.S, Dychkovskyi, R.O. & Tabachenko, R.O. 2010. *Newest technology of coal seam exploration on the bases of underground coal gasification. Ukrainian coal,* 1: 10-14.

Economical evaluation of BUCG for thin coal seams. 1994. NIR Report GP-57: NMU: 57.

Kolokolov, O.V. 2000. *Theory and practice of thermochemical coal process.* Dnipropetrovs'k: NMU Ukraine: 281.

Falshtynskyi, V.S, Dychkovskyi, R.O. & Lozynskyi, V.G. 2010. *Economical justification of effectiveness the sealing rockmass above the gas generator for borehole coal gasification.* Praze naukowe GIG, Gornictwo i srodowisko, kwartalnik, 3. Katowice: GIG: 51-59.

The interaction between dust flows and mist spray in the gravitational field

R. Azamatov
PLC "Pavlogradugol", Pavlograd, Ukraine

ABSTRACT: In analytical calculations taking into account the total area of the dust particles and liquid droplets, cluttering the cross section generation, and given that the absolute relative velocity is the absolute value or magnitude of its vector, and not only the longitudinal component of the velocity of liquid droplets, dust suppression shows the probability of dust particles in the gravitational field in its various diameters.

1 NPRODUCTION

Dust prevention in coal mines as an occupational hazard and a possible explosion source is still not completely solved problem. Existing methods and ways of dedusting of ventilation flows frequently do not provide with creation of sanitary code. Suggested recommendations are contradictory, the range of application of varying measures are not marked out, the mechanism of air hedrodedusting by irrigation are not enough studied.

It is considered that hedrodedusting efficiency is ensured by composite demonstration of inertial, electrostatic and turbulent forces. However, strictly speaking, we should make a separation on dynamic, gravitational and electrostatic forces, as long as the first ones are manifested in the active coverage of the torch irrigation, and the others – outside the active coverage. At the same time turbulent forces function everywhere depending on the speed of liquid droplets' movement and their sizes.

Let's stop at the study of the functioning of turbulent forces. Received results (Saranchuk 1984; Kachan 1986; Mednikov 1980; Kudriashov 1979) of dust particles dynamics and liquid droplets allow us to research how the interaction between mist spray and coal dust is happening when all three forces are manifested. As in a number of works (Saranchuk 1984; Kachan 1986; Mednikov 1980; Kudriashov 1979), let's proceed from the premise that the interaction is possible due to emergence of dust particles relative speed and liquid droplets, as a result of which water drops "comb-out" the cloud of air-borne dust. The more there is time of "combing-out" and concentration of liquid droplets and dust particles, the more the effect is.

2 THE MAJOR PART

Let's consider that air-born dust has already acquired speed far from the source of its formation, coinciding with air velocity and is in suspension state. Concentration of dust at the entrance to the zone of the water curtain will vary according to the proportion to the very dust concentration, the relative velocity of dust particles and liquid droplets, total effective area of the midsection of the dust particles and liquid droplets, as a result we can write

$$V \frac{dC}{dt} = -e \frac{\pi}{4} \left(m d_p^2 + n d_d^2 \right) \sqrt{(u \perp u_0)^2 + v^2 C} , \quad (1)$$

where V – some volume of excavation where the interaction between dust and drop flows takes place, m^3 ; C – dust concentration in a given volume, kg/m^3; e – coefficient of proportionality or coefficient of dust particles capture; d_d – average harmonic diameter of liquid droplets, m; d_p – average harmonic diameter of coal dust particles, m; m – the amount of dust particles in a given volume; n – amount of liquid droplets in a given volume.

Unlike works (Saranchuk 1984 & Kudriashov 1979) they took into account the total area of dust particles and liquid droplets which obstruct the excavation cross-section, and not the sum of their radii squared, which is physically meaningless. In addition, we consider the absolute relative velocity as scalar or value of a vector, and not only the longitudinal component of liquid droplets velocity, so long as the vertical component makes a significant contribution, especially for large diameters of the droplets. Earlier this fact was not taken into account.

Intercepted volume of the excavation can be represented as the product

$$V = V_d \frac{V}{V_d} = \rho \frac{\pi}{6} n d_d^3 / Z , \qquad (2)$$

where V_d – volume of liquid droplets in a given volume, m^3; Z – liquid droplets concentration in a given volume, kg / m^3.

Substituting formula (2) into the equation (1) we have

$$\frac{dC}{dt} = -e \frac{3}{2\rho} \frac{m d_p^2 + n d_d^2}{n d_d^2} \sqrt{(u - u_0)^2 + \upsilon CZ} . \qquad (3)$$

So long as the area ratio of particles and droplets can be expressed in terms of the ratio of their concentrations, it will be equal

$$\frac{m d_p^2}{n d_d^2} = \frac{\rho}{\rho_p} \frac{C}{Z} \frac{d_d}{d_p} ,$$

where ρ – dropping liquid density, kg / m^3; ρ_p – coal particles density, kg / m^3.

As far as $\rho / \rho_p = 1000 / 1300 = 0.77$, the equation (3) will take on form

$$\frac{dC}{dt} = -e \frac{3}{2\rho} \frac{d_p + 0.77 d_d C / Z}{d_p d_d} \sqrt{(u - u_0)^2 + \upsilon^2 CZ} . \qquad (4)$$

The function Z of the concentration of sprayed liquid in the ventilation flow remains unknown. To find it we use the equation of nonstationary impurity transfer taking into account their loss on the excavation wall and on the soil under the influence of gravitation forces.

$$\frac{dZ}{dt} = -\left(\frac{k \upsilon}{h} + \frac{\gamma \Omega}{S} \right) Z , \qquad (5)$$

where k – coefficient of uneven distribution of liquid droplets throughout the height; h – height of the excavation , m; γ – coefficient of liquid droplets losses on the excavation wall, m / s; Ω – perimeter of the excavation, m.

We will seek the solution of the equal (4) and (5) outside the dynamic active coverage of the torch irrigation. Here longitudinal velocities of the liquid droplets are almost equal to the velocities of the ventilation flow, and movement of droplets becomes stationary. The relative absolute velocity of the dispersed liquid will be equal to the constant velocity $\upsilon = \upsilon_2$ of droplets deposition under the influence of gravitation forces. In this case time is expressed through the ratio of the distance to the air velocity:

$t = x / u_0$. Then the solution of the equation (5) can be written as

$$Z(x) = Z_0 \exp\left[-\left(\frac{k \upsilon_2}{h u_0} + \frac{\gamma \Omega}{u_0 S} \right) x \right], \qquad (6)$$

where Z_0 – liquid droplets concentration in the active coverage of the torch irrigation, kg / m^3. The changes in the dust concentration outside the active coverage will also take place not in the process of time but throughout the length of the excavation. Then the equation (4) can be written as

$$\frac{dC}{dx} = -e \frac{3}{2\rho} \frac{d_p + 0.77 d_d C / Z}{d_p d_d} \frac{\upsilon_2}{u_0} CZ . \qquad (7)$$

The entering function of liquid concentration outside the active coverage is determined by the formula (6). Taking the ratio (7) C / Z as a constant C_1 / Z_1 we get the next equation after integration

$$C(x) = C_1 \exp\left[-\frac{3 Z_1 / d_d + 0.77 C_0 / d_p}{2\rho} \times \right.$$

$$\left. \times \frac{e \upsilon_2}{k \upsilon_2 / h + \gamma \Omega / S} \{ 1 - \exp[-(k \upsilon_2 / h + \right.$$

$$\left. + \gamma \Omega / S) / x / u_0] \} \right], \qquad (8)$$

where C_1 – dust concentration on going out of the active coverage, kg / m^3.

Under the function of initial concentration Z_1 we mean total relative flow rate, disregarding its losses in the active coverage:

$$Z_1 = \frac{n q_l}{Q_0} .$$

n – an amount of spray lances; q_l – dispersed water consumption through one spray lance, kg / s; Q_0 – air consumption in the excavation, m^3 / s.

Let's represent capture efficiency e outside the active coverage as a sum of two coefficients:

$$e = e_{gr} + e_e , \qquad (9)$$

where e_{gr} – capture efficiency under the influence of gravitation forces; e_e – capture coefficient under the action of electrostatic forces.

The analysis of calculating data of capture efficiency of dust particles10microns in diameter (or more) shows that we can use the following formula in the gravitation field with sufficient degree of accuracy

$$e_{gr} = \frac{d_p^2}{d_p^2 + 32000/d_d} .$$ (10)

Figure 1 shows dust particles capture efficiency dependence by liquid droplets in the gravitational field at different diameters of particles not taking into account electrostatic forces influence. There are calculated curves of dependence in Figure (10). According to (Saranchuk 1984) average median diameter of suspended coal dust particles in Donbas breakage face is 20-30 km. But it is indicated that fine dust content (less 10 microns) ranges from 12 to 27%. Such particles must be detected by electrostatic methods.

In Figure 2 there is dependence of the dust suppression in the gravitation field

$$P_2 = \frac{C}{C_1}$$

outside the active coverage depending on the concentration of sprayed liquid in the process of 6 spray lances.

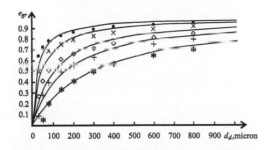

Figure 1. Dust particles capture efficiency dependence on liquid droplets diameters at different diameters of particles in the gravitation field (* – 10 micron, + – 14 micron, ◊ – 17,5 micron, x – 26 micron, ● – 35 micron).

The calculations were made by formulas (8) and (9) for diameters of dust particles 10 microns and 20 microns, the height of excavation $h = 3$ m, irregularity rate of fall coefficient $k = 2$ and in the absence of electrostatic field. Losses on the walls of excavation were taken to be negligibly small ($\gamma = 0$).

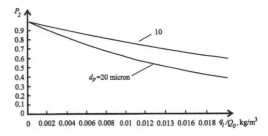

Figure 2. Probability of dedusting in the gravitation field outside the active coverage of the torch irrigation depending on consumption of sprayed liquid in the process of 6 spray conic lances and different diameters of dust particles.

3 CONCLUSIONS

As the results of the calculation illustrate, the probability of dust particles dedusting in the gravitation field is smaller when their diameters are less. Thus, at particles in 10 microns we can decrease their concentration to 60%, and at the middle dust concentration at the particles in 20 microns, it decreases in the gravitation field to 40%.

Thus taken researches shows how important the contribution of the gravitation fields in the processes of hydrodedusting by the irrigation is.

REFERENCES

Saranchuk, V.I., Kachan, V.N. & Rekun, V.V. 1984. *Physicochemical basis of hydrodedusting and prediction of coal dust explosion*. Kyiv: Naukova Dumka: 216.
Kachan, V.N. & Sokolova G.N. 1986. *Choice of effective sprinklers for irrigation*. Actual *questions of physics aerodisperse systems*. Odessa: Point report of XIV all-USSR conf.: 115.
Mednikov, E.P. 1980. *Turbulent transport and deposition of aerosols*. Moscow: Nauka: 176.
Kudriashov, R.R., Voronina, L.D., Shurinova, M.K., Voronina, Y.V. & Bolshakov V.A. 1979. *Wetting of dust and control of particulate air pollution in mines*. Moscow: Nauka: 196.

The stress-strain state of the stepped rubber-rope cable in bobbin of winding

I. Belmas
Dniprodzerzhins'k State Technical University, Dniprodzerzhins'k, Ukraine

D. Kolosov
National Mining University, Dnipropetrovs'k, Ukraine

ABSTRACT: The analysis of tension distribution in the multi layered laying of rubber-rope cable with variable width is executed. Nonuniform pressure of the upper layer of the less width on layers of the rope wound around the drum leads to nonuniform stress-strain state of flat stepped rope. Pressure of more narrow part of rope is provided by the local change of tensions in the wound part of rope.

1 INTRODUCTION

Increment of the mining depths is connected with extension of the lengths of ropes in hoist engines. At considerable length of rope its weight significantly affects the final value (utility) of load. Increment of the end-loading can be achieved by making use of the flat rubber-rope cross-section, which can be realized by introducing of a known design of flat rubber-rope, laying rubber-coated ropes of different lengths. This flat ropes can be wound on reels (Figure 1) of circumflex friction pulley with the conclusion of a free rope special stacker. The introduction of variable width traction requires solving many important scientific and engineering problems including determination the effect of variable width of the rope on stress distribution in the bobbin, formed by its multi-winding on a drum.

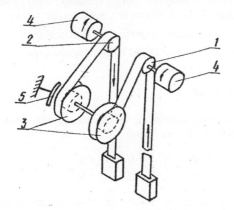

Figure 1. Scheme of the bobbin winder with driving pulleys of friction: 1, 2 – driving pulleys; 3 – bobbins; 4 – electrical motors; 5 – brake.

2 THE STATE OF MATTER

Flat ropes (ribbons) of a constant width are used in handling machines for over 100 years. In such a design, the pressure of the top layer is almost uniformly distributed over the width. The research results of the stress-strain state (SSS) of the pulling rubber-rope are widely covered in the literature (Kolosov 1987 & Belmas 1993). The aim is to define the basic law-governed nature of the SSS of flat rubber-rope wound in several layers under the pressure of the layer of a smaller width.

3 STATEMENT OF THE PROBLEM

Flat stepped rope is wound on the drum in several layers, forming a bobbin (Figure 2). During the traction of a pulling rope of variable width, the layers of each step awarded in previous layer of the smaller width. Let us investigate the SSS of the body formed by multi layer winding of the flat rope under the action of pressure caused by the layers of the smaller width. The adopted design scheme of loading is shown in Figure 3. At the same time let us take into account the following. Cables in rope restrict normal strain of its cross-section. Analytical solution of the problem considering the transverse strain of the cables in ropes is practically impossible. In order to obtain results which would allow to reproduce the mechanism of deformation of the package of layers of a flat rope, consider two cases.

Figure 2. Rubber-rope of variable cross section formed by its multi-winding.

In the first case accept that the cables do not affect the deformation of the rope shell. In the second case consider that layer of cables as incompressible one. At the same time it does not limit the shift in the planes parallel to the rope butts. These two cases lead to a physical model in which the rope is made of isotropic material of its shell. In the first case the thickness of the rope is equal to its actual thickness. In the second – to the rope thickness reduced on cables' diameter.

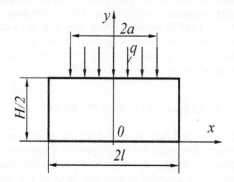

Figure 3. Diagram of loading of the rope with variable cross section.

Radius of the rope bend in a body formed by winding a flat rope considerably exceeds its thickness. This allows neglect the ropes bend and consider each layer as flat one. Determine the SSS of a small number of layers as a solid body, using the Airy function in the form:

$$\varphi = \sum_{n=1}^{\infty} cos(\alpha_n x)\, F(y),$$ (1)

where $F(y)_n = C_{n1} sh(\alpha_n y) + C_{n2} ch(\alpha_n y) +$

$+ C_{n3} y\, sh(\alpha_n y) + C_{n4} y\, ch(\alpha_n y).$

Components of stress for a plane deformation are:

$$\sigma_x = \frac{\partial^2 \varphi}{\partial y^2} = \sum_{n=1}^{\infty} cos(\alpha_n x)\, F(y)_n'';$$

$$\sigma_y = \frac{\partial^2 \varphi}{\partial x^2} = -\sum_{n=1}^{\infty} \alpha_n^2 cos(\alpha_n x)\, F(y)_n;$$ (2)

$$\tau_{xy} = -\frac{\partial^2 \varphi}{\partial x\, \partial y} = \sum_{n=1}^{\infty} \alpha_n\, sin(\alpha_n x)\, F(y)_n'.$$

Relative deformations are:

$$\varepsilon_x = \frac{1+\mu}{E}\left[(1-\mu)\sigma_x - \mu\sigma_y\right];$$

$$\varepsilon_y = \frac{1+\mu}{E}\left[(1-\mu)\sigma_y - \mu\sigma_x\right];$$ (3)

$$\gamma_{xy} = \frac{2(1+\mu)}{E}\tau_{xy} = \frac{\tau_{xy}}{G},$$

where μ, E – Poisson ratio and elastic modulus of rubber.

Displacements along x and y axes can be written in a form:

$$u_x = \int \frac{1+\mu}{E}\left[\sum_{n=1}^{\infty} cos(\alpha_n x)\left[(1-\mu)F(y)_n'' + \mu\alpha_n^2 F(y)_n\right]\right]dx + f(y);$$

$$v_y = -\int \frac{1+\mu}{E}\left[\sum_{n=1}^{\infty} cos(\alpha_n x)\left[(1-\mu)\alpha_n^2 F(y)_n + \mu F(y)_n''\right]\right]dy + f_1(x),$$ (4)

where $f_1(x)$, $f_1(y)$ – unknown functions of integration.

The analysis of relations (4) shows that $f'(y) = 0$, $f_1'(x) = 0$. So, accept $f(x)$ and $f_1(x)$ which describe the displacements of the sample only as a rigid body along the axes.

4 RESULTS

Boundary conditions in accordance with the design scheme (Figure 3) are:

$$x = l,\ \begin{array}{l}\sigma_x = 0,\\ \tau_{xy} = 0;\end{array}\quad y = \frac{H}{2},\ \begin{array}{l}\sigma_y = q,\\ \tau_{xy} = 0;\end{array}\quad y = 0,\ v_y = 0.\ (5)$$

In order to satisfy the first condition accept $\alpha_n = (n+0,5)\pi/l$. The second condition will satisfy approximately by taking $\int_0^{H/2} \tau_{xy} = 0$. Set the pressure by Fourier series $q = A_0 + \sum_{n-1}^{\infty} A_n \cos(\alpha_n x)$.

For obtained set of equations, unknown constants are determined from the following expressions:

$$C_{n1} = \frac{a_n\left(\mu-1+\alpha_n^2\mu\right)}{2\alpha_n^2} \cdot \frac{\left(\alpha_n H + 2ch\left(\dfrac{\alpha_n H}{2}\right)sh\left(\dfrac{\alpha_n H}{2}\right) - ch\left(\dfrac{\alpha_n H}{2}\right)\left(\alpha_n H - 2sh\left(\dfrac{\alpha_n H}{2}\right)\right)\right)}{\left(\mu-1+\alpha_n^2\mu\right)\left[ch\left(\dfrac{\alpha_n H}{2}\right)^2 + \dfrac{1}{4}\alpha_n^2 H_n\right] + \left(1-\mu-\alpha_n^2\mu\right)},$$

$$C_{n3} = \frac{-\dfrac{a_n}{\alpha_n^2}\left(ch\left(\dfrac{\alpha_n H}{2}\right)-1\right) - C_{n1}\left[sh\left(\dfrac{\alpha_n H}{2}\right) - \dfrac{1-\mu+\alpha_n^2\mu}{\mu-1+\alpha_n^2\mu}\alpha_n ch\left(\dfrac{\alpha_n H}{2}\right)\dfrac{H}{2}\right]}{sh\left(\dfrac{\alpha_n H}{2}\right)\dfrac{H}{2}},$$

$$C_{n4} = -C_{n1}\alpha_n\frac{1-\mu+\alpha_n^2\mu}{\mu-1+\alpha_n^2\mu}, \qquad C_{n2} = \frac{a_n}{\alpha_n^2}.$$

Using obtained dependencies the influence of multi-layer winding of pulling rope on the stress state of a formed body is analyzed. In the capacity of a pulling rope the stepped one with the strength of a width unit of core equal to 3150 N / mm was considered. The distributed load is assumed to be unity. The average intensity of pressure is taken equal to unity.

Figures 4-8 shows the surface of the distribution of stress and strain in one layer of pulling rope with six cables in case of pressure on it pulling rope with four cables. Since pulling rope is symmetric, there are only half a sample in figures.

and strains. Extreme normal stresses parallel to x-axis are of the same order of extreme values of the distributed pressure. Tangential stresses arising in the sample are much less than the normal stresses. This shows the absence of sliding of one layer of rope on the other one due to uneven pressure of layers of smaller width. This means that the loss of entry forms as the telescoform bobbin as a result of uneven pressure is not possible. On the other hand it shows the acceptability of the applied layers of rope as a solid body.

However, the strain in the direction of the y axis leads to deviation of forming surface of the winding from a straight line. This deviation may provoke the loss of bobbin shape and a possible change in the kinematics of rope winding, when the rope starts wound not in accordance with Archimedes spiral, but on conic helical line. Such a spooling can take a telescopic form.

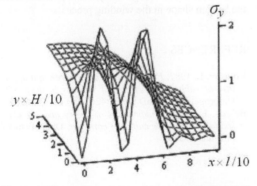

Figure 4. The surface of normal stress distribution in the body of one-layer reel along the y axis.

The surfaces shown in the figures indicate a significant impact of discrete cables of a rope. Nonuniform pressure leads to nonuniform distribution of stresses

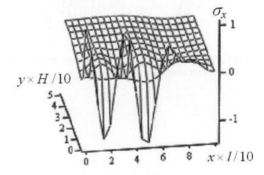

Figure 5. The surface of normal stress distribution in the body of one-layer reel along the x axis.

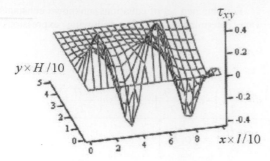

Figure 6. The surface of tangential stress distribution in the body of one-layer reel.

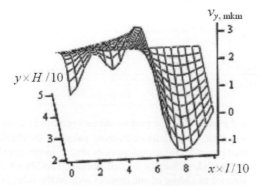

Figure 7. The surface of strain distribution in the body of one-layer reel along the y axis.

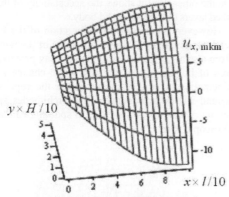

Figure 8. The surface of strain distribution in the body of one-layer reel along the x axis.

Given graphic dependences correspond to the first physical model. In the second model the rope thickness is adopted less than the actual thickness by an amount equal to the diameter of cable. The investigation shows qualitative nature of the proximity of the stress distribution and displacement in the case of decreasing zone of uniform stress distribution. The most conditional reduction of thickness affects on the value of tangential stresses – they have increased several times, but remained considerably less than the normal stresses. The effect of number of cables in the rope on its SSS in the body of reel shows that increasing the number of cables practically no effect on the extreme values of stress and strain in the direction of the y axis. Strains in the direction of the x axis (widening of a sample) grow due to increment of the rope width at the number increasing of cables in it.

5 CONCLUSIONS

Nonuniform pressure of the upper layer of the less width on layers of the rope wound around the drum leads to nonuniform SSS of flat stepped rope in winders for extreme depths. Tangential stresses in the body of bobbin are much less than the normal ones. They can not cause telescoform of the bobbin. Minor tangent tensions indicate admissibility of bobbin body consideration as a continuum body. Ununiform pressure increases the deviation forming surface winding of a straight line. Such deviation may give rise the loss of bobbin shape due to possible change of kinematic process of rope winding when it starts wound not by Archimedes spiral, but on conic helical line. With such a spooling rope can take a telescopic form. Only special rope design can prevent possible loss of the bobbin shape in the winding process.

REFERENCES

Kolosov, L. 1987. *Fundamentals of research and application of rubber-rope cables of winders in deep mines.* Dnipropetrovs'k: PhD thesis: 426.
Belmas, I. 1993. *Fundamentals of theory and calculus of spatial rubber-rope cables of conveyor.* Dnipropetrovs'k: PhD thesis: 360.

Technical and Geoinformational Systems in Mining – Pivnyak, Bondarenko & Kovalevs'ka (eds)
© 2011 Taylor & Francis Group, London, ISBN 978-0-415-68877-2

Advanced method for calculation of deep-water airlifts and the special software development

Y. Kyrychenko, V. Kyrychenko & A. Taturevych
National Mining University, Dnipropetrovs'k, Ukraine

ABSTRACT: The advanced method for calculation of deep-water airlifts is developed. It completely takes into consideration the specifics of solid material transportation through the water and air mixtures flow in the airlifts lifting pipe. The special Exact Calculation software which implements the method was developed. The impact of the structural and discharge parameters onto the energy capacity of the hydraulic lifting process is defined using the software as well as their efficient values for the basic version of the experimental facility at the deposits development depth equals to 6000 m.

Over the past few decades the world community is growing the interest in the development of mineral deposits of the World Ocean. Ocean floor hosts polymetallic ores whose resource is much more than similar land based one. Nowadays, the most challenging in terms of industrial development are deposits of polymetallic nodules, polymetallic sulfide ores, cobalt-manganese crusts, hydrates and phosphorites (Kirichenko 1989). The leading companies of the U.S., UK, Canada and Germany are working on the development of technical tools and methods for extracting ores from the ocean floor. Intensive work in this direction is also being conducted by the National Oceanographic Institute in India, joint-stock corporation of development a deep-ocean resources (DORD) in Japan; Research Institute for Exploitation of the Sea in France, a joint organization Interoceanmetal (Bulgaria, Cuba, Poland, Russian Federation, Slovakia and the Czech Republic); Yuzhmorgeologiya in the Russian Federation, the Union for the Exploration and Development of Ocean Mineral Resources (UEDOMR) in China, the Government of the Republic of Korea.

According to the experts, nowadays the most perspective way of transportation of extracted minerals to basic watercraft is deep-water airlift hydraulic lifting (DWA) due to high reliability in challenging conditions of deep water. Along with this DWA has a significant energy capacity, so the trend of development is the improving the facility efficiency. One of the ways to improve the efficiency of hydraulic hoisting is a selection of their rational design and discharge parameters that minimize energy consumption during transportation.

This article is devoted to solving the actual issue of grounding of the rational design and discharge parameters which provide the hydraulic hoisting exploitation with the highest technical and economic indicators.

Obviously, the accuracy of determining the DWA rational parameters directly depends of the accuracy of the used calculation method.

The analysis of the known methods for DWA calculation has shown that existing methods based on the using of the dimensionless flow rate characteristics and on integrating of the differential equations of motion of the mixture with the various complexity require further improvement. The methods do not take into account the specifics of facilities exploitation, especially transportation of solid material in the composition of the multiphase flow, as well as have many nondescript empirical coefficients. In addition, a common disadvantage of all existing methods is the lack of proper experimental verification. Therefore, the issue of development of the most accurate calculation method and as a consequence, the determining of rational values of design and discharge parameters of deep-water hydraulic hoisting remains open.

The **aim** of this article is the development of an advanced method for the DWA parameters calculation and the special software, as well as establishing the patterns of influence of the design and discharge parameters of hydraulic hoisting to energy capacity of transportation processes.

Existing methods for calculation of deep-water hydraulic hoisting can be divided into two groups. The first group includes methods based on homogeneous models (Kirichenko 2001; Grabow 1977; Grabow 1978, Ueki Syro 1979; Weber 1982) of multicomponent flows and the "energy" interpretation of the occurring processes (the balance between the available and supplied power). Available power – is hydraulic power required for transportation of the pulp, and supplied power – power of supplied air flow. The first group includes Grabow method (Grabow 1977 & Grabow 1978) and Ueki Siro method (Ueki Syro 1979) based on significant assumptions which simplify the mathematical apparatus. However this simplification negatively affects the results accuracy. The advantage of these methods is the calculation speed which allows to calculate a large number of options for establishing a qualitative influence of various factors on the efficiency of hydraulic hoisting.

The second group includes Polyarsky (Polyarsky 1982) and Chaziteodorou (Chaziteodorou 1972) methods, based on the "forced" interpretation of the processes and forces momentum conservation equations separately for each phase. This ultimately adds up to a separate model (Kirichenko 2009) of multiphase flow. Methods of the first group underestimate the required air flow for a given solids based facility capacity, and thus overstate the efficiency value. The main reason for this discrepancy is the fundamental difference between the methods of both groups while determining the total pressure gradient. The calculations performed in a wide range of baseline data shown negligible discrepancy between the results (less than 4...6%) while determining the friction loss using the methods of both groups.

Pressure loss to overcome the weight of the slurry column is not the case. The first group methods assume that these losses, for example at the bottom of the airlift pipe, are the ones required for the lifting of the pulp in the water. However, according to the Polyarsky and Chaziteodorou methods the losses to overcome the weight of the slurry column are the pressure losses based on lifting of solid particulates only in the water.

When solid material densities equal to $\rho_1 = 1100...1400$ kg / m^3 and solid phase concentration equals to $C_1 = 0.03...0.12$ pressure loss to overcome the weight of the slurry column calculated by the methods of the second group exceeds the corresponding loss calculated by the methods of the first group by 60...70%. Obviously, this significant discrepancy affects the pressure in the mixer a lot and as a result, the value of required air flow.

Chaziteodorou method is briefly described above (Chaziteodorou 1972). The method is based on the separate flow model of multiphase flow, where the three-phase flow is considered as a superposition of two-phase flows: water-air mixture and pulp. The essence of the method is solving of equation system which consists of three continuity equations and three motion equations for each phase, supple-

mented by a trailing dependencies and boundary conditions.

The pressure and specific gravity of air, velocity of the phases and cross-sectional areas occupied by each phase as a function of the coordinates of the pipeline are determined as the result of the solution of this system. The initial value of the specific gravity of air depends on the pressure in the mixer. The method of successive approximations is used to find the value of the pressure and compressed air flow in the mixer the way that the pressure in the upper section of pipe equals to the atmospheric pressure while solving equations of motion in the lifting pipe. Required initial values of velocities and areas of the phases are determined by solving of a system of two continuity equations and two motion equations for the pulp in the feed tube. This system is obtained by zeroing of an air phase in the equations which describes the three-phase flow in a lifting pipe. Methods of the second group are more reasonable and accurate than the methods of the first group (see Figure 1), mainly due to basing on a universal separate models of multiphase flow, taking into account the effect of interfacial forces. Nevertheless, it has a number of disadvantages, mostly related to incorrect interpretation of certain physical processes and the neglect of several factors which significantly affect the efficiency of transportation.

For example, the assumption that solid particles in three-phase mixture are transported by fluid only does not reflect the physics of the formation of airlift "traction", which ultimately affects the calculated speed of solids lifting. It is also mistakenly assumed that the drag coefficient of the pipeline remains constant along the entire length of the lifting pipe and does not depend on the specific structure of the flow. In addition, the bottleneck of the method is the lack of sustainable transportation of solid particles control while fluctuations of speed and much more.

The authors believe that the improving of the calculation method accuracy associated with the requirement of the complete accounting of all main factors which determine the physics of the solid particles transportation processes. Based on years of experience in the industry, the authors identify the following key factors and mechanisms:

1) Transportation of solid particles in the lifting pipe of deep-water hydraulic hoisting is performed by air/water mixture but not water. I.e., the basic flow characteristics are determined by the similarity criteria, depending on the parameters of the mixture rather than a liquid.

2) The drag coefficient of the lifting pipe is not constant along the length of the pipeline and depends of the basic parameters of the flow, expressed in terms of characteristics of the mixture and above

all the specific flow structure. Each flow structure corresponds to a different mathematical model. The dependencies for determining the drag coefficient for different mixture flow structures are given in (Kirichenko 2009). The expressions for determining of the boundaries of flow structures in the lifting pipe of deep-water hydraulic hoisting depending on the defining criteria of the flow are also presented in that article. As a result of adaptation of these characteristics to the base hydraulic hoisting from a depth of 6000 m, we have firstly obtained the following values of the stability limits of mixture flow structures, expressed by the discharging gas content variation range β.

- the boundary between bubble and slug structures $\beta = 0.25...0.3$;
- the boundary between slug and annular structures $\beta = 0.65...0.8$;
- the boundary between the annular and dispersed structures $\beta = 0.92...0.94$;

3) The condition for sustainable transportation of solid particles, according to which transportation flow rate must exceed the critical speed up to 15...20% (Adamov 1982) should be considered at each step of integration of the motion equations along the airlift pipeline. The following formula (Adamov 1982 & Skorynin 1984) is used to determine the critical speed of the pulp in the feed tube taking into account the cojoint fall of particles group:

$$V_{cr.p} = \left(1 - 0.35 C_{vol}\right)\left(1 - \left(\frac{d_n^a}{D_p}\right)^2\right) \times$$

$$\times \left(1 - C\right)^2 \left(\frac{4 \cdot g \cdot \left(d_n^a\right)^2 \left(\rho_1 - \rho_0\right)}{3 C_x \cdot \rho_0}\right)^{0.5}, \qquad (1)$$

where C – throttling concentration of solids in the pulp; C_x – drag coefficient of solid particle and water; ρ_0 – fluid density, ρ_1 – solid material density; C_{vol} – volumetric consistency of the slurry in the inlet pipe; d_n^a – average diameter of the nodules; D_p – pipeline diameter; g – acceleration of gravity.

The following dependence is used accordingly for the airlift lifting pipe (Adamov 1982 & Skorynin 1984):

$$V_{cr.m} = \left(1 - 0.42 C_{vol.m}\right)\left(1 - \left(\frac{d_n^a}{D_p}\right)^2\right) \times$$

$$\times \left(\frac{4 \cdot g \cdot d_n^a}{3 C_{xm}}\left(\frac{\rho_1}{\rho_m}\left(1 + q_z\right) - 1\right)\right), \qquad (2)$$

where $V_{cr.m}$ – three-phase mixture critical speed in the lifting pipe of deep-water hydraulic hoisting; $C_{vol.m}$ – mixture volume consistency in the lifting pipe; q_z – the average specific air consumption; C_{xm} – drag coefficient of solid particles in the mixture; ρ_m – mixture density.

The presence of a downward phase of the solid particles movement in gas shells in the slug flow structure should be additionally taken into account. The slug flow structure is the most dangerous possibility of failure (crisis) of sustainable solids transportation. This may cause the backing of the pipeline (Kirichenko, Evteev & Romaniukov 2007).

4) It is necessary to correct design scheme of the method by adjusting the discharge parameters to guaranteeing the required transportation velocity in cases of solids sustainable transportation breaches.

5) The opportunity of sound "locking" in the lifting pipe (critical flow) which limits the effectiveness of hydraulic hoisting (Kirichenko 1989) as well as the possibility of "flooding" of the flow at the annular and dispersed flow pattern mixture (Walys 1972) should be taken into consideration.

6) The air solubility and the transportation pipeline angle influence to the value of the real volumetric gas content (Kirichenko & Sdvizhkova 1990, Kirichenko, Samusia, Avrahov & Ivanchenko 1991) must be provided.

7) The impact of the supplying air system characteristics (compressor station + air line) to the operational modes of the facility (Kirichenko, Avrahov & Samusia 1989) must be taken into account.

Considering these factors an advanced method for calculating the hydraulic hoisting (AMCHH) which improves the accuracy of the results has been developed by authors.

The laboratory experiment (Samusia, Evteev & Kirichenko 2008) has been performed in order to approve AMCHH. Experimental investigations have been conducted in the hydraulics and hydraulic drive laboratory of the Mining Mechanics Department of the National Mining University. The experiment was based on an integrated experimental hydraulic stand that allows to perform the physical modeling of one-, two-and three-component flows in the pump, airlift and airlift pumping facilities (Samusia & Evteev & Kirichenko 2008). Moreover, the hydraulic stand design provides the variation possibility of the mixer relative dynamic immersion value in the lifting pipe at the ranges equal to

217

0.4...0.95. This allows to simulate processes of the short mining facilities and deep-water facilities with considerable length.

The calculated values obtained by different methods (adapted for the liquid lifting) and experimental data comparison results are shown on Figure 1.

The chart shows that the Polyarsky and Chaziteodorou methods which use heterogeneous models give more accurate results than the Grabow and Ueki Siro methods, based on a homogenous model. However, the most accurate calculation method is AMCHH developed by authors, which has been chosen as a base one for further investigations.

The comparison of calculated data and experimental data of different researches carried out within the marine and mining conditions was performed for the purpose of bringing the scale of experimental facilities to full-scale mining-sea airlifts. In particular, the Donetsk Polytechnic Institute in conjunction with the "VNIIProzoloto" institute have tested the marine airlift systems for lifting water in the Baltic Sea near the port of Liepaja. Experimental researches were conducted on the basis of the research vessel "Shelf-1" at the depths up to 90 meters (Adamov 1982).

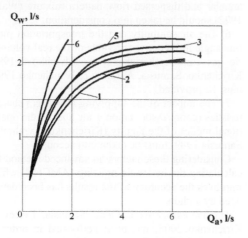

Figure 1. Water volume-flow (Q_w) air volume-flow (Q_a) variation relation under standard conditions; 1 – own experimental data; 2 – AMCHH; 3 – Chaziteodorou method; 4 – Polyarsky method; 5 – Grabow method; 6 – Ueki Syro method.

Figure 2 shows the comparison of experimental data (solid line) and calculated using the AMCHH data (dotted line) for the specified parameters of facility:

$1 - d = 0.1\,\text{m},\quad L = 5.0\,\text{m},\quad h = 47.0\,\text{m}$
$2 - d = 0.15\,\text{m},\quad L = 8.5\,\text{m},\quad h = 47.0\,\text{m}$

$3 - d = 0.1\,\text{m},\quad L = 0.5\,\text{m},\quad h = 59.3\,\text{m}$
$4 - d = 0.1\,\text{m},\quad L = 21.0\,\text{m},\quad h = 59.3\,\text{m}$

where d – diameter of the pipeline; L – length of the inlet pipe; h – mixer penetration.

As follows from the graphs the calculation accuracy increases in proportion to airflow. The maximum accuracy does not exceed 20%.

Researchers from the "Karlsruhe" University (Germany) performed the experimental investigation on the recovery of lignite, sand and gravel using the airlift method (Weber 1976, Weber 1982) in the laboratory and the existing quarry facilities.

The diameter and the length of the pipeline in laboratory facility were equal to 0.1 m and 7.8 m correspondingly. At air flow equals to 0.027 m³ / s the maximum flow for the solids at this facility were equal to 3.45 kg / s, and the solids volume concentrations in the flow tube reaches the 33% (Weber 1976).

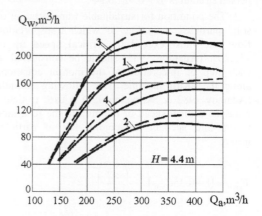

Figure 2. Experimental and calculated throttling characteristic of the marine airlift system comparison.

The airlift system (Weber 1982) with the total length of 441 m has been tested in the "Rheinische joint-stock company" lignite quarry.

The following characteristics of the facility were used: diameter of the pipe 300 mm; mixer penetration 42...248 m, the air volume flow 0.22...0.713 m³ / s; solid material maximum flow 115 t / h, solid material obtained concentration 0...8%. Experimental investigations were conducted using the short supplying tube ($L_n = 5...6$ m) as well as long one ($L_n = 101...341$ m).

Table 1 selectively shows the experimental data and the air flow values Q_{ac}, which were calculated for the parameters listed in the table using the AMCHH. The maximum discrepancy of these experimental and calculated data does not exceed 24%.

In order to implement the proposed method the "Exact Calculation" software was developed by the authors. It is the C++ console application compatible with win32/64 platforms. The software expects the input data as an XML file and outputs the results in XML format. In addition, the results can be exported to Microsoft Excel (xls) spreadsheet.

Systematic numerical investigations for the basic variant of deep-water hydraulic hoisting (capacity equals to 100000 tons / year for "dry" raw material extracted from the depth of 6000 m) were conducted in order to establish the regularities of design and output parameters of hydraulic hoisting influence on energy capacity of the transportation processes.

The main variable parameters have the following ranges: the real volume concentration of solid material $C_1 = 0.02...0.15$; mixer penetration $H_{mp} = 1500...3500$ m; air mass flow $M_2 = 2.5...65$ kg / s; solid material mass flow $M_1 = 5...12$ kg / s.

The most representative results are selectively shown on Figures 3-8.

Table 1. The comparison of calculated data and experimental data.

| Lifting material | Design parameters | | | | Discharge parameters | | | | | |
	h	L_n	H	$\alpha = \dfrac{h}{h+H}$	Q_{ae}, m³ / s	Volume concentration C_s, %	Solid flow rate Q_s, m³ / s	Water flow rate Q_w, m³ / s	Q_{ac}, m³ / s	δ, %
Gravel $\rho_s = 2575$ kg / m³ $d_n = 5$ mm	171	101	7	0.96	0.187	1.13	0.002	0.177	0.168	10.2
	174	101	7	0.961	0.256	2.29	0.0045	0.191	0.214	16.4
	177	101	7	0.962	0.384	3.39	0.0057	0.162	0.313	18.6
	180	101	7	0.962	0.405	3.95	0.0095	0.232	0.335	17.3
	186	101	7	0.964	0.260	1.89	0.0038	0.195	0.230	11.5
	216	101	7	0.969	0.249	2.17	0.0039	0.175	0.205	17.6
	222	101	6.9	0.97	0.329	3.7	0.0077	0.201	0.288	12.6
	225	101	6.9	0.97	0.240	2.51	0.0041	0.160	0.194	19.3
	69	290	6.6	0.912	0.570	2.06	0.004	0.191	0.490	14.0
	111	290	6.6	0.944	0.374	4.13	0.0056	0.124	0.306	18.1
	152	290	7.7	0.952	0.262	1.49	0.0026	0.169	0.211	19.3
	104	341	6.3	0.943	0.544	3.22	0.0053	0.158	0.475	12.6
	246	197	6.8	0.973	0.510	4.46	0.00935	0.200	0.439	13.9
	246	197	7.3	0.971	0.367	2.58	0.0054	0.205	0.306	16.7
	42	6.2	7.2	0.853	0.575	4.74	0.0127	0.255	0.470	18.2
	42	6.2	7.2	0.853	0.390	2.67	0.0068	0.248	0.324	17.0
	42	6.2	7.2	0.853	0.233	2.69	0.00535	0.193	0.201	13.8
Sand $\rho_s = 2610$ kg / m³ $d_n = 0.6$ mm	245	197	7.4	0.97	0.484	3.86	0.0075	0.186	0.432	10.7
	246	4.9	6.4	0.971	0.252	2.64	0.0055	0.203	0.223	11.5
	248	4.9	8.4	0.97	0.390	7.1	0.127	0.166	0.354	9.3
	248	4.9	8.4	0.97	0.456	6.4	0.0121	0.178	0.401	12.1
	148	101	8.4	0.946	0.488	5.89	0.0107	0.172	0.455	6.7
	148	101	8.9	0.946	0.220	3.25	0.0052	0.154	0.200	9.1
	148	101	8.4	0.946	0.355	6.01	0.0113	0.176	0.327	8.0
Lignite $\rho_s = 1143$ kg / m³ $d_n = 50$ mm	103	341	7	0.936	0.584	6	0.0169	0.263	0.488	16.4
	103	341	7	0.936	0.713	7.5	0.0201	0.249	0.581	18.5
	103	341	7	0.936	0.412	4.8	0.0116	0.232	0.328	20.3
	153	290	6.8	0.957	0.691	8.6	0.0254	0.270	0.535	22.6
	146	296	7.3	0.956	0.527	7.8	0.0211	0.251	0.408	22.5
	245	197	7.4	0.97	0.505	5.4	0.0157	0.274	0.389	23.0
	245	197	7.3	0.97	0.497	4.7	0.0147	0.300	0.383	22.9
	245	197	7.3	0.97	0.388	4.7	0.0127	0.259	0.296	23.8

Figure 3 shows the relationship between discharge parameters and energy parameters of airlift, which provides the required capacity in a fixed position of the mixer. There is an inflection point (extremum) on the curves shown on graph, which indicates the existence of a rational concentration of solid material which corresponds to the minimum air consumption at selected geometric parameters of the facility. As shown on the graph, the maximum value of efficiency corresponds to the minimum values of specific power and air mass flow.

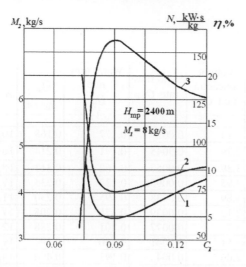

Figure 3. The dependence between the basic parameters of airlift and the real volume concentration of solid phase; 1 – the specific power (N); 2 – airflow (M_2); 3 – Efficiency (η).

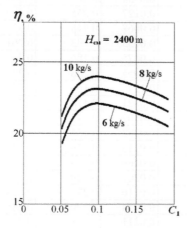

Figure 4. Dependence between the efficiency and the real volume concentration of solid phase for various solids capacities.

Figure 4 shows the influence of the pulp concentration and the capacity of the deep-water solids based hydraulic hoisting to the efficiency of the facility at the fixed mixer penetration.

As follows from the graphs the increasing of the solids mass flow in the investigating range leads to an efficiency increasing. Furthermore each value of the solids capacity corresponds to a rational value of the pulp concentration.

Figure 5 shows the dependence between the air mass flow as well as the airlift efficiency and the real volume of solid phase concentration for a given solids capacity and fixed position of mixer and different pipe diameters.

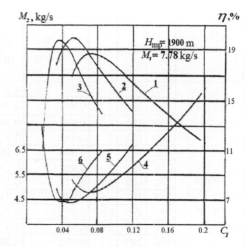

Figure 5. Dependence between the air mass flow as well as the airlift efficiency and the real volume of solid phase concentration at different pipe diameters: 1, 2, 3 – efficiency values for the pipeline diameters equal to 0.20, 0.22, 0.24 m, respectively; 4, 5, 6 – M_2 values for the pipeline diameters equal to 0.20, 0.22, 0.24 m, respectively.

As it is shown on the graphs the maximum efficiency value is achieved with a diameter of pipeline equals to 0.227 m, which nearly coincides with similar calculations for the base version of deep-water hydraulic hoisting provided by VNIPI "Okeanmash".

Figure 6 shows the dependence between the main discharge parameters of the deep-water hydraulic hoisting and the pressure in the mixer at a given solids capacity and fixed mixer penetration for the same pipe diameters. As follows from the graphs decreasing of air flow, providing a given facility capacity at the selected mixer penetration increases the pressure in the mixer as well as the efficiency of the facility.

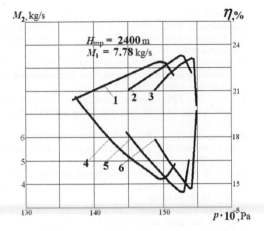

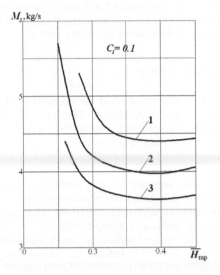

Figure 6. Dependence between the air mass flow as well as the airlift efficiency and the mixer pressure at different pipe diameters. 1, 2, 3 – Efficiency values for the pipeline diameters equal to 0.22, 0.27, 0.32 m, respectively; 4, 5, 6 – M_2 values for the pipeline diameters equal to 0.22, 0.27, 0.32 m, respectively.

As follows from the Figure 7, the minimum air flow rate that provides a given output corresponds to the dimensionless mixer penetration $\overline{H}_{mp} = 0.41$, equivalent to $H_{mp} = 2380$ m. It should be noted that the value of mixer penetration equals to 1900 m has been used in calculations provided by VNIPI "Okeanmash".

Figure 8 shows the dependence between the airlift efficiency and the dimensionless depth of mixer immersion at various solids capacities. As follows from the graphs each capacity of the facility for solids corresponds to the rational mixer penetration which minimizes energy capacity. At the same time in the investigated range of parameters mixer penetration affects the value of airlift efficiency no less (Figure 8) than the diameter of the pipe (Figure 5).

An analysis of the results of systematic numerical experiments established a **new scientific result**. The rational dimensionless mixer immersion of airlift $\overline{H}_{mp}$ which provides the maximum efficiency of facility in the range of airlift solids capacity variation equals to 7...10 kg / s and depths of ferromanganese nodules mining equal to 3500...7000 m is

$$\overline{H}_{mp} = 0.36...0.42.$$

Based on comprehensive analysis and the comparison of the dependences (Figures 3-8) which show the mutual influence of design, discharge and energy parameters of deep-water hydraulic hoisting, for the basic variant of the system, the following rational parameters which minimize power consump-

tion of the facility were determined: $C_1 = 0.09$; $H_{mp} = 2380$ m, $M_2 = 4.12$ kg / s.

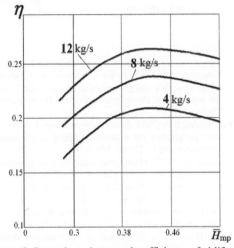

Figure 7. Dependence between the air mass flow and the dimensionless depth of mixer immersion for various solid flow rates; 1, 2, 3 – solid flow rates equal to $M_1 = 10$, 8, 6 kg / s, respectively.

Figure 8. Dependence between the efficiency of airlift and dimensionless mixer penetration for different solid materials flow rates.

The method considered as invention for launching and operating of the DWA with the mixer, which immersion depth exceeds the maximum pressure produced by the compressor has been developed by the authors of article.

The further stage of research is the development of a graphical version of the software, based on cross-platform Qt library, which would visualize the deep airlift processes. There are three variation are planned to be implemented: standalone application, add-in for SolidWorks CAD and the component for the CAE-solution for modeling complex deep-sea mining systems developed by the authors.

CONCLUSIONS

1. The advanced method for calculation of DWA parameters is developed. The method provides highly accurate results due to the consideration of the full facilities operation specifics and features of solid material transportation composed of a heterogeneous mixture.

2. The accuracy of the method is confirmed through own laboratory experiments and comparison with experimental data from other investigations in marine and mining conditions.

3. The "Exact Calculation" application written in C++ for win32/64 platforms has been developed as the software for advanced calculation method.

4. Regularities of influence of design parameters and discharge parameters to the energy capacity of hydraulic lifting processes and their rational values for the basic version of the experimental facility at a depth of 6000 m were determined using the "Exact Calculation" software.

REFERENCES

Kirichenko, E.A. 1989. *Rational parameters of deep-water airlift selection and grounding taking into consideration the influence of the supply air system.* Dnipropetrovs'k: Thesis: 172.

Kirichenko, E.A. 2001. *Scientific grounding of the pipe systems parameters for hydraulical lifting of the minerals.* Dnipropetrovs'k: Thesis: 181.

Grabow, G. 1977. *Hydro-pneumatische und hydraulische Ferderung von Feststoffen aus groben Meerestiefen, Pumpen und Verdichterin formationen.* Halle, 2: 39-45.

Grabow, G. 1978. *Optimierung Hydraulischer Forderuer fahren sur submarinen. Gewinnung mineralischer fohstoffe. Frieberger Forschung-shefte*, 2: 111.

Ueki Syro. 1979. *Airlift feature examination.* Saiko to

Hoan, 8.

Polyarsky, U. 1982. *Mining of manganese nodules from the seabed and ocean floor using pneumohydraulic transportation.* Preglad gorniczy: 152-161.

Chaziteodorou, G. 1972. *Entwurf eines Abbanschemas Zur Gewinnung und Fv're teremg von Mankankollen aus der Tiefsee.* Meerstechnik, 2.

Adamov, B.I. 1982. *Research and development of deep-water airlift for solid material lifting.* Donetsk: Thesis: 192.

Skorynin, N.I. 1984. *Research and development of deep-water multimixing airlift facilities for rock lifting.* Donetsk: Thesis: 205.

Weber, M. 1976. *Das Airlift-Verfahren und seine Einsetzbarkeit zur Furderung von Mineralien aus der Jiefsee.* Dusseldorf: 141-162.

Weber, M. 1982. *Vertical hydraulic conveying of Solids by air-lift.* International journal of miner, 3: 137-152.

Kirichenko, V.E. 2009. *Grounding of deep-water airlift parameters taking into consideration the transient processes*: 176.

Kirichenko, E.A. & Evteev, V.V. & Romaniukov, A.V. 2007. *Investigation of the slug flow structure parameters in the lifting pipe of deep-water airlift.* Scientific bulletin of NMUU, 9: 66-72.

Walys, G. 1972. *One-dimensional two-phase flows.* Moscow: Mir: 440.

Kirichenko, E.A. & Sdvizhkova, E.A. 1990. *Issue of the air solubility influence on the airlift flow rate characteristics.* "Mining technology" researches and specialists reports thesis. Dnipropetrovs'k.

Kirichenko, E.A. & Samusia, V.I. & Avrahov F.I. & Ivanchenko O.A. 1991. *Deep-water airlift operational specifics.* Collected reports thesis of technical methods of ocean exploration. V. 1. Oceanology institute of USSR.

Kirichenko, E.A. & Avrahov F.I. & Samusia, V.I. 1989. *Deep-water airlift design parameters optimization taking into consideration the supply air system characteristics.* Reports thesis of "Technical methods of oceans and seas exploration". Moscow.

Samusia, V.I. & Evteev, V.V. & Kirichenko, V.E. 2008. *Experimental research of the vertical two-component flow parameters in airlift.* Scientific bulletin of NMUU, 12: 68-74.

Pivniak, G.G. & Kirichenko, E.A. & Evteev, V.V. & Shvorak, V.G. & Kirichenko, V.E. 2009. *Airlift launching and stopping methods and the system for its implementation.* Patent 2346161 Russian Federation, E21C50/00, E21C45/00, F04F1/20.

Perspectives of innovation diffusion in Ukrainian mining industry

T. Reshetilova &V. Nikolayeva
National Mining University, Dnipropetrovs'k, Ukraine

ABSTRACT: Effective usage of considerable potential of Ukrainian mining enterprises is being hampered because innovative ideas are not implemented.

As Ukrainian international policy is EU oriented, the country's industries should be developed on the basis of innovative potential priority. Based on the established tendencies of innovative activity in the EU countries, the position of Ukraine in this process was determined.

A scientific idea of using assessment of impact of natural factors on the mine performance as an instrument of implementing innovation diffusion into mining industry is realized in the paper.

Some results of economic assessment of natural factors impact are given, which for the first time enabled quantitative determination of the degree of complexity of mining and geological conditions at every single enterprise. This enables determination of perspectives of attracting innovations in different mining and geological conditions.

1 INTRODUCTION

A country's social and economic growth can be achieved largely due to active implementation of innovative ideas into scientific potential of enterprises.

Numerous studies determined (The official site of the President of Ukraine) the following characteristics of scientific and technical development of Ukraine on the present stage:

• technical and technological lagging behind the developed countries of the world;

• high level of resource-intensiveness (energy, material, fund and labor consumption) of the production process;

• low level of labor productivity;

• scientific work largely aimed at military needs;

• certain isolation in terms of international scientific and technological exchange;

• non-rational use of country's scientific and technical potential etc.

As for the main branches of Ukrainian industry is concerned, scientific and technical policy should facilitate the transformation of technological and production structures, usage of advanced technologies for structural reconstruction of the economy in general and solving social and economic problems of the society.

Fuel and energy complex, metallurgical complex, machine building, chemical and building industries and forestry should be referred to as the main branches of industry in Ukraine. They account for 87% of basic production assets, 78% of employment and almost 78% of gross industrial product (about 58% of gross national product). This determines the importance of these industries for the country's economic development.

Mining complex accounting for over 36% takes special place among the main industries.

Taking into account the fact that Ukrainian international policy is aimed at EU membership, we can say that further development of the country should be intensified according to clear objectives and priorities of development based on the innovative potential of the country (Chuchno 2008).

In this respect analysis of the tendencies of innovative activity in the countries of the European Union in comparison with the current situation in Ukraine is of the most immediate interest for optimization the ways of raising innovative potential of the main branches of Ukrainian economy as a whole and mining industry in particular.

2 ANALYSIS OF TENDENCIES OF INNOVATIVE ACTIVITY IN THE EU COUNTRIES AND IN UKRAINE

European Union is an integrated association of 27 countries with different level of economic development. This difference allows for common tendencies of development in the sphere of innovations to be combined with individual way of development in every single country.

General trend of encouraging innovations is manifested in creating a common European research and innovation environment, expansion of vertical and horizontal coordination of innovative policies, strengthening local innovation policy and constant increase in financing scientific research and development (Surinach 2009). All abovementioned shows that the EU countries have a coordinated strategy in the sphere of innovative development. This has enabled the appearance of common internal market of innovations and, consequently, quicker diffusion of innovations between countries and companies in the EU.

Within the common strategy every country has its own innovative systems which differ in their ways, methods and means of governmental control of innovative activity, including intensity and ways of supporting interaction between scientific, industrial and educational components (Surinach 2009).

Well developed countries (Germany, the Netherlands, Belgium, Austria and others) have taken up the most rapidly growing and therefore the most perspective segments of innovative technologies and world market products. This allows them to maintain a high level of innovation activity.

Innovation activity is a complex characteristic of business which includes its intensity, timeliness and ability to raise enough potential of necessary quality and quantity. It is defined as percentage of enterprises engaged in innovation activity to the total number of enterprises in the country.

The level of innovation activity of enterprises in the EU was calculated on the basis of statistical data for 2009 (Surinach 2009 & Eurostat 2010) (Figure 1). The same strategy was used for Ukraine.

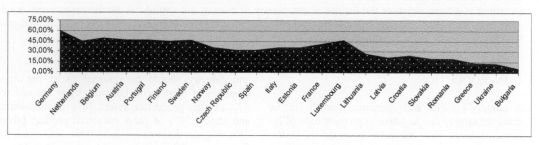

Figure 1. Level of innovation activity in the countries of the European Union and in Ukraine.

As the result of the research all countries were divided into 5 groups according to the level of their innovation activity:

• Group 1 is represented by Germany which level of innovation activity (60%) is the highest in the EU.

• Group 2 consists of the Netherlands, Belgium, Austria, Portugal, Finland, Sweden and Luxembourg. The level of innovation activity in these countries is in the range of 40-50%.

• Group 3 unites countries which level of innovation activity is between 30 and 40%: Norway, the Czech Republic, Spain, Italy, Estonia, France and Lithuania.

• Latvia, Hungary, Slovakia and Romania belong to Group 4. Innovation activity in these countries is 20-30%.

• Countries with the lowest level of innovation activity (less than 20%) were included into Group 5 (Greece – 13%, Ukraine – 12.8% and Bulgaria – 5%).

As we can see, Ukraine is one of the least developed countries as for the level of innovation activity.

A detailed analysis of innovation activity of Ukrainian enterprises showed that the total number of industrial enterprises that put innovations into practice during 2004-2009 went up by 23.2 (from 958 to 1180). In the beginning of 2010 the level of innovation activity of Ukrainian industrial enterprises is still low at 12.8% or 1411 enterprises. In 2004-2009 there was a decrease of implementation of new technological processes by 32.5% and a slight rise in the number of commercial productions of innovative goods (by 8.5%). The dynamics of implementing innovations at Ukrainian industrial enterprises in 2004-2009 is shown in the Figure 2.

The analysis of the innovations spheres Ukrainian business entities developed showed that the majority of enterprises spent money on purchasing automobiles, facilities, equipment, other fixed assets and covered other capital costs associated with implementation of innovations (7% of total industrial enterprises or 767 units). Only 0.8% of all industrial enterprises or 90 companies spent money on buying new technologies and 2.9% of enterprises or 324 business entities carried out research and development (http://mpe.kmu.gov.ua).

The structural analysis of implementing innovations in manufacturing sector showed that such industries as chemical, petrochemical and machine building were the most active in this process. The greatest number of innovative products was produced by machine building industry (average of 43.4%), food industry (19.5%), chemical and petrochemical industries (12.4%) and the smallest – in mining and extraction industries (1.5%) (ukrstat.gov.ua).

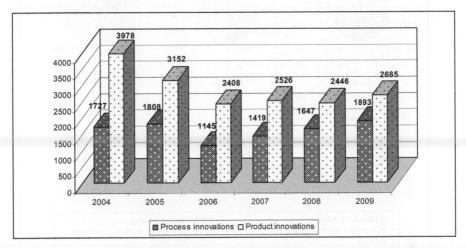

Figure 2. Dynamics of implementing innovations at Ukrainian industrial enterprises in 2004-2009, units (based on Sobkevich 2010).

According to the report made by the experts of The Directorate-General for Economic and Financial Affairs of European Commission, the level of innovative activity of a country depends on its industrial structure (Surinach 2009 & Eurostat 2010).

These data were used to develop the diagrams of industrial structure of each of the countries under consideration. The diagrams show the industrial structures of the countries which were earlier grouped according to the level of innovation activity (as shown in Figure 2). The branches of industry were classified according to the EU standards (Surinach 2009 & Eurostat 2010):

Mining industry which includes mineral resource industry, oil and gas industries;

Manufacturing industry: machine building, metal-processing industry, construction, chemical, light and food industries;

• Transport and communication services;
• Computer and other business services;
• Other industries.

To conduct comparative research, a similar classification of Ukrainian industrial structure was made based on the data given by the State Committee of Statistics for 2009 (Figure 3).

The results of the research showed that in Germany manufacturing industry accounts for 42.57%, while mining – for only 0.43%, computer and other business services are rather developed (15.95%) as well as transport and communications (17.72%). In Austria manufacturing industry accounts for 48.86%,

while mining for only 0.8%. In Spain mining industry has a share of 1.09%, manufacturing accounts for 60.93%, transport and communications – 10.36%, computer and other business services – 3.79%. In Ukraine the share of mining industry in the industrial structure of the country is 11.39%, manufacturing accounts for 64.37%, transport and communications – for 5.58%, computer and other business services – for only 1.08%. As we can see, the share of mining industry in Ukraine is 10-15 times exceeds this level in other countries, which directly affects the lower level of innovation activity in the country.

Prior to studying the innovation activity of Ukrainian mining industry, the structure of innovation diffusion methods should be analyzed. Business enterprises can act in two ways: generate innovations or adopt them from outside.

The analysis of levels of generation and adoption of innovations in the countries of the EU and Ukraine was carried out (Figure 3). The innovation generation coefficient was calculated by determining the share of enterprises that generate innovations independently in the total number of businesses performing the innovation activity in the country.

The adoption coefficient was determined by the number of enterprises that adopt innovations in the total number of businesses performing the innovation activity in the country. It was taken into consideration that innovation adoption occurs if it was designed either in cooperation with other enterprises or by other enterprises independently.

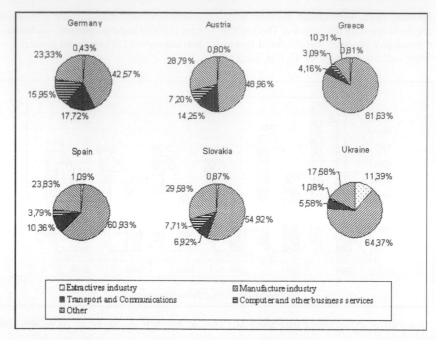

Figure 3. Industrial structure of the economy of European countries (based on Surinach 2009 & Eurostat 2010).

The countries of the EU have a common tendency to "generate" innovations; their index of generation largely prevails over that of adoption (Figure 4).

In Germany (Group 1) business enterprises aim mainly at "generating" ideas than at adopting (level of generation is 37.2%, level of adoption is 22.8%).

Maximum value of the level of innovation generation in the countries of Group 2 ranges from 17.1 to 37.8%.

In Group 3 France has the highest level of generating innovations – 27.4% (adoption level is 13.6%). In the countries of the fourth group generation level is 11-12.5%, adoption level is 5-6%.

The countries of the fifth group have the lowest level of generating and adopting innovations (2.8% and 2.2% correspondingly).

In Ukraine percentage of enterprises generating innovations is 7.15%, percentage of enterprises that adopt innovations is 3.85% (Ukrstat 2010).

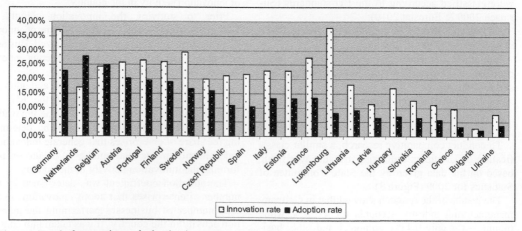

Figure 4. Level of generating and adopting innovations in the countries of the EU (based on Surinach 2009; Eurostat 2010; Ukrstat 2010).

3 PECULIARITIES OF INNOVATION ACTIVITY OF MINING INDUSTRY OF UKRAINE

Nowadays branches of the mining industry develop under complex and sometimes contradictory influence of socio-economic and natural factors. The product of the mining industries is mineral raw materials which are non-renewable unlike other natural resources. This natural characteristic is functionally important. It predetermines the necessity for each mining enterprise to expand in space extending the period of its operation. The level of economic and technological progress determines the degree and boundaries of enterprise expansion in the area.

Non-renewable nature of mineral raw materials dictates the necessity for each enterprise to deepen their development. But at certain depths it becomes technically difficult and makes no economic sense to continue working by both open-cut and underground method of extraction. So the enterprise has to be closed or restructured in the way that enables its future operation.

Real situation is much more complicated than the principal scheme offered above. Not only natural but also social, economic and environmental factors affect the state of affairs.

Ukrainian mining industry is characterized by a number of features some of which are favorable for further development, some have a hampering effect.

Among favorable features the following can be named: large deposits of coal of different ranks suitable for generating electric and thermal energy for technological needs; close proximity of coal, iron ore and other mineral deposits; proximity of large fuel- and energy intensive manufacturing plants and consumers of metal (metallurgical, chemical, machine-building and other enterprises); developed infrastructure; large industrial potential of working mining enterprises.

Unfavorable features are the following: difficult and extremely difficult mining and geological conditions; low coal content; the necessity to extract black coal and anthracite only by underground method; poor condition of mines that work for a long time without reconstruction which resulted in deterioration of their engineering characteristics; large-scale wearing of main assets, especially equipment; low level of mechanization in cargo-handling, transportation and auxiliary processes.

The general characteristic indicating the condition of mining complex is its structure in respect of the operation period since construction and after last reconstruction. The increase of this period increases the complexity of extraction and, consequently, the cost of reconstruction.

Nowadays there exist 167 working mines and 4 quarries in Ukraine. 102 of them have production capacity of 600.000 t\y, 32 – from 600 to 1 million t\y, 33 – more than 1 million t\y. Operation period of 50% of all mines reaches 50 years, out of which 74 mines (33%) are being exploited more than 50 years, 38 (17%) – for more than 70 years. In the last 30 years only 34 mines were opened (about 15% of the general number of working mines). For the 50 years' period only 30.1% of all working mines were reconstructed; over the last 25 years – only 10%. This means that 90% of mines that were reconstructed work for more than 30 years after the reconstruction. We can say that during such long period the reconstruction has become obsolete and the mines need a new one.

At the present stage of the mining industry development mining enterprises whose work was considered economically non-feasible are being closed down. At the same time in the course of economic reforms in Ukraine some decisions are being taken about transformations in the mining industry. Hence, starting from 2011, only the mines that prove their feasibility will continue the operation (the Ministry of Coal Industry, 2010). On the basis of the analysis of all factors influencing the work of mining industry, the reconstruction of 40 mines of 18 mining corporations is considered to be a top priority. Among them there are 9 mines which produce the most valuable and scarce ranks of coal (coking, fat, non-baking coals). They are situated mainly in Donetsk-Makeevskyi and Central regions, which are characterized by extremely difficult mining and geological conditions due to the great depth of working.

In this respect an important decision should be made on whether generation or adoption is the best way to attract innovations. It is largely determined by the specific character of a coal mine as a complex manufacturing system. This issue should be addressed in detail.

Ukrainian coal extraction enterprises are concentrated in three regions: Donetsk, Lviv-Volynian and Dnepr coal basins. In the first two basins black coal and anthracites are extracted, in the Dnepr coal basin – brown coal. There are 21 geological industrial areas in the Donetsk coal basin but mines operate in only 14 of them. The regions are characterized by similar mining and geological conditions and coals of definite ranks. All mines are divided in certain connections to the boundaries of geological industrial areas, so some mines are characterized by great diversity of mining and geological conditions. The thickness of seams range from 0.8 m to 1.8 m; the seams dip at different angles – low, shallow, sloping and steep; temperature conditions range from difficult to very difficult; the level of danger is very high; gas presence is excessive.

Mining and geological conditions of the Lviv-Volynian basin are difficult due to the small and unstable thickness of seams, instability of rock, high methane content of coals and rocks and necessity of specific shaft sinking methods. The thickness of industrial seams ranges from 1.5m to 6m. Their area and thickness are unstable; the depth varies from 10 to 150 m, which enables open-cast mining of a considerable amount of deposits. Mining and geological conditions of operation are rather complex due to high level of watering of loose water-bearing sands.

Technological coal deposits are mainly concentrated in 5 regions: Donetsk-Makeevskyi (44% of all industrial deposits), Krasnoarmeyskyi (16.3%), Central (10%), Almazno-Maryevskyi (8.3%) and Lugansk-Krasnodonskyi (9.8%). The most valuable and scarce ranks of coal (coking, fat, non-baking coals) are mainly situated in Donetsk-Makeevskyi and Central regions, which are characterized by extremely difficult mining and geological conditions due to the great depth of working.

Energy coals are extracted in all regions, while anthracites can be found in two of them, Chistyakovo-Snezhnyanskyi (Donetsk region) and Krasnoluchsko-Antracitovskyi (Lugansk region).

A number of natural factors affecting the choice of methods of mining and means of mechanization and their specific character doesn't allow to directly adopt innovations generated for certain deposits in Ukraine and other countries.

The degree of natural factors impact on the production process of some mines allows for a forecast to be made about the expediency of generating innovations for their development. If the enterprises are unable to considerably improve their economic results due to difficulty of mining and geological conditions, it is rational not to waste investment funds on generating or adapting innovations.

On the basis of the developed method of assessing the degree of influence, quantitative estimates for determining the directions of Ukrainian mine restructuring were obtained.

The expenses associated with the most serious natural factors affecting the increase of production costs were calculated for each mining enterprise. The ratio of these expenses to a ton of production gives us the portion of the production cost that depends on natural factors. The calculations showed that this figure is different for every mining enterprise. On the mines of the Donetsk region, for example, due to great depth and excessive gas presence, the portion of the production costs is high for such processes as whole mine ventilation, mine workings ventilation and excavated rock tempera-

ture. On the contrary, due to the relatively big thickness of the seams and low watering, the portion of the cost of these processes is low. The situation on the Pavlogradugol mines is almost opposite: the part of the cost connected with the seam thickness and watering is high, the part of the cost depending on the depth of extraction, gas presence and geodynamic aspects is low. So, to get the unbiased assessment of the natural factors impact, that is to obtain the part of the production costs that depends on the natural factors, we have to compare a single mine production costs with industry average figure.

This method enables us to draw a fairly grounded conclusion that on the Pavlogradugol mines, for example, average 16% of overall production costs are connected with the impact of natural factors, so, to raise the efficiency of operation, innovations can be implemented for 84% of the costs. Whereas on the Dobropolieugol mines as much as 30% of the production costs are connected with natural conditions, on the Donetskugol mines this figure is even bigger – 45%.

If the part of the production costs connected with unfavorable natural conditions of mining is very high, the economic expediency of additional investment in innovation projects is questionable. On the other hand, such approach is not universal and requires additional expert evaluation of innovation projects offered for generation or adoption. This is explained by the fact that there are some technical solutions aimed at decreasing the impact of natural factors on the production processes which can improve the operating efficiency of the mining enterprise. As a result, it can join the list of mines recommended for restructuring.

4 CONCLUSIONS

The analysis of innovation activity as a comprehensive characteristic of industrial enterprises of the EU and Ukraine showed that Ukraine lags behind in this process.

The mining sector that takes a significant position in the structure of Ukrainian industries (36%) is especially appropriate for the diffusion of innovations.

A specific feature of Ukrainian mining industry in comparison with other countries is its complex mining and geological conditions. The expert evaluation of these conditions should influence the decisions of economic expediency of innovations diffusion.

The suggested method of assessing the degree of natural factors' impact on the production process allowed giving quantitative estimates of the complexity of mining and geological conditions. The estimates could be used in analyzing the perspectives of attracting innovations.

REFERENCES

Chuchno, A. 2008. *Scientific and technological development as the object of Evolutionary Economics Theory.* Economy of Ukraine. Issue 1: 12-22.

Eurostat (European Statistics) Access: http://epp.eurostat.ec.europa.eu/portal/page/portal/statistics/themes.

Geiec, V. 2010. *Liberal democratic foundations: a course on modernization of Ukraine.* Issue 3: 4-20.

Petrovska, T. 2009. *World experience of coal sector restructuring.* Building a market economy in Ukraine. Issue 19: 408-412.

Sobkevich, O., Sukhorukov, A., Savenko, V. & Zhalilo, Y. 2010. *Development of the industrial potential of Ukraine during the post-crisis recovery.* Kiev: Nishi: 48.

State Statistics Committee of Ukraine Access: http://www.ukrstat.gov.ua/

Surinach, J., Autant-Bernard, C., Manca, F., Massard, N. & Moreno, S. 2009. *The Diffusion/Adoption of Innovation In The Internal Market (Economic Papers).* European Communities: 340.

The program of economic reforms in 2010-2014. The Committee on Economic Reforms under the President of Ukraine. Access: www.president.gov.ua/docs/Programa_reform_FINAL_1.pdf.

Statistic Information of Ministry of Coal Industry. Access: www.mvp.gov.ua.

Technical and Geoinformational Systems in Mining – Pivnyak, Bondarenko & Kovalevs'ka (eds)
© 2011 Taylor & Francis Group, London, ISBN 978-0-415-68877-2

Use of dust masks at coal enterprises

S. Cheberyachko, Y. Cheberyachko & M. Naumov
National Mining University, Dnipropetrovs'k, Ukraine

ABSTRACT: There are given recommendations concerning choice of dust masks for metal mining enterprises depending upon protective efficiency, resistance to breath, particle size distribution, work type, operating time, and climate. Based on the experimental researches is received, that protective efficiency of the respirators depends on the filters quality and time of the working. Improvement of the protective efficiency is the result of the increasing airflow resistance at first, and to decreasing of it then. Shown, that raising of temperature and air humidity is the result of the respirator quality parameters decreasing.

1 INTRODUCTION

Dust is the most common disadvantage of environment in mining industry. Dust impact on organism of workers may result in progress of development dust diseases and dust bronchitis. Today dust mask is one of efficient means of respiratory protection. Therefore, mistakes made while selecting it greatly negate total protective effect. On the other hand, adverse weather conditions as well as work hardness influence too. Besides it should be taken into consideration that mismatched means of individual protection of respiratory organs (MIPRO) result in unjustified supertension of body functional system, and hence in sharp work decrement. That's why regulatory actions on dust mask application are of great importance as they help minimize effect on human vital function, and maximize protection against damage effect.

To justify choice of required type of protective means for specific labour conditions with account of securing maximum possible human performance it is necessary to take into consideration design features of dust masks, protective efficiency, composition and quantity of harmful agents within environment, term of protective effect, operating regime, and climatic conditions. Assess each factor characterizing MIPRO, and influencing its attribute characteristics.

2 DISCUSSION AND RESULTS OF RESEARCH

Design Features. MIPRO can be expendable and non-expendable. Expendable means are used on the basis that their expected duration is one shift when level of harmful agents is not more 50 mg / m³. If the condition is breached non-expendable means should be used.

Protective Efficiency of a Dust Mask. The factor is assessed with the help of protection coefficient K_3.

All filtering MIPRO are divided into the groups with different protective efficiency: low ($K_3 < 10$), average ($K_3 = 10...100$), and high ($K_3 > 100$). European Standards mark low protection coefficient as 1, average protection coefficient as 2, and high protection coefficient as 3 (Basmanov 2002).

To determine protection efficient K_3, penetration coefficient K_3 is determined experimentally. The latter is concentration ratio of harmful substance within MIPRO to concentration of the substance within environment (Basmanov 2002)

$$K_3 = 100 / K_p.\qquad(1)$$

Table 1 shows penetration coefficient determined on standard test-aerosol method of oil spray with 0.28-0.32 microns diameter, and protection coefficient for the most commonly used domestic MIPRO. As the Table 1 demonstrates Puls and RPA-TD have optimum protection level among non-expendable masks.

Pressure difference determining by means of filter resistance to airflow is significant parameter of MIPRO. As a physiological characteristic, respiratory resistance is a value which on one hand is connected with resistance to pulmonary ventilation rate, structure of respiratory cycle, and work hardness. On the other hand, it depends on environmental conditions, and design features of respirators. But if parameters of human external respiration and environment are stipulated then pressure difference depends only on filtering material properties (Basmanov 2002).

$$\Delta p = \frac{4\nu\mu\beta H}{r^2(-\lambda - 0.5\ln\beta)}, \qquad (2)$$

where ν – linear velocity of gas flow, m / s; β – fiber packing density; H – filtering layer thickness, m; r – fiber radius, m; μ – dynamic viscosity of gas, N·s / m^2; λ – compensation factor depending on filtering material type.

Connection between pressure difference and penetration coefficient of test-aerosol method can be shown as (Basmanov 2002).

$$K_n = exp(-\alpha[\Delta p]), \qquad (3)$$

where Δp – pressure difference on MIPRO, Pa; α – filtration coefficient.

$$\alpha = \frac{2r\eta_\Sigma}{4\pi(-\lambda - 0.5\ln\beta)\nu\mu},$$

where η_Σ – cumulative rate of capture of particles depending upon all filtering mechanisms (electro-static, inertial, engagement, diffusion).

Table 1. Characteristics of dust masks.

Type of mask	Pressure difference under 30 l / m air consumption, Pa		Penetration coefficient on test-aerosol method of oil spray, K_3, %
	Intake of breath	Breathing-out	
RPA-TD-1	55	30	0.5-2
RPA-TD-2	25	30	0.5-2.5
Puls-K	55	15	0.5-2
Puls-M	25	15	0.5-2.5
Klyon-P	30	30	0.5-3.5

Studies of expression (3) help to conclude that improvement of protective efficiency is possible if filtering layer thickness is increased, and fiber diameter is decreased. In turn it will result in growth of pressure difference on MIPRO in Figure 1, and load on respiratory organs. It is not expedient to use MIPRO with high protection level working with non-toxic matters as production activities performed in dust respirators and all-purpose respirators having 40-60 Pa resistance should belong to classes which severity is a category higher to compare with similar activities performed without respirators (Basmanov 2002).

On the other hand, increase in pressure difference not always results in improvement of protective properties of respirator. In some cases due to imperfect design of leakage seal increase in resistance of filtering elements factors into growth of impure air inflow to undermask space. Paper (Golinko 2008) obtains dependence of protective efficiency of respirator on protective efficiency of filters agreed upon value of their respiratory resistance

$$K_n^p = 10^{-\alpha(R_{f.r}S)} + 0.8\frac{\rho d^2}{18\eta}\sqrt{\frac{4\pi\nu^3}{Q_1 - \Delta p / R_{f.r}}},$$

where $R_{f.r}$ – filter resistance, N·s / m^5; S – filter area, m^2; ρ is density of aerosol particles, kg / m^3; d – diameter of aerosol particle, m; η – kinematic viscosity of air, m^2 / s; Q – consumption through

respirator, m^3 / h; ν – particle velocity, m / s.

Some optimum is available under which the greatest protective efficiency of MIPRO is ensured in Figure 2.

With it, further improvement of protective properties of filter results in deterioration of respirator protective efficiency. It depends on the fact that due to increase in filter resistance to airflow (to be unavoidable with its filtration quality improvement) redistribution of airflows takes place, and inflow of polluted air through gas mask leakages increases.

Respiratory resistance also influences human physiological state. When specific value is achieved, functional shifts in respiratory system (breath phase prolongs, inspiratory volume increases, intra-alveolar pressure increases, and breathing rate cuts) originate. It results in performance decrement of human.

Largely, protection coefficient of respirator as well as its resistance depends on characteristics of filtering material applied. The material is used to produce one of the MIPRO key elements – changeable dust filter. FPP, elephlen are mostly used. Table 2 shows key characteristics of filters (Golinko 2008).

It would seem that application of elephlen having comparatively high penetration coefficient should be less efficient to compare with FPP materials. Nevertheless, owing to slight resistance of elephlen filter total protective efficiency to compare with MIPRO having with FPP 15-1.5 high-efficiency filters (Table 2). It confirms idea on redistribution of airflows within respirator when filter resistance increases.

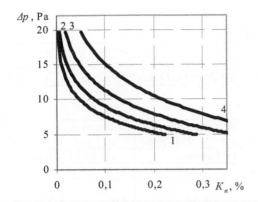

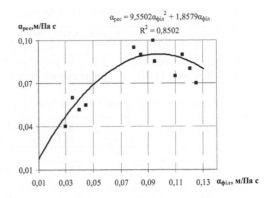

Figure 1. Dependence of penetration coefficient on pressure difference on MIPRO under different diameters of filtering material fiber: 1 – 1.5 microns; 2 – 2 microns; 3 – 2.5 microns; 4 – 3 microns.

Figure 2. Dependence of coefficient of filtering action of respirator on coefficient of filtering action of filter.

Table 2. Results of laboratory tests of RPA-TD respirator filters.

Products under test	Pressure difference if air consumption is 30 l / min, ΔP, Pa	Penetration coefficient on MT if air consumption is 30 l / min, K_n, %	Penetration coefficient of microslice powder M-5, K_n, %
Elephlen filters	54.5±2.7	0.05±0.005	0.005±0.005
FPP 15-1,5 filters	73.4±3.5	0.01±0.005	0.0001±0.005
FPP 15-0,6 Filters	61.3 ± 2.0	0.3±0.03	0.01±0.005
RPA-TD respirator with FPP 15-1.5 Filters	39±3.2	1.2±0.02	–
RPA-TD respirator with elephlen filters	25+2,4	0.9+0.05	–

It is impossible to make MIPRO knowing nothing of *composition and quantity of harmful agents within environment*. For example, coal dust concentration in breathing zone of worker is 300 mg / m³, MPC for coal dust with SiO_2 content in it up to 10% will be 4 mg / m³. Hence, the worker should use MIPRO which protective coefficient is no less than 75 (300/4 = 75). Table 3 shows types of MIPRO recommended to be used at coal enterprises depending upon both composition and quantity of harmful agents in air (Kaminski 1999).

Protective Effect Term. Under the conditions of high dust level native to coal enterprises the factor is one of the most important. Operating time of respirators depends on period of terminal resistance achievement. According to GOST 12.4.041-89 it is 100 Pa under given air consumption 30 l/min through MIPRO.

Results of numerous experimental research connected with studies of mechanisms of deposition of dust particles on fibers of filtering materials helped to obtain empiric expression to determine operating time of respirator (Cheberyachko 2007).

$$t = \frac{\left(F(\Delta p) - F_B^2\right) d \rho_n \varphi F_0}{4 F_B Q C},$$ (4)

where

$$F(\Delta p) = \left\{\left[\frac{6(100 - \Delta p_0)\pi^2 L}{k_n \rho_n \varphi F_0}\right] + F_B^3\right\}^{\frac{2}{3}},$$

and Δp_0 – initial pressure difference on filter, Pa; k_n – proportion factor depending on filtration velocity, m⁴ / s² (5-8); L – total length of fibers; F_0 – total area of filter, m²; $F_B = \frac{2\beta H}{a}$ – summary surface of filter fibers; P – mass of dust particles deposited on filter, kg; ρ_n – bulk density of dust particles, kg / m³; φ – ratio of non uniformity of dust distribution within filter area (1.3-1.5).

Table 3. Recommendations on MIPRO choice depending on labour conditions.

Recommended types of MIPRO if MAC is exceeded		
Up to 10 times	10 to 100 times	More than 100 times
Filtering Half-masks Lepestok-5, Rostok-3, U-2K	Filtering Half-masks: Lepestok-40, Rostok-2, Snezhok-P, and Snezhok-F. Cartridge Respirators with Rubber Half-masks: RPA-TD, Puls, Klyon-P	

Table 4. Recommended time of use filters for RPA-TD respirator.

Dust Concentration, mg / m^3	Operating period of filters according to air consumption					
	30 l / min	95 l / min	30 l / min	95 l / min	30 l / min	95 l / min
	Of FPP 15-1.5		Of Elephlen		Of Meltblown	
50	27.8	16.7	38.9	23.3	34.4	20.7
100	13.9	8.3	19.4	11.7	17.2	10.3
200	6.9	4.2	9.7	5.8	8.6	5.2

Using formula (4) there was calculated operating time of filters for RPA-TD respirator with critical 100 Pa resistance in Table 4. If the filters are used longer then it will result in deterioration of protective efficiency due to infiltrations along obturation.

Job Description. Some MIPRO designs may be inapplicable due to sharp increase in respiratory resistance when it is referred to activities taking great efforts, and following by considerable volume of lung ventilation. For example, peak volumes of lung ventilation when activities are intensive and very intensive are 200 l / min and 250 l / min accordingly. With it according to GOST 12.4.041-89 initial resistance to respirator breath is determined under 30 l / min consumption, and it can not reflect real situation with energy consumption of organism.

Figure 3 shows values of resistance to breath for respirators having two canisters (RPA-TD, Puls, Astra-2) with filters made of "FPP 15-1.5" and elephlen under different consumptions of air.

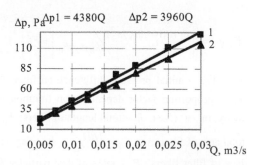

Figure 3. Dependence of pressure difference within filters of respirator (1) and respirator RPA-TD (2) on air consumption.

Weather Conditions. Job conditions differ greatly from standard ones ($t = 20$ °C; $\varphi = 50\%$; $P = 101.1$ kPa) under which evaluation of respiratory quality takes place according to GOST 12.4.041-89. In mine workings air temperature is 26 to 31 °C, air humidity is 90-100%, and atmospheric pressure is high.

Research conducted in Figures 4 and Figure 5 show that raise of temperature and raise of relative air humidity result in pressure difference increase. It can be explained by the fact that resistance to breath is directly proportional to gas viscosity varying under the effect of temperature. Besides, humid air passing through filtering material forms light film on the surface of fibers. With time the film cracks, and drops originate there where crossed fibers contacted. Gradually, they cover filter pores. As a result, resistance to MIPRO breath increases.

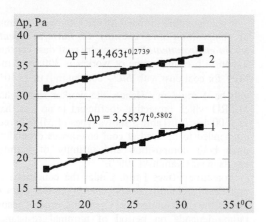

Figure 4. Graphs of MIPRO pressure difference on environmental temperature: 1 – Elephlen Filters; 2 – FPP 15-1.5 Filters.

Extra increase to pressure difference on MIPRO through elevated temperature and air humidity may be considered as

$$\Delta p = k_1 k_2 R_1 Q_1,$$

234

where k_1 – extra pressure difference resulting from temperature increase; k_2 – extra pressure difference resulting from increase in relative air humidity (they are determined experimentally for each type of filtering respirator) (Tables 5 and 6).

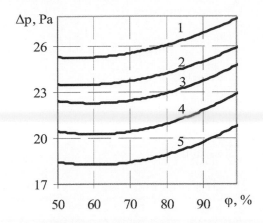

Figure 5. Dependence curves of pressure difference of respirators on humidity of air passing through filter under different environment temperatures, ⁰C: 1 – 30; 2 – 26; 3 – 24; 4 – 20; 5 – 16.

Thus, choice of respirator should cover a number of factors not only its filtration efficiency and adequate protection index. It can drastically degrade protective properties even of high-quality MIPRO if working regime and weather conditions a re not taken into account.

Table 5.Compensation value for temperature.

Temperature, ⁰C	+16	+20	+24	+28	+30
Coefficient, k_1	0.9	1.0	1.1	1.2	1.2

4 CONCLUSIONS

As the research demonstrates, it is very important not to exceed recommended operating time of filters. The matter is that dirty filter is characterized by increased resistance to breath which growth may factor in originating infiltrations along leakage seal. It also should be taken into consideration that each respirator type corresponds to its filter type not to decrease indices of the latter. It is of great importance for half-mask to fit a face being in accordance with standard size. Besides, design of respirator should provide minimum "dead zone" within undermask space as well as maximum visibility while working.

Table 6. Compensation value for air humidity.

Relative air humidity, φ %	Air flow temperature, ⁰C				
	16	20	24	26	30
70	0.9	1.0	1.1	1.2	1.3
80	0.9	1,0	1.1	1.2	1.3
90	1.0	1.1	1.2	1.2	1.3
100	1.1	1.2	1.3	1.3	1.4

REFERENCES

Basmanov, P.I., Kaminski, S.L., Korobeynikova, A.V. & Trubitsina, M.E. 2002. *Means of Individual Protection of Respiratory Organs*. Reference Manual. – SBR: SPH "Iskusstvo Rossiyi": 399.

Golinko, V.I., Cheberyachko, S.I. & Cheberyachko, Yu.I. 2008. *Use of Respirators at Coal and Ore Mining Enterprises*. Monograph. – Dnipropetrovs'k: NMU: 99.

Kaminski, S.L. & Korobeynikova, A.V. 1999. *Means of Individual Protection of Respiratory Organs. Choice, Application. Working Regime. Recommended Practice.* SBR: "Krismas+'': 96.

Cheberyachko, Yu.I. 2007. *Estimation of Particle Size Distribution Influence on Protective Efficiency of Dust Respirators*. Learned Bulletin of NMU, 8: 72-74.

Analysis of the tendency of modern economics development influence on the potential of Ukraine's coal industry reformation

Y. Demchenko, V. Chernyak & S. Salli
National Mining University, Dnipropetrovs'k, Ukraine

ABSTRACT: Possibilities of negative influence of modern macroeconomic processes of world economics on Ukraine's coal industry reformation potential are defined in this paper and the ways of restricting this influence are proposed.

Any section of the national economy of each country is, in fact, is a mini-state or "state within a state".

Coal industry in the national economy, unlike, for example, metallurgy which is a mini-state of the production-finance type (since the section is the basic exporter and currency provider) or machinery construction which can be considered as a base for innovation-economic development (under condition of the development and introduction of new technologies) represents a "state" of social type because:

– firstly, its production is a strategic element of the Ukraine's energy independence guarantee in the world economic and political systems;

– secondly, its personnel is territorially based basically (except for mines of Lvivs'ko-Volyns'kyy basin) in eastern and partially central regions of Ukraine that at present are one of the counterbalances of political system, unbalancing of which is capable to cause some changes of any social-economic regime.

During many dozens of years coal industry has been, is and will be unprofitable and this is a fact that cannot be ignored, because it is a fund consumer but not a fund provider in the state. And evaluation of its activity is only based on financial-economic analysis, that is, by its financial result does not have any sense, and from some perspective is criminal. It is the same as to demand profitable activity from the Army and the State Pension Fund. Trying to provide balance between incomes and expenditures is correct, but to demand incomes from the industry enterprises and increase of labor productivity by any means is not correct at all.

It is necessary to face the truth and acknowledge that to change the industry balance essence "incomes ≅ expenditures" to "incomes ≡ expenditures + profit" means nothing else as to change social-political condition of the whole state.

Save for this, trend of the industry development in terms of positive economic result and high productivity is low-perspective by the following factors.

The first factor – consumer's, or low availability of labor productivity increase.

The experts state: "Labor productivity of the Ukrainian miners remain to be very low compared to foreign indices: twice as low than in Poland, 5 times lower than in average in the countries of western Europe, 20 times lower than in the USA. That is, to extract 1 mln. tons of coal a year it is needed 6 thousand of our miners compared to 3 thousand in Poland, 1.2 thousand in western Europe, 300 – in the USA" (Zvolyns'ka: http://www.dkrs.gov.ua/kru/uk/publish/article/35202).

But let us recalculate it in terms of a manpower cost. Let us look what the official public informational sources tell us concerning this matter.

Head of the Russian independent labor union of coal industry – I. Mokhnachyuk (Russia, 2006): "Today, if to take miners of Germany where salary is within 2-2.5 thousand euro, and salary of miners in Russia which ranges in 350-400 dollars, it is hard to even compare them. I am not even talking about the salaries in Ukraine, where they are even lower" (Korablev).

Ex-prime minister of Ukraine Y. Tymoshenko (2008): "Average salary of the miners all over the world is practically three times greater than in our country"(Tymoshenko: http://news.finance.ua/ru/~/1/0/all/2008/01/16/115938.).

The then deputy of coal industry minister Y. Yashenko (2008): "Minugleprom is going to approximate the salary of domestic miners to the world-average – up to 1 thousand US dollars. Till the end of 2008, the average salary within the industry will reach 3066 UAH" (http://www.inoread.ru/new_4_16046.html).

The head of the coal department labor union AT

"Mittal Steel Temirtau" V. Sidorov (Kazakhstan 2005): "If to talk about ratio of labor productivity and salary, then Karagandian miners earn (300-400 US dollars – note of the authors) 12 times less than their German colleagues, and 22 times less than American miners" (Chekanova: http://articles.gazeta.kz/art.asp?aid =59380.).

So, if to take into consideration the above-mentioned numbers, then with average price of 1 t of coal being 70 US dollars, monthly wage of German miner "costs" around 45 tons of coal or 16% of his productivity, US miner – 80 tons or 8%, for Ukrainian miner – 5 tons or 9% of the amount he earns (if to calculate for 1 month – around 550 tons).

Or such calculation: Ukraine spends 25 mln. US dollars for the salary of 6 thousand workers from 1 mln. tons of sold coal; Germany – for 1200 workers – 43 mln. US dollars, USA – for 300 miners – 22 mln. US dollars.

And we are quite seriously, from the viewpoint of modern economics, considering the following scheme: to find several dozens (maybe even hundreds) billions of dollars for reconstruction of perspective mines, to increase production volume for each working mine (for example, in two times), to reduce personnel amount (for example, in 1.5 times), to increase the salary (in 3 times) proportionally to the total increase of labor productivity (in 3 times), and to finally receive the same 25 mln. US dollars of expenditure for labor remuneration per 1 ton of coal. So what is the effect of such increase in? Is that because number of specialists for some time will be engaged in searching for necessary incredible amounts of investments and more than 30% of miners will be "kicked out" on labor market? Or because, those miners that have remained will have increase in salary by 3 times? Decent but incredible: "or maybe, so what"?

For the period needed for restructuring and, this will take not less than 5-7 years even with sufficient investments, prices and inflation will "eat" this triple increase minimum by half. And this is only by the conditions of "consumer economics". But there are other causes of this "influence factor".

Significant increase of labor productivity will be "faithfully" followed by a considerable increase of workers' incomes. Increase of workers' incomes in one section will certainly stimulate the necessity of incomes increase in other sections and, as a sequence, – general increase of consumption volumes. From viewpoint of theoretical economics, this is a good effect and it testifies about progressive development of the given economic system. But this is a good effect only under conditions of its unlimited resources presence. With unlimited resources, at every step of such development, the pressure from

sources of their provision intensifies: both internal and external.

And that means "development" from viewpoint of modern economic realia.

1) Development that shows an experience of the "developed countries", is, first of all, the movement of gravity center of a business activity from the real economics sector into services sector of commodity and money circulation, and also active formation of "virtual economics" sector. In the 70-80's of the previous century, the "real (initial + secondary) sector" made up 60-70% of the world economy volume, now its part makes up not more than 30-40%. At this, such propositions are composed predominantly "thanks to" GDP structure of the "developed countries" (Table 1 based on the materials (Yul'chieva 2008 & Golomolzina 2010)). The "third world" countries, on the contrary, remain such that are functioning due to the secondary sector (Table 2, Gorkin).

So, in order to develop coal sector in a balanced manner and develop itself as a solid country, Ukraine will have to increase tertiary sector of economics by faster rates than in industry that is quite problematic in conditions of the limited role of Ukraine's financial component in the world economic system.

2) Development is a definite redistribution of the incomes. For the last thirty years the economics has been increasing by 4-5% annually, considering the fact the earlier this parameter did not exceed 0.5-1.0%. Thanks to this increase – out of almost 7 billion people that live on this planet, only 1 billion live according to the incomes level "higher than average", taking into account that near 2 billion people have incomes rates lower than 2 dollars a day. At the beginning of the 60's the ratio in terms of incomes between higher and lower layers of the world social pyramid made up 30:1. Up to the 90's this number had increased twice and at the beginning of the present century it reached 82:1, and during the first decade the experts note its rapid increase up to 500:1 (Karaganov: http://www.wpec.ru/text/200802 010827.htm.).

That is, race for the high production rates and, as a result, for high income of the coal sector workers, with high likelihood will end in a such way that in an absolute value (not in relative – in n-times or by m-percent) only the incomes of a definite quite limited number of people will increase.

3) economic development is also an increase of all types of resources consumption. According to the «Global Footprint Network» (GFN) association, that defines what kind of ecologic trace will remain on the planet, the consumption of Earth's resources for the last 10 years is going very fast: 1986 – the

humanity used annual resources till the 31st of December; 1995 – till November, 25; 2005 – till October, 2. In 2010 – until September, 21. If the resources consumption rates are increasing at the same rates, then in the middle of 2030 to maintain such way of life it will be needed an equivalent of two planets Earth.

Thus, coal industry has very big reserves of coal but it consumes not only this type of resources. Coal enterprises need metal, wood and, again, special place for them is taken by the financial resources.

Table 1. Dynamics of the national economy sectors development in the developed world countries at the end of the XX and beginning of the XXI centuries.

| Country | Year | Economics sector, % | | | Total |
		Primary (agriculture)	Secondary (industry)	Tertiary (services sector)	
France	1975	22	38	40	100
	1990	4	50	46	100
	2008	2	21	77	100
Germany	1976	5	48	47	100
	1990	4	43	53	100
	2008	1	30	69	100
Japan	1976	28	32	40	100
	1990	17	30	53	100
	2008	1	31	68	100
Great Britain	1985	33	32	35	100
	2000	17	38	45	100
	2008	1	24	75	100
United States of America	1975	26	45	29	100
	1990	15	44	41	100
	2008	1	22	77	100

Sources: international statistic data of UNCTAD, The Global Competitiveness Report 2009-2010/ World Economic Forum, Geneva, 2009.

Table 2. Countries with the highest share of the secondary sector in GDP in 2005.

| World rank based on the secondary sector share | Country | Total, % | including, % | |
			Processing industry	Extracting industry, building, communal services
1	Equatorial Guinea	91.2	4.2	87.0
2	Quatar	74.1	8.3	65.8
3	Iraq	72.3	1.3	71.0
4	Lybia	68.1	2.6	65.5
5	Angola	67.2	3.6	63.6
6	Congo (Brazzaville)	67.0	5.4	61.6
7	Azerbaijan	61.9	7.8	54.1
8	Gabon	56.2	8.2	48.0
9	Kuwait	53.9	7.1	46.8
10	Oman	53.6	7.6	46.0
11	Saudi Arabia	53.5	10.1	43.4
12	Algeria	53.3	6.1	47.2
13	OAE	51.5	13.1	38.4
14	Malaysia	49.9	29.5	20.4
	...			
	World	28.3	17.8	10.5

Second factor is a financial but low-perspective way in finding additional investments into coal sector.

During the last 30-40 years the world services market has increased twice and in the tertiary sector the strict tendency of financial markets progressive increase has formed – currency, fund and credit. Business intercourse of the international financial organizations, banks and insurance companies has increased by 30 times compared to 1964 reaching 10 trillion US dollars, and a share of financial component in this sector of economics of the developed countries has reached 20-27%.

World financial sector is becoming stronger and getting more influence on the primary and secondary sectors of economics. The bank investor becomes the basic industry development investor of any country. But what is bank finances today, what are they backed by?

The fact that money can be made by way of extra charges, interests, currency exchange rates – was known 20 years ago but not in such volumes. Indices, ratings, evaluation results – this is what the riches are accumulated on and corporative empires are destroyed. When someone says "The world bank has decreased credit rating of such-and-such Ukrainian bank from B++ to C-", it means that any future treaty of our bank with a foreign one, if it takes place at all, will be way more expensive for our compatriots then earlier and this money will be "sought" by the bank in real sector of economics.

The market of shares has taken center place among the financial banks. The shares have become the basic method of capitals formation. In 1970 the stock segment of the world capital market had a turnover equal to 3% of the global GDP. In 1995 it made up 136% at the turnover index of an international currency market increase by 80 times (from 1973 to 1995). The turnovers at the market of shares make up 800 billion dollars a day, and at the bond market – 950 billion dollars. Total turnovers of the stock market have almost equalized with the turnovers of the currency market (1.9 billion dollars a day) (Mikhaylov 2000.).

Not the production volumes, nor its quality, not even the capability for the equipment renewal and the production assortments become main factors that form incomes of the economic subjects. The heads of serious business more and more often wake up thinking not about the volume of production that their enterprises have sold but how "the Dow Jones index has changed during the night, or the New York stock exchange, currency exchange rates and so on".

But not only the dynamics itself is important, or the value of these indices, ratings, interest rates and currency exchange rates, but what share in these incomes the real production makes or ensured by it money supply. What does stock or bank capital without real sector of economics really mean? Up to 95% of this capital consists of the debts - state, corporative and private ones. And the value of these debts is defined by the financiers themselves by the way of their subjective distribution into the risk levels.

Industrial enterprises, receiving investments of such capital, virtually became hostages of the "game at the risks field". This visually has been demonstrated by the world financial crisis, the beginning of which was the "unsuccessful games" of the financial investors on the USA mortgage market. As a consequence, even such finance-industrial giants as SKM and "Mittal Steel" had to shut down their investment projects.

Thus, none of the institutions even of a state level will give any guarantees for Ukraine's coal sector strategy of restructuring realization based on financial investments attraction.

The third factor – shadow or low-perspective way of economic relations criminalization.

Annually the volume of "shadow business" increases which is a weapon, drugs, gambling games and so on, and its influence on "legal economics increases as well". The specialists state that (Tuzova 2009): "The received results as a consequence of multiple expert estimations generalization from 1990 to 2005 testify about the significant size of the world economics shadow sector and about an increase of shadow economics in life of the world society at modern stage of the social-economic development. As a result of the expert data processing in 145 countries of the world it has been established that as of 2005 the volume of shadow economics made up 16% of its total official GDP (gross domestic product). This number can be considered as final for the world economy as whole.

In the group of the economically developed countries producing 73% of the world gross product within the period of 1990-2005, in 19 countries from 21,

economics shadow sector increase with respect to the official registered GDP is observed. In one case "zero" growth is registered and in another one – insignificant decrease. The biggest shadow economics of the world is the USA's shadow economics where the added value produced in the shadow sector has reached almost 1 trillion dollars (2005)"

Ukraine at present is among the countries with the highest level of corruption and criminalization in the economics, and to hope that the reforms of the domestic coal industry will be carried out without "shadowers" is impossible. Thus, it is necessary to consider that up to 30-40% of the "found" investments simply "will be used not for their intended purpose"

All above analyzed tendencies of the modern economics development testify that the realization of the large-scale investment ideas on the level of separate states and their industrial sectors without "support" of the "leading players" that have some influence on these tendencies is factually impossible.

Solving development problems of the "countries that develop" at the expense of the same countries resources is only an unrealizable in modern economic conditions dream.

As for the Ukraine's coal industry problems – the reality of their solving can be provided only in case of initiating and achieving real changes by the Government based on the following areas of activities:

– first of all, establishment of the rigid requirements for rational use of resources in the systems of all economic levels. "1 ton more or one ton less – it does not really matter" – such approach must not be used anymore.

– secondly, on the macroeconomic level: definition and adherence to the proved consumers, structural and integration propositions of economics development. Establishment of optimal ratios of real and financial (including shadow finances) components of the sectors of economic and asset-productive systems of various scale.

– thirdly, regulative restriction of the financial institutions "game field" in the investment projects of economics real sector. It is necessary to "break the situation" in which the prime cost and production cost are not formed based on the interrelation of "mini-states" of customer and seller, but by a "super state" – financial system.

– fourthly, improvement and provision of the legislative base adherence concerning struggle against the economics shadowing.

– fifthly, legislative determination and formation of a social thought regarding the social-economic status of Ukraine's coal sector.

REFERENCES

Zvolyns'ka, O.V., Zhuk, I.I. & Novychenko, A.P. *"Black gold": resource potential or a headache of the state.* – [Electronic resource]. Materials of the official site of the Main Control-revision management of Ukraine – Regime of access. http://www.dkrs.gov.ua/kru/uk/publish/article/35202.

Korablev V. *Makhnachyuk summarizes of the regional council of trade unions.* [Electronic resource]. Materials of the portal "Mayak" – Regime of access: *http://old.radiomayak.rfn.ru/schedules/57/22403.html.*

Tymoshenko: *Ukrainian miners are the poorest miners in the world* – [Electronic resource]. Materials of the portal "Finance.ua" – Regime of access: http://news.finance.ua/ru/~/1/0/all/2008/01/16/115938.

Salary of Ukrainian "will approach" to the world's average – [Electronic resource]. Materials of the portal "InfoReed". Regime of access: http://www.inforead.ru/new_4_16046.html.

Chekanova, S. *Will foreign countries help them?* – [Electronic resource]. Materials of the portal "Gazeta.KZ" – regime of access: *http://articles.gazeta.kz/art.asp?aid=59380.*

Yul'chieva, G.N. 2008. *Management of hotels services: theory, practice, perspectives for Kazakhstan. Author's abstract of dissertation of a PhD*: 08.00.05. Almaty: 47.

Golomolzina, N.V. 2010. *Service as a form of economic appropriation of useful properties of consumer's benefits and factors of production. Author's abstract of dissertation of a PhD*: 08.00.01. Ekaterenburg: 30.

Gorkin, A.P. *Geography of a processing industry of the world at the beginning of the XXI century* – [Electronic resource]. Materials of a portal "Problematic and sector issues of economic geography". Mode of access: http://geo.1september.ru/2008/01/17-t.htm.

Karaganov, S. *Russia and "gold billion"* – [Electronic resource]. Materials of a portal "Club of the world politic economics" – Mode of access: *http://www.wpec.ru/text/200802010827.htm.*

Mikhaylov, D.M. 2000. *World financial market: tendencies and instruments*. Moscow: Test: 19-25.

Tuzova, A.A. 2009. *Geography of shadow sector of the modern world economy. Author's abstract of candidate dissertation*. Moscow: 26.

The results of the convergence researching in the longwall

S. Vlasov & A. Sidelnikov

National Mining University, Dnipropetrovs'k, Ukraine

ABSTRACT: The results of the convergence researching in the longwall, that were obtained applying three-dimensional computer model of the stratified transversely isotropic rock mass, are presented in the article. The wallrocks convergence distribution regularity along a longwall depending on stope attitude along an extraction pillar allows to assess a reserve of hydrocylinder deformation capacity and to predict incidents in the longwalls caused by fitting of powered support onto rigid base. Parameters and results of the three-dimensional simulation of the rock mass are valid for Western Donbas conditions.

1 INTRODUCTION

The state programs of Ukrainian mining industry development "Ukrainian Coal" and "Energy Strategy of Ukraine for Period till 2030" stipulate considerable increasing value of coal output due to employment and implementation at the operating enterprises highly efficiency coal mining techniques and technologies. However the intensification of the coal mining process has defined a wide range of the mining problems caused by negative underground pressure consequences in development faces and longwalls. It extremely decreases coal-mining output and extremely increases costs of manufacture and industrial accident rate. Therefore in current economical conditions coal mining enterprises make high demands to accuracy of calculation parameters the rock mass stress and strain state because applying of effective approaches to rock mass state management is integral part of employment highly efficiency coal mining technologies and up-to-date highly productive mining equipment. Information about wallrocks convergence distribution regularities along a longwall depending on stope position along an extraction pillar allows employing a powered support more efficiently.

2 FORMULATING THE PROBLEM

As a result of a stope movement and increasing the size of a goaf there is a redistribution of the stress and strain state in a rock mass. One of the main manifestations of underground pressure in the longwall is the wallrocks convergence. A value of the convergence on a line of hydrocylinders' location at a border of the working and mined-out space is an important technological parameter because it is the most probable place of fitting powered support onto rigid base what by-turn makes impossible the movement sections of powered support and creates an emergency in the longwall.

Comparison the convergence distribution regularity on the line of hydrocylinders' location with permissible according to powered support specification compliance allows to draw conclusion about possibility of applying certain powered support type or dimension-type in specific mining and geological conditions and to assess a reserve of hydrocylinder deformation capacity. It makes possible to predict incidents in the longwalls caused by fitting of powered support onto rigid base.

3 PURPOSE OF THE ARTICLE

Researching the regularity of wallrocks convergence distribution on the line of powered support's hydrocylinders location depending on longwall attitude along an extraction pillar on the basis of results step-by-step simulation of stope movement in the three-dimensional computer model of the extracting block that is located in the stratified transversely-isotropic rock mass.

4 SIMULATION TECHNIQUE

By means of software package that realizes computation by finite element analysis, computer 3-D model for conditions of longwall #874 located on the seam C_8^H at the mine "Zapadnodonbaskaya" of OJSC "Pavlogradugol" has been designed. Initial technological and mining-and-geological data are presented in Table 1.

Table 1. Initial data to model construction.

Parameter	Value
Longwall length, m	190
Powered support	КД-80
Angle of dip, deg	0-2
Extraction thickness, m	0.95-1.05
Support of the drift	КШПУ-13.2
Name of the coal seam	C_8^H
Depth of mining, m	460

A finite element mesh is formed as 8-node tetra-hedrons under the scheme Delano-Voronova. The finite element's maximal linear dimension not overrides 10 m. Take into account that model has one symmetry plane, which is normal to stope's middle the mathematical computation performs only for one symmetric part. At computation all necessary boundary conditions in edge model parts and on the symmetry plane are fulfilled. The powered support's affect is simulated by distributed load onto seam's roof and foot that specifies in the powered support location. The geometry parameters of distributed load zone correspond to geometry parameters of powered support КД-80 roof and base plate. Distributed load zone locates at distance 2.5 m from the stope's plane and has rectangular shape with area extend 1.5×190 m. Enforcement onto roof is specified about 2.5 MPa, onto foot – 3.0 MPa.

The simulation is performed from the editing room point till point of longwall attitude along an extraction pillar 235 m with step of 10 m. At simulation every next stope position along an extraction pillar as initial data the simulation results of previous position were taken. Such approach to simulation allows consider step-by-step the redistribution processes of stress and strain state in the rock mass what appreciably increases scientific significance and practical importance of obtained results.

The model's geometry parameters and rock mass physical and mechanical parameters were defined according to (Sidelnikov 2009). More particularly the simulation parameters are specified in (Sidelnikov 2009 & Vlasov 2010).

5 RESULTS

The value of convergence in the longwall depending on stope position along the extraction pillar is an important technological parameter that allows to assess a reserve of hydrocylinder deformation capacity and to predict incidents in longwalls caused by fitting of powered support onto rigid base.

Generally the value of convergence is possible to present in the following form

$$K = f(z) + \Pi(z), \qquad (1)$$

where $f(z)$ – function of roof subsidence along the longwall length in certain stope position; $\Pi(z)$ – function of ground swelling.

Table 2. The value of a, b, C depending on the stope position.

Stope position, m	a, m^{-3}	b, m^{-1}	C, m	Z, m
5	$3 \cdot 10^{-10}$	$-2 \cdot 10^{-6}$	0.006	
15	$8 \cdot 10^{-10}$	$-5 \cdot 10^{-6}$	0.110	
25	$4 \cdot 10^{-10}$	$-5 \cdot 10^{-6}$	0.168	
35	$-3 \cdot 10^{-10}$	$8 \cdot 10^{-7}$	0.217	
45				
55	$1.45 \cdot 10^{-9}$	$-2 \cdot 10^{-5}$	0.296±13%	
65				[-95; +95]
$^*S = 0$	$2 \cdot 10^{-9}$	-10^{-5}	0.162±10%	
$S = 1$	$6 \cdot 10^{-10}$	$-1.2 \cdot 10^{-5}$	0.271±10%	
$S = 2$	$6 \cdot 10^{-10}$	$-1.8 \cdot 10^{-5}$	0.361±4%	
$S = 3$	$1.45 \cdot 10^{-9}$	$-2 \cdot 10^{-5}$	0.305±3%	
$S = 4$	$1.45 \cdot 10^{-9}$	$-2 \cdot 10^{-5}$	0.317±3%	

*Value of the parameter S is defined according to formula (3).

According to analogy of plate and shell bending, which bend on the curved surface of fourth order if they are influenced by the distributed loads, to describe the convergence distribution along a longwall were used fourth-order equation. So the value of wallrocks convergence along the longwall length in

244

certain stope position along an extracting pillar can be found from the expression

$$K = aZ^4 + bZ^2 + C,\qquad(2)$$

where a, b, C – empirical coefficients that depend on stope position along an extraction pillar; Z – co-ordinate along a longwall length.

The empirical coefficients a, b, C depending on the stope position are presented at the Table 2.

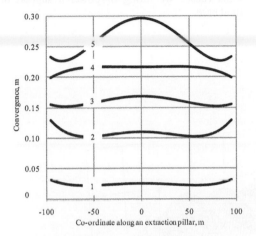

Figure 1. The convergence distribution along the longwall length: 1 – 5 m; 2 – 15 m; 3 – 25 m; 4 – 35 m; 5 – 45 – 65 m.

From the Table 2 evidently that till point of stope position 45 m from the editing room the convergence distribution regularity in every event has different coefficients. It is necessary note that indeed because coefficients a and b have bigger infinitesimal order the finite value will be little vary from the value coefficient C (Figure 1).

Starting from the stope position 75 m obviously expressed periodicity is observed. Therefore to de-

fine the convergence distribution regularity in the longwall in any stope position along an extraction pillar was suggested to identify a parameter S, which characterizes the stope position within one period irrespective of a longwall attitude along an extraction pillar. Grouping the convergence distribution regularity according to quality and quantity characteristics within one step of periodicity underground pressure consequences, it was determined five typical stope's positions in which were determined previously mentioned empirical coefficients. On the assumption of the above-stated, this parameter should satisfy to an inequality (3)

$$0 \le S = \frac{L - 75 - 50n}{10} \le 4,\qquad(3)$$

where L – any stope position, m, $L \in \left[75;\ L_{ex.p}\right]$;

$L_{ex.p}$ – extraction pillar length; n – any multiplier, which is selected, so that the inequality (3) is carried out, $n = 0, 1, 2, ..., N$. The obtained value of the parameter S is rounded to the nearest integer and then according to Table 2 the suitable empirical coefficients are selected. As appropriate to obtain more exact value in intermediate stope positions (step less than 10 m) it is necessary to apply a method of proportional parts (Figure 2).

Dependences of the wallrocks convergence distribution along the longwall length on the line of hydrocylinders' location for stope position in which the parameter $S = 0, ..., 4$ is presented at Figure 3.

The three-dimensional diagram of variation wallrocks convergence on the line of the powered support hydrocylinders' location along an extraction pillar is presented at Figure 4.

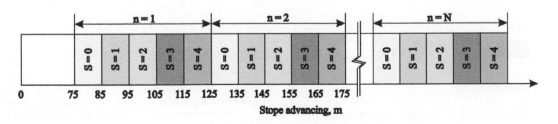

Figure 2. Periodicity underground pressure consequences along an extraction pillar.

245

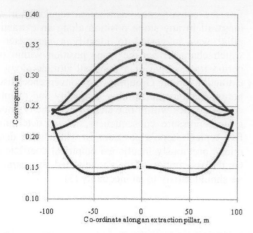

Figure 3. The convergence distribution along the longwall length: $1 - S = 0$; $2 - S = 1$; $3 - S = 3$; $4 - S = 4$; $5 - S = 2$.

6 CONCLUSIONS

Wallrocks convergence distribution regularity on the line of powered support hydrocylinders' location is described by the fourth-order equation of form $K = aZ^4 + bZ^2 + C$ and depends on stope position along an extraction pillar that is characterized by variable coefficients a, b, C. Stated regularity allows to assess a reserve of hydrocylinder deformation capacity and to predict incidents in the longwalls caused by fitting of powered support onto rigid base.

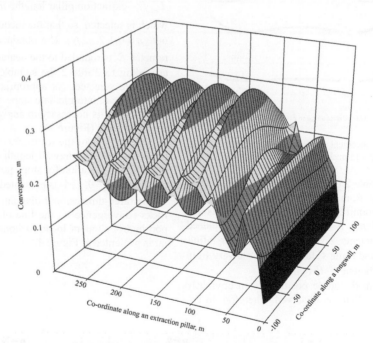

Figure 4. Wallrocks convergence distribution regularity along a longwall and along an extraction pillar:

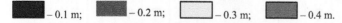 – 0.1 m; – 0.2 m; – 0.3 m; – 0.4 m.

REFERENCES

Sidelnikov, A.A. 2009. *Substantiation parameters of the rock mass 3-D simulation around the longwalls and development faces.* Dnipropetrovs'k: The geotechnical mechanics, 82: 77-85.

Vlasov, S.F. & Sidelnikov, A.A. 2010. *Research of the roof fall mechanism applying three-dimensional model of the stratified transversely-isotropic rock mass at a longwall advance.* Dnipropetrovs'k: Naukovii visnik NMU, 2: 14-17.

Financial conditions of mining enterprises activities in Poland, years 2003-2009

M. Turek
Central mining institute, Katowice, Poland

I. Jonek-Kowalska
Silesian university of technology, Faculty of organization and management, Zabrze, Poland

ABSTRACT: Mining enterprises in Poland have to cope with numerous problems. Since the beginning of 1990s in this industry restructuring processes are being conducted, with the aim of improving the efficiency of coal mine functioning as well as their adjustment to the conditions of free market economy. In this article, an attempt was made to assess the financial state of one of the mining enterprises operating in Poland. The research period covered years 2003-2009. Liquidity, debt, profitability and working efficiency were analyzed. It is worth mentioning that presented results of the research constitute only one of the stages from the Ministry of Science and Higher Education research project on the topic of *Alternative models of financing operational activities of mining enterprises* conducted in the Central Mining Institute.

1 INTRODUCTION

A large number of conditions, both of internal and external type, influence the financial state of Polish mining enterprises. Among the most important and most recent external factors is the situation in the international coal market and European legal regulations concerning carbon dioxide emission. The industrial restructuring that is being carried out since 1990s and the permanent undercapitalization of the Polish coal mining industry are also not meaningless. Among the internal factors, influencing in a substantial way the financial state of Polish mining enterprises are such factors as low cost competitiveness, connected with a fixed production cost structure and a 50% share of salary costs in general costs. A strong position of trade unions, preventing rationalization of employment and the introduction of salary system aimed at improving efficiency is also a major problem of Polish enterprises.

The conditions mentioned above influence the state of Polish mining enterprises. The assessment of this state is the subject of the consideration in this article. The research was conducted in one of the largest mining enterprise operating in Poland. The research period covered years 2003-2009. The research included:

❑ identification of financial liquidity using the indicators of current, quick, capital and instant liquidity.

❑ determination of general debt condition as well as short and long term.

❑ assessment of capability to service debt.

❑ verification of operational efficiency in the area of current assets and short-term liabilities management.

❑ assessment of performance effectiveness

2 ASSESSMENT OF FINANCIAL LIQUIDITY IN THE RESEARCHED MINING ENTERPRISE

In the first stage of research, an assessment of current and quick financial liquidity was made as well as indicators of capital and instant liquidity were calculated for the examined enterprise. The indicators of current, quick, capital and instant liquidity are presented in table 1 (Turek & Jonek-Kowalska 2009).

In the whole examined period, financial liquidity is considerably below the lower standard value and up to year 2006 it shows a constant decreasing trend. A small improvement of the indicator occurred in years 2007-2009. The enterprise in the analyzed period does not possess the capability to regulate the current liabilities from current assets in the light of accepted standards. Even if industry characteristics are taken into account, the continuous decrease of capability to cover current liabilities is disturbing. In 2006 it was the lowest in the examined period and allowed only to regulate 40% of current liabilities from current assets.

It is worth mentioning that both in the domestic and foreign industrial analyses the value of current liquidity indicator for mining is about 1.5 (Brigham

2005; Pomykalska & Pomykalski 2009). According to the above, it can be stated that current liquidity of a researched enterprise is much lower than the industrial average. However, it can be observed that the given value corresponds to the overall of mining activity, not only coal mining, in which the value of current liquidity indicators is usually lower than the one given. Such low value of the indicator stems from the relatively small share of current assets in the enterprise's property and a high level of current liabilities in the sources of funding.

Table 1. Indicators of current, quick, capital and instant liquidity in researched enterprise, years 2003-2009.

Years	2003	2004	2005	2006	2007	2008	2009
Current liquidity	0.62	0.55	0.50	0.40	0.46	0.46	0.61
Quick liquidity	0.45	0.40	0.30	0.21	0.29	0.29	0.37
Capital	12.5%	17.52%	7.25%	3.78%	14.19%	13.55%	18.78%
Instant	3.18%	4.50%	9.04%	6.91%	1.63%	10.77%	9.94%

Source: Based on given enterprise's financial statements.

The indicators of quick liquidity are also shaped similarly to indicators of current liquidity – they differ considerably from the accepted general standards and have a falling trend up to year 2006, then in years 2007-2009 they are slightly increasing (Włoszczowski 2008). It is worth mentioning, that the average for mining is about 0.9, so the examined enterprise does not fulfill also this slightly lowered industrial value (which, however, is calculated for whole mining industry, not only coal mining). In the researched period, from the most liquid assets, the examined enterprise is able to regulate only from 21% (2006) to 45% (2003) of current liabilities.

The capital indicators included in Table 1 are characterized by a large variability within the time from 3.78% in 2006 up to 17.78% in 2009, which means that the value of current liabilities covered by short-term investments is subjected to strong fluctuations. It stems from a strong level of variability in time of short-term investments (Skowronek 2000 & Dębski 2005).

The last of the calculated indicators is the instant liquidity indicator, that gives information about the capability of the enterprises to instantly regulate to most due current liabilities. The indicators of instant liquidity are also subjected to strong variability in time, reflecting different levels of capital in the examined period. It confirms the statement about the unstable situation in the area of capital management. The highest coverage of the liabilities most due was characteristic for year 2006 and amounted 10.78% while the lowest one was noted in 2007 and amounted 1.63%.

In the light of results of the analyses carried out it may be stated that the examined enterprise does not have financial liquidity. Such results suggest the necessity of examining the reasons for indicators' deviation from standards and the need to make managerial decisions aimed at the improvement of the situation in the area of financial liquidity. Below, few suggestions related to changes in current asset and current liabilities management are presented.

The problem of the examined enterprises is mainly the level of current liabilities, which puts the enterprise at the risk of losing financial liquidity and the capability to regulate these liabilities (Sierpińska & Jachna 2007). The share of current liabilities in total liabilities is presented in Table 2.

Table 2. The share of current liabilities in total liabilities and the structure of those liabilities in the examined enterprise in years 2003-2009 (in %).

Years	2003	2004	2005	2006	2007	2008	2009
The share of current liabilities in total liabilities	38.02%	45.98%	31.85%	35.15%	40.68%	45.62%	40.04%

Source: Based on given enterprise's financial statements.

Current liabilities in the whole examined period exceed 30% of total liabilities. It is worth emphasizing that 70-80% of current liabilities in the examined enterprise are overdue. Therefore both the level and the internal structure of current liabilities influence in a very negative way the financial liquidity of the examined enterprise.

Furthermore, the examined enterprise is coping with the problem of underfunding, which also influences in a negative way present and past capital structure. Therefore in this area, it is advised to:
❑ examine the structure of short-term liabilities including their type and maturity date
❑ estimate the cost of current liabilities

□ decide to reduce the most costly short-term sources of financing

□ monitor repayment of obligations in order to ensure they are regulated on time

In the area of current liabilities management the most import ant problem is high variability of short-term investment, including capital. Therefore, one should revise the policy of capital management and determine specific reasons and circumstances of these variable states. If the instability and irregularity in this area stems from the lack of management rationality it is necessary to determine and implement guidelines related to capital management. In this area it is suggested to:

1. Identify cash needs based on the historical levels of cash inflows and outflows.

2. Determine the optimal level of cash in the enterprise, by using the previously mentioned Miller-Orr, Stone and Baumoll-Allais-Tobin models or Monte Carlo method (Machała 2001).

3. Determine the minimal and maximum capital limits.

4. Create cash estimates.

5. Invest surplus capital and complement capital deficiency based on present needs in this area.

Suggested action would allow to stabilize instant liquidity and capital. Furthermore, they would enable rational capital management aimed at both pro-

viding payment security as well as minimizing the costs related to keeping cash (Nawacki 2006).

According to the above, the analysis of financial liquidity would enable getting information about the capability of the enterprise to regulate its current liabilities. Any irregularities in this area made it possible to indicate potential areas, in which managerial decisions with the use of previously mentioned financial management instruments could be made (Dudycz 2008 & Wysłocka 2008).

3 ASSESMENT OF DEBT IN EXAMINED MINING ENTERPRISE

During the process of debt identification, which is an another research stage, the indicators used were:

□ debt ratio

□ equity ratio

□ debt-equity ratio

□ long-term debt ratio

□ short-term debt ratio

□ covering long-term liabilities by fixed assets (Jonek-Kowlaska 2009).

Information about the indicators mentioned above, for the examined enterprises in years 2003-2009, are presented in Table 3.

Table 3. Debt indicators in examined enterprises, years 2003-2009.

Years	2003	2004	2005	2006	2007	2008	2009
debt ratio	96.07%	87.88%	82.52%	85.33%	85.37%	85.56%	85.76%
equity ratio	3.93%	12.12%	17.48%	14.67%	14.63%	14.44%	14.24%
debt-equity ratio	24.43	7.25	4.72	5.82	5.84	5.92	6.02
long-term debt ratio	5.67	1.49	0.90	0.99	0.71	0.72	0.65
short-term debt ratio	8.47	2.37	1.46	2.00	2.29	2.23	2.33
long-term liabilities cover by fixed assets	3.37	3.82	4.38	4.79	6.67	6.39	6.68

Source: Based on given enterprise's financial statements.

In the examined period, the level of debt is generally very high. In 2003 more than 96% of property was funded by foreign capital. In the whole examined period the debt exceeds 80%. The enterprise is exposed to a large financial risk, as it does not fulfill the security standards created for debt. The property of the enterprise is financed by own capital in a very small extent, from 3 to 17%.

A very high level of financial risk is also showed by the level of own capital's debt. In 2003 the liabilities exceeded the value of own capital 24 times. Although in following years the value of the indica-

tor is subjected to a significant decrease to a level of 5-6, but is still too high in relation to the security standard, the value of 1.

The long-term debt ratio in two first years of the analysis is also on the very high level and exceeds the optimal standard that is in the range <0.5; 1.0>. In years 2005-2009 the value of the examined indicator lowers to the values perceived as safe. It is the result of both systematic decrease of long-term liabilities as well as the rise of own capital.

The short-term debt ratio is also subjected to fall. In 2003 short-term liabilities exceeded more than eight times the own capital. In following years the value of the indicator fluctuates in the range 1.46-2.37.

This presented slow but steady improvement of the situation in the area of debt comes from the decreasing level of liabilities, both short- and long-term. However the pace of debt repayment is faster. The property is encumbered above everything by the short-term liabilities coming from taxes, duties, social insurance and salaries.

According to the data in Table 3, the security of long-term liabilities by using tangible assets is periodically rising, which is also a proof of the improvement of situation in the area of debt and its security. In 2003 tangible assets covered long-term liabilities more than thrice. In year 2007 the security in tangible assets is almost sevenfold.

According to the data above, in the examined period there has been an improvement of the structure of financing sources in the enterprise (Michalak & Turek 2009). Systematically the level of short- and long-term debt was decreased, therefore decreasing the financial risk as well. However, both the general debt and the level of short- and long-term liabilities are still high in comparison to standards that are considered secure. It is worth reminding that a substantial part of those liabilities are overdue, which makes the assessment of the level of debt in the examined enterprise worse (Michalak 2009).

Apart from the debt level and structure in the examined enterprises, the capability to service debt was also assessed, using the indicators of times interests earned and fund flows to debt ratio. In table 4 the value of indicators used to assess capability to service debt of the examined enterprise in years 2003-2009 was presented.

Table 4. Indicators used to assess capability to service debt of the examined enterprise in years 2003-2009.

Years	2003	2004	2005	2006	2007	2008	2009
times interests earned	-*	4.22	4.37	-*	0.4	-*	1.53
fund flows to debt ratio	-*	10.68%	9.74%	6.60%	9.28%	12.19%	11.47%

* – negative indicator values caused by loss.
Source: Based on given enterprise's financial statements.

According to data in table 4, the enterprise only in years 2005 and 2004 had the required capability to cover the interests with profits. The indicator exceeded the lower range of the standard to a small degree. In 2007 the interests were covered by profit only in 40% and in 2009 in 153%, in 2003, 2006 and 2008 the enterprise showed a gross loss which caused a total lack of capability to regulate interests from the financial result achieved.

The indicator of fund flows to debt ratio in the examined period fluctuated in the range of 6-12%, showing irregular variations in time. Such variations were connected mostly with the rise of depreciation, high value of which allowed to maintain a high value of the indicator (Michalak, Turek 2009). Furthermore, the improvement of the indicator should also be credited to the decreasing debt. According to the indicator values received, the lowest fund flows to debt ratio featured in 2003, where no surplus was achieved and year 2006 when the from the surplus only 6.6% of enterprise's liabilities could be regulated. The highest value was noted in 2008, the enterprise was able to regulate 12.19% of liabilities from the surplus. With such level of the indicator the repayment of the whole debt would have taken 10 years for the enterprise. Nevertheless, it is worth noting that the improvement of this relation is not connected with the improvement of activity effectiveness for the enterprise, because the financial result in the examined period was subjected to very irregular variations.

Table 5. Net financial result of the examined enterprise in years 2003-2007 (in mil PLN).

Years	2003	2004	2005	2006	2007	2008	2009
Net profit (loss)	-599.912	450.850	254.726	-96.096	11.085	24.710	25.046

Source: Based on given enterprise's financial statements.

4 THE ASSESSMENT OF OPERATIONAL EFFECTIVENESS IN THE EXAMINED MINING ENTERPRISE

The operational effectiveness was determined using the return on assets, equity and sales indicators. The values of the profitability for the enterprise in years 2003-2009 are presented in Table 6 (Turek & Jonek-Kowalska 2010).

In the examined period the return on sale was subjected to strong fluctuation. The highest profit margin on sales was achieved by the enterprise in 2004 and 2005. In that period 1 PLN of net income from sales generated respectively 0.05 and 0.03 PLN of net profit. Because of the negative financial result the return on sale was the lowest in 2003 and 2006. In that period each 1 PLN of sale income brought respectively 0.09 and 0.01 PLN of loss.

Table 6. Profitability indicators in the examined enterprise in years 2003-2009.

Years	2003	2004	2005	2006	2007	2008	2009
return on sale	-9.41%	5.07%	2.96%	-1.16%	0.14%	0.24%	0.24%
return on assets	-6.04%	4.16%	2.34%	-0.94%	0.11%	0.32%	0.34%
return on assets with operational profit	-5.46%	3.34%	0.34%	-1.7%	-8.83%	4.25%	-2.25%
return on equity	-153.47%	34.34%	13.41%	-6.41%	0.77%	1.79%	1.76%

Source: Based on given enterprise's financial statements.

A similar situation was formed in case of general return on assets regarding net profit. The best indicators are in years 2004-2005 and the worst in 2003 and 2006. It is worth emphasizing that after taking operational profit into consideration the situation in the area of general return on assets is even worse as only in 2004, 2005 and 2008 the return on assets in positive, but lower than the one measured by using net profit. It means that there is ineffectiveness in the enterprise's operational sphere.

Changes of profitability with the highest amplitude of fluctuation concerned return on equity. In 2004 1 PLN of equity brought 0.34 PLN of profit while in 2003 1.53 PLN of loss. The fact of clear instability in the area of enterprise's operation effec-

tiveness is also worth attention. The examined indicators are characterized by high variability in time and lack of clear, single trend of changes.

5 ASSESSMENT OF OPERATIONAL EFFICIENCY IN EXAMINED MINING ENTERPRISE

In a following stage of the research an operational efficiency assessment was made. In table 7, the values of indicators in this area in years 2003-2009 are presented.

Table 7. Operational efficiency indicators in the examined enterprise in years 2003-2009.

Years	2003	2004	2005	2006	2007	2008	2009
global assets turnover	064	0.82	0.79	0.81	0.83	1.10	1.04
fixed assets turnover	0.84	1.03	0.94	0.94	1.02	1.36	1.42
current assets turnover	2.72	4.11	4.95	5.72	4.37	5.70	2.83
receivables turnover	7	9	10	13	13	14	13
days' sales in receivables	55	39	36	27	27	26	28
inventory turnover	22	28	20	21	26	52	15
days' supply in inventory	16	13	18	17	14	7	24
own equity turnover	16.32	6.77	4.53	5.51	5.65	7.63	7.29
current liabilities turnover	2	3	3	3	2	2	3
days' sales in current liabilities	149	140	125	135	148	158	141

Source: Based on given enterprise's financial statements.

According to the data in Table 7, income from sales forms about 79-110% of the value of enterprise's assets. The lowest value of this indicator is 64% and was noted in the first year of enterprise's operation. A relatively low value comes from the capital intensity of the examined industry.

The indicators of fixed assets turnover amount from 0.84 in 2003 to 1.42 in 2009. Their slow increasing trend is also worth noting, which is a proof of an improvement in the effectiveness of fixed asset use in generating sales income.

The indicators of current assets turnover were also improved over time. In 2003 the enterprise recreates the state of these assets almost thrice in a year, while in 2006 and 2008 almost six times. The fall in the indicator comes in 2009 when it amounts 2.83.

Short-term receivables turnover was also more effective. The enterprise collected short-term receivables in 7-13 cycles in a year. This cycles were systematically shortened. In 2003 exaction took the enterprise 55 days while in 2006 and 2007 a single cycle lasted only 27 days and in 2008 26 days. The policy of managing receivables also was substantially improved.

According to the data in table 7, the enterprise realized in the examined period from 2 to 3 cycles of receivables repayment yearly. A single cycle lasted from 125 to 158 days.

A larger diversity in the examined period was showed by the supply turnover. Supply turnover went most smoothly in 2008 (52 cycles) and was the slowest in 2009 (15 cycles). A single cycle was completed within 15-52 days.

To finish the operational activity assessment, one should also refer to the indicator of own equity turnover, which was subjected to a gradual decrease. In 2003 1 PLN of equity generated 16.32 PLN of sales income, while in 2009 only 7.29 PLN, and the lowest value was in 2005 – 4.53 PLN. Such situation was connected with a very low share, at the beginning of own equity in financing structure and its gradual growth which was not accompanied by a proportional increase of sales income.

6 CONCLUSIONS

To sum up, the worst financial indicators in the examined enterprise concern the area of liquidity. The enterprise in whole examined period does not have financial liquidity. High level of current liabilities in years 2003-2009 affects in a negative way the level of examined indicators. The situation is further complicated by a very high share of overdue liabilities in the structure of current liabilities.

In the area of profitability it is impossible to determine clear trends. The best level of all profitability indicators the enterprise experienced in 2004, 2005 and 2008. Unfortunately, in 2006 and 2007 there was a fall of sales income and an increase of operational costs, which resulted in the profitability indicators falling down again.

The debt of the enterprise in years 2003-2009 decreased slightly, both in the area of short- and long-term liabilities. The financial risk was also decreased. Positive changes in the area of debt should be attributed to effectively performed financial restructuring process and gradual debt reduction. However, it should be emphasized that 85% of debt level is still very high and the capability to cover interests by earnings is unstable and the fund flows to debt ratio has desired values only due to high depreciation.

The presented research was conducted in the frames of the project by Ministry of Science and Higher Education entitled *Investments in coal mining industry in terms of their financing* (N N524 464836) conducted by Silesian University of Technology.

REFERENCES

Analiza ekonomiczno-finansowa przedsiębiorstwa, pod red. Cz. Skowronka, Wydawnictwo Uniwersytetu M. Curie-Skłodowskiej, Lublin 2000.

Brigham, E.F. & Huston, J.F. 2005. *Podstawy zarządzania finansami*, PWE, Warszawa.

Dębski, W. 2005. *Teoretyczne i praktyczne aspekty zarządzania finansami przedsiębiorstwa*, PWN, Warszawa.

Dudycz, T. 2008. *Wykorzystanie podejścia dynamicznego do pomiaru płynności finansowej*, Studia i Prace Wydziału Nauk Ekonomicznych i Zarządzania Uniwersytetu Szczecińskiego, 7. Zarządzanie przedsiębiorstwem.

Jonek-Kowalska, I. 2009. *Ocena kondycji finansowej przedsiębiorstwa*, [w:] *Start. Przedsiębiorczość akademicka*, pod red. A.Pradeli, Wydawnictwo J. Skalmierskiego, Gliwice.

Machała, R. 2001. *Praktyczne zarządzanie finansami firmy*. Warszawa: PWN.

Michalak, A. & Turek, M. 2009. *Analiza struktury kapitału w kontekście źródeł finansowania przedsiębiorstw górniczych*. Warszawa: Wyd. Sigmie PAN.

Michalak, A. & Turek, M. 2009. *Ocena struktury i dynamiki zmian kosztów w przedsiębiorstwie górniczym*. Krakow: Publikacje naukowe AGH „Przegląd Górniczy".

Michalak, A. 2009. *Ryzyko i koszt kapitału jako determinanty modeli finansowania inwestycji*, [w:] *Dylematy kształtowania struktury kapitału w przedsiębiorstwie*, opracowanie naukowe z serii "Przedsiębiorczość", Warszawa: Wyd. Szkoły Głównej Handlowej.

Nawacki, G. 2006. *Koszty nie odstraszają* "Puls Biznesu".

Pomykalska, B. & Pomykalski. P. 2007. *Analiza*

finansowa przedsiębiorstwa. Warszawa: PWN.

Sierpińska, M. & Jachna, T. 2007. *Metody podejmowania decyzji finansowych. Analiza przykładów i przypadków.* Warszawa: PWN.

Turek, M. & Jonek-Kowalska I. 2010. *Koncentracja przedsiębiorstw przemysłowych – przyczyny, przebieg, efekty.* Gliwice: Wydawnictwo Politechniki Śląskiej.

Turek, M. & Jonek-Kowalska, I. 2009. *Ocena płynności finansowej jako kryterium podejmowania decyzji zarządczych w przedsiębiorstwach górniczych.* [w:] Kraków: Szkoła Eksploatacji Podziemnej. Instytut Gospodarki Surowcami Mineralnymi i Energią Polskiej Akademii Nauk, Sympozja i Konferencje, 74.

Włoszczowski, B. 2008. *O komparystyce finansowej,* [w:]

Finanse we współczesnych procesach kreowania wartości, pod red. W.Caputy i D.Szwajcy. Warszawa: CeDeWu.

Włoszczowski, B. 2008. *Dyskusyjne problemy interpretacji wskaźników finansowych,* Studia i Prace Wydziału Nauk Ekonomicznych i Zarządzania Uniwersytetu Szczecińskiego, 5. *Budżetowanie działalności jednostek gospodarczych. Teoria i praktyka.*

Wysłocka, E. 2008. *Potrzeba zmian w przedsiębiorstwie a ocena jego wyników finansowych,* [w:] *Doskonalenie procesu zarządzania przedsiębiorstwem w obliczu globalizacji,* pod red. R. Borowieckiego i A. Jaki. Kraków: Wydawnictwo Uniwersytetu Ekonomicznego.

Coordinating program of cargo traffic control in coal mines in the process of disturbed land reclamation

L. Mescheryakov, A. Shyrin & T. Morozova
National Mining University, Dnipropetrovs'k, Ukraine

ABSTRACT: The model of automation control system of disturbed land reclamation adapted to real production conditions is recommended taking into account the analysis of surface restoration stripped by underground mining operations.

Specialized Land Reclamation Administration (SLRA) in Western Donbass is engaged in surface restoration stripped by mining operations. It develops individual program of reclamation operations and corresponding package of measures concerning technical and biological land reclamation for each mine.

Using restored areas for forestation is a traditional trend of stripped land reclamation in the region. It is caused by the fact that while performing package of measures on restoration and preparation of stripped areas for forestation mining waste is maximally used and utilized for land reclamation. (Grebenkin 2002).

Rock is one of the types of coal mine waste. Annual rock output in the region is more than 2500 thousand of tons. At the same time 55% from the total rock output is utilizes and used for reclamation and 45% is located in special places – rock piles.

Technical plan on land reclamation for forestation (Figure 1) involves gradual control of clearing reclaimed areas, transporting and heaping layers of burned out rock from rock dump, shifting and stocking rock delivered to the surface directly from the mine, clay and fertile soil layer coating, as well as surface layout, driveway and drainage network installation, etc. (Shyrin 2010).

Mines, open pits, vehicular enterprises and other region production units which should form cargo traffic as well as coordinate it in time and space were engaged to perform a package of measures on technical land reclamation.

It is necessary to point out that in conditions of stimulating mining operations and further development of explored mineral deposit the majority of Western Donbass mines has sufficiently changed traditional technology of coal recovery. Accordingly, the volume of mining development operations has been changed. Therefore, operating system on supplying rock recovered by underground mining method to restored areas has failed. Moreover, in conditions of mining enterprise reconstruction problems of land restoration became non-essential for regional operating mines. (Shyrin 2006) Analysis of estimating ecological safety of regional land technical reclamation system showed the absence of clear program of cargo traffic control.

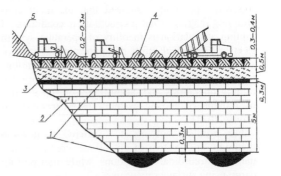

Figure 1. Technological plan of subsided area mining reclamation: 1 – burned-out terricone rock and dump layers; 2 – rock pile layers; 3 – clay protective layer; 4 – potentially fertile layer, removed before dumping; 5 – plant cover.

According to the results of factor analysis of the reasons of land restoration program failure stripped by mining operations it is determined that the closed link in the total system of controlling traffic flow of rock used for technical reclamation is the subsystem of secondary mine transport maintaining technological plan of driving development workings. Data of shift-average losses of development face productivity can confirm it (Figure 2).

Surface ropeway as a single transport means is predominantly used at the mines of Western Donbass while driving development workings using power-loader in conditions of active soil rock swelling. However, disruption of secondary traffic subsystem was noted even while using surface ways of high technical quality (DKNP-1.6). As a consequence, it led to the failure of planned volumes of development mining operations and rock delivery to the surface for technical land reclamation.

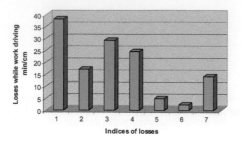

Figure 2. Structure of shift-average losses of development face productive operation: 1 – empty train expectation; 2 – problems connected with organization; 3 – transport service accidents; 4 – face equipment breakage; 5 – accidents on the sites of mining transport; 6 – conveyor transport service accidents; 7 – other operations.

Studying the reasons of unrhythmical surface ropeway operations showed that the main constraining factor of their productivity in conditions of mining stimulation is inability of operating system to ensure technical safety of equipment and prognosis methods of extraordinary (accident) situations taking place in extreme conditions of development mining operations and, accordingly, in the absence of reliable diagnostic systems. (Mescheryakov & Shyrin 2011).

For problem solving it is offered to develop methodic and system of operational control of dynamic extraordinary situations while transporting cargo using surface ropeways.

Technological outlines of driving workings by cutter-loader with different components of transport equipment were subjected to mine study of operating parameters.

Analysis of productive operation of 27 development workings equipped by surface ropeways was made to determine reasons holding down the rate of driving development workings due to the transport fault.

While performing mine study of non-standard production situations the main reasons of decreasing productive operations of surface ropeways were classified as standard, non-standard, critical and accidental. This approach is especially of current interest for making prognosis concerning possible consequence of mine transport operation. To avoid mistakes while developing effective control methods of technological operating surface ropeways in non-standard situations due to the reasons of productive operation decrease transient modes from

one state to another one were introduced.

It was supposed that such approach to operating system of secondary transport control will enable to take into account the most important dynamic features of surface ropeway function in complicated mining conditions.

The main problem for people controlling processes of cargo traffic by surface ropeways is that they often have to make decisions not having the full information about the mine and they cannot imagine possible consequences of their decisions.

In such situation surface ropeway operator should know how to decrease uncertainty, how much it is admitted and what actions are preferable.

In typical mining conditions the process of surface ropeway control starts with informing operator about the problem he should make a decision. For example, there is an accident in development working connected with runaway car. If this situation is typical and well studied the decision is made at once. If the situation is characterized by the high level of uncertainty as well as information lack then subjective experience is not enough for making decisions.

To decrease the uncertainty level connected with equipment it is required to have information about system operation in transient periods ensuring more based forecast of expected results. According to the experience of such problem solving it is known that such additional information enables to get data about possible failure of surface ropeway operation and to change initial intuitional variant of decision making.

REFERENCES

Grebenkin, S., Yermakov, V., Shyrin, A. and others. 2002. *Environmental protection measures against rock dump.* In the book: Geomechanical and technological problems of mine closure in Donbass. Donetsk: DonNTU.

Shyrin, A., Rastsvetaev, V., Dyatlenko, M. & Posunko, L. 2010. *Complex estimation of transport and technological plans of driving development workings by cutter-loader in Western Donbass mines.* Materials of the 5th International research practical conference "Mining and ecological problems". Dnipropetrovs'k: National Mining University, 1: 41-48.

Shyrin, A., Posunko, L. & Piskunova, O. 2006. *Social and economical peculiarities of forming model of mine closure finishing the development of mineral deposits.* Geotechnical mechanics: Interdepartment collection of scientific papers. Dnipropetrovs'k: IGM SFS of Ukraine, 6: 65-71.

Mescheryakov, L. & Shyrin, A. 2011. *Perculiarities of controlling disturbed land reclamation processes at the stage of further coal development.* Materials of Underground Mining School. Krakow: PATRIA.

The perspectives of bioindication methods using in the assessment of toxicity of industrial waste

A. Gorova, A. Pavlychenko & E. Borisovskaya
National Mining University, Dnipropetrovs'k, Ukraine

ABSTRACT: The principles of bioindication methods using in environmental objects control are considered. The necessity of ecological danger assessment of industrial wastes with bioindication methods is grounded. The known biotesting methods of environmental objects quality are improved considering the specificity of industrial wastes. The assessment algorithm of potential ecological danger of industrial wastes is offered with the use of bioindication methods.

1 INTRODUCTION

Rate setting of environmental contamination is based presently on sanitary and hygienic principles, scilicet on the necessity of primary protection of man. Values of maximum permissible concentrations of different substances in water, air and soil, which are usually taken for rate setting of industrial wastes receipt to the environment, are set exactly on the assumption of danger degree for a man.

There is no doubt that it is justly, but it is not enough on the score of few reasons. Firstly, in some circumstances the observance of hygienical norms of pollutants' content in the environment does not ensure the safety for a man, because plants and animals, which inhabit contaminating territory, can concentrate contaminations in their organisms. So they can get to the organism of man by trophic chains in amounts exceeding of maximum permissible concentrations in dozens of times. Secondly, there is no particular reason univocally to claim that sanitary and hygienic norms, which are set relatively to a man, ensure the safety of other biotic components and ecosystems generally.

For example, upon the conditions of the continuous influence of sulphur dioxide at its concentrations in the atmosphere air which not exceeding maximum permissible concentrations for a man, the damage of such contamination sensitive objects as lichens and conifers happens. Consequently, the norms of content of different contaminants in the environmental objects do not ensure the equal degree of protection of wild-life objects from all types of these substances. Scilicet it is incorrect to rate the state of different components of ecosystem only by

the pollution danger degree for a man.

One of possible way to resolve this problem is the application of bioindication methods of the anthropogenic influence on ecosystems and their components. A bioindication is a discovery and determination of the biologically meaningful anthropogenic pressures on the basis of reactions of living organisms and their associations to these pressures (Krivolutskiy 1988).

Bioindication is used with the purpose of exposure of self-potential of ecotope and permissible pressures of exogenous substances. It is used for the control of the populations' state in the early diagnostics and warning of contamination consequences as well as for the creation of the complex system of the ecological monitoring with the exposure of negative changes in natural environment state et cetera.

Organisms can be bioindicators when their presence and state depends on the conditions of environment. As all living objects are the open systems which the stream of energy, information and matter goes through, all of them, including a man, in a different degree are suitable for the decision of bioindication tasks.

Bioindicators' advantages are:

1) in the conditions of the chronic anthropogenic pressures these indicators can react even on relatively weak influences because of the dose accumulation effect, these reactions show up at the accumulation of some critical values of the total dose pressures;

2) they summarize an action of all and singular biologically important anthropogenic factors in an environment and reflect the environment state on the whole, including its contamination and other

changes;

3) the necessity of registration of physical and chemical parameters, which characterize the environment state is diminishing;

4) they specify the ways and places of accumulation of different sort of contaminations in the ecological systems as well as the ways of hit of these agents in the organism of a man;

5) reactions of bioindicators can be base for the estimation of the harmfulness degree for wild-life of any substances synthesized by a man.

In accordance with the organizational levels of the biological systems it is possible to select the different levels of bioindication: biochemical, cytological, histological, morphological, organism level and population level.

There is more and more attention is paid to the question of assessment of wastes danger by bioindication methods in modern scientific literature. So, in the paper (Sedyh 2000) wastes of the boring drilling were tested on the seeds of four different higher plants (a common pine-tree, a Siberian fir-tree, a balsamic poplar, a five-stamen willow). The laboratory germination of seeds and length of plants sprout were taken into account in experiments. Both depression of seeds germination and sprout length and their stimulation were marked, depending on the concentration of wastes of the boring drilling. A Siberian fir-tree was proved as the most sensitive culture. The results of the conducted researches allowed to recommend this method for a receiving enough informing material about phytotoxicity of the boring drilling wastes, that allows to use it for the exposure of the most toxic varieties of the boring drilling wastes. It can be used also for the screening of tolerant types of plants as well as for a study of ecological factors influence on technogenic material toxicity.

In the next article (Borisenkova 2000) the testing of ash and slag wastes of thermal power plant was conducted on laboratory animals – white rats and mice. The results of experiments showed the absence of animal death (at the different ways of single introduction) as well as the absence of action symptoms of ash and it's water extraction even at the maximally entered doses. By the results of biotests the investigated wastes were attributed to untoxic.

The protracted biotesting (during 6 months) was conducted on white rats for the determination of combined toxic action of heavy metals, which are containing in galvanic sludge (Rusakov 1998). The results of experiment testify to the unfavorable affecting of wastes to the bioindicators' organisms. It shows up mainly in disturbance of morphological composition of peripheral blood, dysfunction of metabolic processes and decline of organism resis-

tance. The system of biochemical indexes (permeability of biological membranes, basic processes of vital functions of cell, state of the enzyme systems) used in this work allowed to expose the most vulnerable links in the chain of intoxication on cellular, organ and organism levels at the influence of industrial wastes. It can be used as base for the exposure of informing indexes of biological action at the assessment of toxicity of industrial wastes which contain heavy metals.

The assessment of mutagenic and carcinogenic danger of industrial wastes is needed for the comprehensive toxicological estimation. The next work (Zhurkov 1998) is devoted the decision of this task. In this paper wastes of aviation industry were tested on bacteria (Eyms test) and white mice (micro nucleus test). The amount of revertant colonies of bacteria and amount of marrow erythrocytes with micro nucleuses for mice were served as indexes of mutagenic and carcinogenic effect. The research results indicated that wastes of aviation industry have carcinogenic and mutagenic potential and the level of this potential depends on the type of wastes (connective, glass cloth, carbon cloth, different glues and other). The data given can be examined as a first attempt of complex assessment of mutagenic and carcinogenic danger of industrial wastes.

It is necessary to mark that not only development of effective bioindication methods but also their standardization with the purpose of introduction in environmental protection practice has a large value. So, in Russian Federation were developed and ratified normative recommendations «Criteria of rating of wastes to the class of danger for a natural environment» (Criteria 2001). The classification of industrial wastes to five danger classes is brought in these recommendations: the 1st class – extraordinarily dangerous, the 2nd class – high-dangerous, the 3rd class – moderato dangerous, the 4th class – low-hazard and the 5th class – practically not dangerous.

Rating of wastes to the class of danger for a natural environment according to this method can be carried out by calculation or experimental way. In the case of rating of wastes by calculation method to the 5th danger class it is necessarily to confirm this level by an experimental method. In the case of confirmation default of the 5th danger class by experimental method wastes must be attributed to the 4th class of danger. An experimental method is based on the biotesting of water extraction of wastes by two test-objects from different systematic groups (daphnia and infusoria, ceriodaphnia and bacteria or water-plants etc.). The class of danger, exposed on a test-object that showed the highest sensitiveness to analysable wastes, sets to the final result.

It should be noted that in the Ukrainian legislative

base it was not succeeded to find out normative documents, containing recommendation on application of biotests for determination of danger of wastes.

2 FORMULATING THE PROBLEM

Coming from the above-mentioned analysis, it is obviously that objective necessity presently exists for development and introduction of adequate methods of ecological danger assessment of industrial wastes. Using of bioindicators for the decision of this task is possibly only at implementation of certain conditions:

1. Test-organisms must not perish at the sharp changes of testing terms.

2. Bioindicators must be presented on possibility in great numbers and they might to possess genetically homogeneous properties.

3. Test-objects must provide possibility of the simple sampling and have the parameters that can be estimated by sight.

4. Testing on these organisms must not require the heavy spending of time and resource.

5. Results of indication must be enough representative and stable, the range of errors of measuring on comparison with other testing methods must not exceed 20%.

Plants fully conform to these requirements. Among all living organisms, which are used for the aims of bioindication, one of the first places belongs to them legally.

They are stable enough to the sharp changes of natural ecological factors and at the same time sufficient sensitive to influence of anthropogenic factors. In the conditions of one experiment plants can be presented by tens and hundreds of genetically homogeneous organisms. Changes in growth and development of plants are easily identified by sight; there is no necessity for difficult expensive equipment during carrying out tests et cetera.

Thus, the methods of bioindication by plants satisfy the conditions of assessment determination of ecological danger of wastes and can be used for the decision of this task. It is necessary to mark that bioindication researches do not eliminate but organically complement the methods of analysis that generally accepted presently. They allow to determine total complex influence of all of the components of wastes on living organisms, and, consequently, to determine a danger for ecosystems and man.

With the purpose of reception of more reliable information about the danger degree of wastes testing must be conducted on a few levels of bioindication. Among all variety of biotests for these researches a "growth test" and "Allium-test", related accordingly to the organism and cellular levels of bioindication, were chosen.

3 RESEARCH METHODS

3.1 *Features of growth test at the biotesting of wastes*

Content of growth test consists of accounting of germination, intensity of seeds sprouting and changes of growth indexes of sprouts of indicator culture, which are grown on the investigated samples of soil, water, water extractions of soils, silts, wastes et cetera.

This test differs in extraordinary simplicity of execution, in availability and it does not require plenty of time. The similar speed-up testing is irreplaceable at the initial stage of research, when the large number of variants (factors, doses, types of plants) is "sifted" with the purpose of selection among them the most perspective one for further, more deep researches. Taking into account that the return reactions of plants in this test are fixed on the earliest stages of their ontogenesis, which are characterized by heightened sensibility to negative influences, growth test is the reliable instrument for the assessment of toxic effects of not only strong but also weak, potentially dangerous toxicants.

Practically any type of plants can be a bioindicator in this test. However a preference is ordinary given to quickly sprouting test-cultures which are typical for this region.

The using of growth test for the determination of toxicity of solid industrial wastes has certain features. Unlike soil, water, ground deposits and other natural objects, wastes are not natural, but artificial formation, the degree of toxicity of them, as a rule, considerably exceeds the toxicity of environmental objects. Therefore under research of toxic properties of wastes it is expedient to conduct the biotesting of both wastes and water extraction from wastes, especially having regarded the traditional conditions of their storage and elutriating water-soluble toxicants by atmospheric sinking.

It is necessary to note that clear requirements of preparation of water extraction from wastes are absent in the studied normative literature. The only exception is "The Method of Composition and Properties Determination of Industrial Wastes with the Purpose of Establishment of Possibility and Conditions of Their Reception to the City Landfills of Municipal Solid Wastes" (ND #3897-85, 1985). In accordance with this document extraction is got by dilution of wastes with water of drinkable quality in correlation 1:1. Necessary time of waste contact with water is 15 days. However this method foresees determination of physical and chemical properties of industrial wastes. Obviously, in the case of determination of their potential toxicity time of water extraction preparation can be another depending on the type of test-culture and other conditions.

Therefore with the purpose of adaptation of growth test to the features of industrial waste testing and determination of rational time of water extraction preparation, this parameter was varied from a few minutes to fifteen days during experiments.

For brief experiments with the purpose of determination of sharp phytotoxicity as the test-objects a few types of plants usually are used, which are differed in the rapid germination of seeds. In this case three classic for a growth test and typical for climatic conditions test-cultures were chosen: *Allium cepa L.* (onions), *Raphanus sativus L.* (sowing radish) and *Triticum aestivum L.* (soft wheat). The using of three different types of plants allowed exposing the culture which is the most sensitive to the toxic influence of industrial waste.

For the assessment of industrial waste toxicity by growth test a water extraction was used that was made from a slag and fly ash of the Dnepropetrovsk incineration plant (DIP).

The statistically reliable inhibition of considered marks (amount of germinating seed, length of root system, mass of root system, length of above-ground part et cetera) in comparison with analogical indexes in control served as an index of phytotoxicity of water extraction of wastes.

Phytotoxic effect (PE), namely degree of oppressing of growth processes was determined in percents in relation to control (by mass and by sprouts or roots length of test-culture) according to the formula (Bilyavskiy 2004):

$$PE = \frac{m_o - m_x}{m_o} \cdot 100\% ,\qquad (1)$$

where m_o – mass or length of sprouts (root or above-ground part) in control; m_x – mass or length of sprouts in variants of research.

Thus, the growth test is the perspective method of express assessment of phytotoxicity of industrial wastes, but it has to be improved to appreciate the specific of incineration products of municipal solid wastes (MSW), namely it needs the validation of rational time of water extraction preparation, continuance of experiment and type of sensitive test-culture.

3.2 Features of Allium-test at the biotesting of wastes

Among the great number of test-systems the cells of onion apikal meristem (*Allium-test*) are one of the most comfortable models for the assessment of cytotoxic and genotoxic effects of pollutants: the large sizes of cells and morphology of chromosomes let neatly to describe chromosomal aberrations and other defection of mitosis.

Besides, onion is the eukaryote as well as all higher plants and the structural organization of its chromosomes is very close to the chromosomal apparatus of both man and animals. Moreover all existent in onion cells processes are identical to the processes which existent in animal cells.

Successful long-term experience of assessment of environmental objects quality (soil, water, ground deposits) by *Allium-test* allows to recommend it for the assessment of toxicity and mutagenicity degree of industrial wastes and, in particular, MSW incineration products. One of basic advantages of this method is the possibility of simultaneous research of both cytotoxicity and mutagenicity of investigated objects using one preparation, scilicet to get two assessments of ecological danger of wastes instead of one. But considering that the *Allium-test* is traditionally used for research of toxic and mutagenic features of environmental objects, it needs adaptation to the features of testing of solid industrial wastes. Therefore the time of water extraction preparation from a slag and a fly ash was also varied during experiments with the purpose to choose its rational value.

For testing of wastes of the Dnepropetrovsk incineration plant onion seeds germinating was conducted on a filtration paper in double-dishes. A fly ash, a slag and also water extraction from these wastes were taken as investigated objects. Every variant of research was presented by three replications for authenticity of experiment.

Cytological preparations were prepared from 1 mm of root tips (meristems) of onion. All figures of mitosis were considered at cytological preparations: interphase, prophase, metaphase, anaphase and telophase, which were met among the 5-6 thousands of reviewed meristem cells.

The value of mitotic index (*MI*) was determined as a relation of dividing cells amount to the general amount of the reviewed cells and it was figured per thousand (Pausheva 1988):

$$MI = \frac{m'}{n} \cdot 1000 , ‰ \qquad (2)$$

where m' – the amount of the divided cells; n – the amount of investigated cells.

Reduction of mitotic index value in comparison with the control testifies to the cytotoxic action of pollutants which are present in the probed samples.

Cells with aberration chromosomes were considered at the same preparations: bridges and fragments in anaphase and telophase as well as clinging, spraying and fragments of chromosomes in metaphase. The frequency of reoccurrence of mitosis pathological figures was figured in percents from the amount of the divided cells according to the formula:

$$A^x = \frac{G}{m'} \cdot 100 \, , \% \qquad (3)$$

where G – the amount of aberration cells; m' – the amount of the divided cells.

Increase of the number of pathological figures of mitosis in comparison with the control testifies to the increase of mutagenicity of the investigated samples.

Thus, the *Allium-test* is the informing and perspective method of express assessment of cytotoxicity and mutagenicity of industrial wastes too. The purpose of implementation of this test was to find out the level of cytotoxic and mutagenic effect, which is rendered by the investigated wastes to a test-object, and also to improve this method by the validation of rational time of water extraction preparation from the MSW incineration products.

4 RESULTS DISCUSSION

4.1 Research of phytotoxic properties of wastes by bioindication method

As a result of the conducted biotesting the correlation between the value of phytotoxic effect (PE , %) and time of leaching of the probed wastes (t , *days*) was obtained. It has the polynomial character and looks this way ($R^2 = 0.95 - 0.99$):

– for *Raphanus sativus L.* –

$$PE(t) = 0.0044t^5 - 0.168t^4 + 2.313t^3 - \\ -13.836t^2 + 34.21t - 1.799 \qquad (4)$$

– for *Allium cepa L.* –

$$PE(t) = 0.0046t^5 - 0.189t^4 + 2.826t^3 - \\ -18.11t^2 + 46.4t \qquad (5)$$

– for *Triticum aestivum L.* –

$$PE(t) = 0.004t^5 - 0.158t^4 + 2.318t^3 - \\ -15.39t^2 + 46.41t - 0.26 \qquad (6)$$

Received equalizations were checked for adequacy by the Fisher criterion F at the set level of authenticity α . All received equalizations have $F_b > F_{k_1 ; k_2 ; \alpha}$ that is the chosen correlations describe the results of experiments adequately.

The results of comparison of sensitiveness of the used test-cultures in relation to toxic influence of the investigated wastes are given in a Table 1, where the values of phytotoxic effect, exposed at the different types of plants in variants with 15 day's extraction are showed (PE_{15}).

Table 1. Comparison of test-cultures sensitiveness by the phytotoxic effect.

Test-cultures	PE_{15} , %	
	slag	fly ash
Raphanus sativus L. (sowing radish)	40.9	95.7
Allium cepa L. (onions)	52.4	100.0
Triticum aestivum L. (soft wheat)	77.0	100.0

Data given in this table evidently show that slag and fly ash of Dnepropetrovsk incineration plant possess the obviously expressed phytotoxicity. This index for different test-objects varies within the limits of 41-100%, besides a fly ash is two times more toxic in the average than a slag.

It is also set that the time of preparation of waste extraction must amount no less than 15 days for the exposure of maximal value of PE at the determination of phytotoxicity of industrial wastes by growth test.

For the matter of sensitiveness of the used cultures, namely degree of their reaction to the rendered influence, in order of sensitiveness growth they can be ranged as follows: *Raphanus sativus L.* → *Allium cepa L.* → *Triticum aestivum L.* (Figure 1). The continuance of experiment which is necessary and sufficient for finding out phytotoxic properties of MSW incineration products amounts 72 hours.

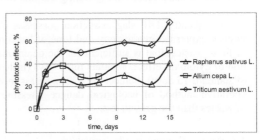

Figure 1. Determination of optimum extraction time from the MSW incineration products and the determination of the most sensitive test-culture.

4.2 Research of cytotoxic properties of wastes by bioindication method

Results of determination of mitotical index value are presented at the Figure 2. The statistically reliable reduction of mitotic activity with comparing to control was marked in all researched variants. It testifies to the presence at the slag extraction of Dnepropetrovsk incineration plant of toxic effect at cellular level (cytotoxicity). The most depressing of growth and development processes of bioobject was marked in a variant with 13-day's and 15-day's extraction – the mitotic index was equal 44.75‰ and 46.00‰, that is below than in control accordingly on 66.9% and 66.0%.

The mathematical treatment of experiment results was executed to educe the correlation between general mitotic index in the cells of *Allium cepa L.* and the time of slag leaching.

Polynomial equalization of regression describes the correlation between intensity of cellular division and the time of leaching most adequately. It has the next view:

$$MI(t) = -0.0024t^5 + 0.109t^4 - 1.81t^3 + \\ + 13.33t^2 - 41.9t + 97.99 \qquad (7)$$

The authenticity of approximation is 72% here ($R^2 = 0.72$). The chart of the obtained function is also presented at the Figure 2.

The analysis of the received polynomial equalization of $MI(t)$ by the 2nd derivative showed that at the interval [0, 1] this function had three extremums, besides the first minimum is in the point (3; 52). Consequently, the general mitotic index in the cells of phytoindicator goes down during the first three days, arriving at the minimum value of 52‰ (that is below than in control on 62%).). Distinction between the maximal value of intensity of cellular division of $MI(0) = 98$ ‰ and minimum value $MI(15) = 47$ ‰ amounts about 51‰, and greater part of these defections (90%) takes place exactly during the first three days.

Consequently, DIP slag extraction possesses obviously expressed cytotoxicity and time of leaching of these wastes, that is sufficient for the determination of cytotoxic effect, amounts three days (instead of standard 15 days). Scilicet the *Allium-test* is more operative method of ecological danger assessment of wastes than the growth test.

4.3 Research of mutagenic properties of wastes by bioindication method

Results of determination of reoccurrence frequency of aberration chromosomes are presented at the

Figure 3. The analysis of results of *Allium-test* conducted on wastes of Dnepropetrovsk incineration plant showed that the complex of slag pollutants, besides a cytotoxic effect, possessed a strong mutagenic action. In all investigated variants the amount of chromosomal aberrations is in the onion cells grown on slag extraction considerably exceeded control.

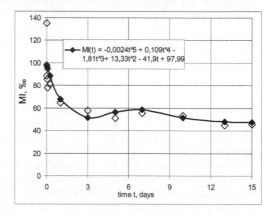

Figure 2. Correlation between mitotic index value in the meristem cells of *Allium cepa L.* and the time of DIP slag leaching.

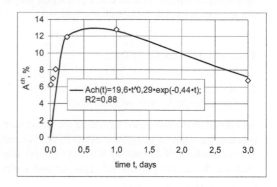

Figure 3. Correlation between the amount of aberration figures of mitosis in the meristem cells of *Allium cepa L.* and the time of slag leaching.

There were various chromosomal pathologies in meristems of phytoindicator: bridges and fragments in anaphases and telophases, and also clinging of chromosomes in metaphases. Even in the case when time of water contact with a slag made 5 minutes, the amount of figures of mitosis with chromosomal anomalies exceeded the level of aberrations in control more than in three times. It was also marked that the increase of time of contact of slag with water caused the sharply increase of mutagenic activity of extraction.

The mathematical treatment of experiment results was executed for the exposure of correlation between the amount of chromosomal pathologies in the cells of *Allium cepa L.* and the time of slag leaching $A^{ch}(t)$. In connection with that fact that the time of slag leaching, which is necessary and sufficient for determination of cytotoxic effect by the *Allium-test*, as marked before, amounts three days, for mathematical treatment experimental information was used in a time range of first three days.

The most coefficient of determination ($R^2 = 0.88$) was obtained for exponential equalization of regression of the next view:

$$A^{ch}(t) = a \cdot t^b \cdot exp(-c,t); \text{ at } a = 19.6 ; \; b = 0.29$$
and $c = 0.44$. \hfill (8)

Verification of the received model on the F – criterion of Fisher confirmed its adequacy to the results of experiment, as Fisher's coefficient is $F_{6;5;0.05} = 4.95$ for this case $F_b = 13.88 / 1.98 = 6.99$; $F_b > F_{k_1,k_2,\alpha}$.

The received dependence of chromosomal aberrations amount in meristems of test-object from the time of slag leaching is presented at the Figure 3. Analysis of the obtained exponential equalization $A^{ch}(t)$ by the 2nd derivative showed that this function had maksimum in the point (0.7; 13). That is that the amount of chromosomal pathologies in the bioindicator cells at the leaching of the investigated wastes for a day long arrives to the a maximal value – 13%, that in 7.6 time exceeds this index in control. Thus, it is elicited that the necessary and sufficient time of extraction preparation for determination of mutagenicity level of these wastes amounts one day.

As a result of the conducted researches the effect of cytotoxic and mutagenic influence of slag of Dnepropetrovsk incineration plant is set on the test-culture of *Allium cepa L.* This effect shows up in depressing of cellular division intensity and in increasing of chromosomal pathologies amount in meristems of indicator. Thus the display of cytotoxic and mutagenic features of slag has irregular character and depends on the time of leaching of wastes extraction – 90% of defections in the onion cells take place at duration of extraction preparation from 24 to 72 hours.

Therefore the obtained dependences of mitotic index value and amount of aberration chromosomes in the phytoindicator cells allowed defining the time of wastes leaching which is necessary and sufficient for finding out their potential ecological danger: for determination of cytotoxic features of slag this time amounts three days, for determination of mutagenic properties it amounts one day. Consequently, the *Allium-test* is the informing and operative method of ecological danger assessment for not only environmental objects but also for industrial wastes, besibes this method allows to obtain the information about both toxic and mutagenic features of different wastes after relatively short time of duration.

5 CONCLUSIONS

A biotesting of industrial wastes at cellular and organism level of bioindication is the necessary condition of determination of their potential ecological danger. The biotesting of wastes at cellular level allows exposing early defections in functioning of test-object, while research of wastes at organism level of bioindication reflects the accumulation of the proper pathological changes on previous levels. Scilicet it allows estimating the consequences of changes in cells for the organism on the whole.

Considering foregoing, the next algorithm of assessment of potential danger of industrial wastes for a natural environment is offered.

1. Assessment of danger class of wastes by the calculation method.

2. Assessment of danger classf of wastes by the experimental method:
 – estimation of phytotoxicity of wastes by the growth test;
 – estimation of cytotoxicity and mutagenicity of wastes by the *Allium-test*.

3. Comparison of calculation results with the results of experimental researches and more accurate definition of final danger class of wastes on this base.

Thus, the implementation of experimental bioindication researches will allow defining the level of ecological danger of industrial wastes and comparing it to the results, obtained by a calculation way. It will give the opportunity to produce adjustment of danger class of the investigated wastes for a natural environment. It will also allow to correct the understated sums of collection in a budget for contamination of natural environment and to elevate level of ecological safety in industrial wastes management.

REFERENCES

Krivolutskiy, D.A., Stepanov, A.M., Tikhomirov, F.A. & Fedorov, E.A. 1988. *Ecological rate setting on the example of radio-active and chemical contamination of ecosystem.* Methods of bioindication of environment are in the districts of APS. Moscow: 4-16.

Sedyh, V.N. & Tarakanov, V.V. 2000. *Influence of wastes of oil and gas digging on the germination of seeds of arboreal plants: raising of problem.* Dendrology, 4: 51-55.

Borisenkova, R.V. and others. 2000. *Assessment of danger of fly ash and slag wastes of Ulan-Ude Thermal Power Plant.* Medicine of labour and industrial ecology, 4: 8-13.

Rusakov, N.V. and others. 1998. *Assessment of danger of industrial wastes, containing heavy metals.* Hygiene and sanitation, 4: 27-29.

Zhurkov, V.S. and others. 1998. *A hygienical assessment of mutagene potential of industrial wastess.* Hygiene and sanitation, 4: 30-32.

Criteria of rating of wastes to the class of danger for a natural environment. Ratified by the order of Russian Ministry of Natural Resources from. 2001: 511.

The maximum amount of toxic industrial wastes, assumed for storage at the city landfills of municipal solid wastes: ND #3897-85. [Valid from 1985-05-30].

Bilyavskiy, G.O. & Butchenko, L.I. 2004. *Elements of ecology: the theory and the practice. School-book.* Kyiv: Libra: 268-269.

Pausheva, Z.P. 1988. *Praktikum on plants cytology.* Moscow: Agropromizdat: 271.

Increased gas recovery from the wells in coal-rock massif due to applied interval hydraulic fracturing and fracture filling with gas-conductive materials

V. Perepelitsa, L. Shmatovskiy & A. Kolomiets
M.S. Polyakov's Institute of Geotechnical Mechanics, Dnipropetrovs'k, Ukraine

ABSTRACT: This article presents research results of the well gas-recovery dynamics received after interval hydraulic fracturing of the coal-rock massif and thickening of the fractures by filling the fractures with loose gas-conductive materials were employed in A.F. Zasyadko mine with the help of an experimental plant.

To intensify a coal-bed methane drainage, Institute of Geotechnical Mechanics of National Academy of Science of Ukraine (IGTM of NASU) has worked out and patented a method to create a net of gas-conductive fractures (Patent 67863 Ukraine). The method is assigned for the wells drilled in the gas-contained coal-rock massifs and assumes application of interval hydraulic fracturing and fracture thickening and filling with a loose gas-conductive material.

We also designed and constructed a special experimental plant for water-sand mixture preparation (WSMP plant) that consists of two-chamber thick-wall cylinder, system of hydrovalves and hydraulic regulators (Figure 1).

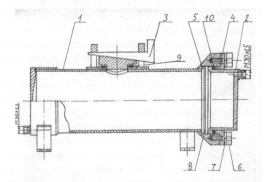

Figure 1. Structural layout of the WSMP plant: 1 – sand chamber; 2 – water chamber; 3 – wedge to close the sand chamber; 4 – ring casing to fasten the water chamber; 5 – sieve # 1; 6 – screws to fasten the water chamber; 7 – lock washer; 8 – sieve № 2; 9, 10 – rubber seals.

Capacity of the sand chamber is 0.037 m³, and capacity of the water chamber is 0.005 m³. Water pressure for water-sand mixture preparation should be within 1.0-3.0 MPa.

Figure 2 shows a set of equipment used in specially-drilled wells in order to produce an interval local hydraulic fracturing (LHF) in a coal-rock massif, create a net of gas-conductive fractures and fill the fractures with a sand-water mixture.

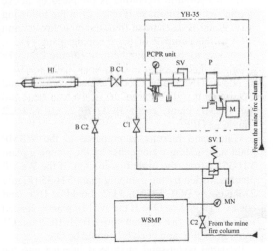

Figure 2. Hydraulic circuit of equipment that is used to make LHF in the wells and fill net of fractures with loose gas-conductive material.

The set of equipment (Figure 2) consists of a pumping plant УН-35 that discharges water through the hydraulic lock HL into the well subject to the hydraulic fracturing and includes a plunger pump P with motor M, safety valve SV for pressure up to 30 MPa and PCPR (Pressure Control–Pressure Relief) unit that controls pressure level. The pump capacity is 35 l / min., pressure output is 30 MPa. Besides, the set (Figure 2) also includes a WSMP plant for water-sand mixture preparation that is equipped with manometer MN to control pressure in the water chamber. Water, to produce a local hydraulic fracturing (LHF), is discharged from the mine fire column to the УН-35 plant and then runs through the hydraulic lock HL. In order to prepare a water-sand mixture and inject the mixture into the drilled well, water to the WSMP plant can be either pumped by the pump УН-35 with the help of the ball cock BC-1 or discharged from the mine fire column (FC) with the help of the ball cock BC-2. Cocks C1 and C2 provide operation of all above circuits, and safety valve SV-1 is used to protect the WSMP plant against any overloads.

Operations of the system to produce the local hydraulic fracturing with water-sand mixture injected into the well (Figure 2) are shown in Table 1.

Table 1. List of operations to produce the LHF and inject a water-sand mixture into the well.

Operations	Position of the cocks			
	BC1	BC2	C1	C2
Sand load	OFF	OFF	OFF	OFF
Water-sand mixture preparation:				
– by the pumping plant	OFF	OFF	ON	ON
– from the mine fire column	OFF	OFF	OFF	ON
Injection of the water-sand mixture to the well:				
– by the pumping plant	OFF	ON	ON	OFF
– from the mine fire column	OFF	ON	OFF	ON
Local hydraulic fracturing	ON	OFF	OFF	OFF

Local hydraulic fracturing is produced by above special set of equipment (Figure 2) in the following way. Water is discharged from the mine fire column (FC) to the pump P of the pumping plant УН-35. According to the hydraulic circuit (Figure 2), cocks should be in the following positions: ball cock BC1 should be ON, and ball cock BC2 and cocks C1 and C2 should be OFF (see Table 1). To load sand into the WSMP plant, cocks BC1, BC2, C1 and C2 should be OFF.

Water-sand mixture is prepared in the WSMP plant with adding water either by the pumping plant УН-35 or from the mine fire column. If the WSMP plant is loaded by the pumping plant УН-35 (Figure 2) the cocks BC1, BC2 and C2 should be OFF, and cock C1 should be ON. If the WSMP plant is loaded from the mine fire column cocks BC1, BC2 and C1 should be OFF, and cock C2 should be ON.

Prepared water-sand mixture is injected into the well by one of the two circuits (Figure 2), i.e. when the WSMP plant is loaded either by pumping plant УН-35 or from the mine fire column.

For the first circuit, the cocks BC1 and C2 should be OFF, and BC2 and C1 should be ON.

If the WSMP plant is loaded from the mine fire column (FC) (Figure 2) cocks BC1 and C1 should be OFF, and cocks BC2 and C2 should be ON.

Bench tests of the WSMP plant were conducted in hydraulics workshop of A.F. Zasyadko mine in a bench of special-purpose design. The plant was tested in two modes:

Mode I – water was fed to the WSMP plant from the surface column at working pressure 1.0 MPa;

Mode II – water was fed to the WSMP plant by the pump УН-35 at working pressure 3.0 MPa.

In both modes, sand chamber was fully ($V = 0.037$ m^3) filled with sand of particle size 0.00025 m. There were 10 cycles for the chamber filling per each mode. Average time duration for the sand fully discharged from the chamber was 2.5 minutes (mode I) and 1.0 minute (mode II).

These bench tests successfully proved viability of the WSMP plant. The set of equipment (УН-35 pumping plant and WSMP plant) for the local hydraulic fracturing satisfied all technological requirements and was recommended for experimental validation in the A.F. Zasyadko mine.

Below is an explanation of how to prepare the wells for the mine experimental studies of dynamics of gas-recovery efficiency resulting from employing the interval hydraulic fracturing in the coal-rock massifs and thickening of the fractures by filling the fractures with loose gas-conductive material. Injection wells of the given diameter length (0.0046÷0.0078) m (subject to further reaming up to 0.112 m) are drilled by rigs ЭБГП-1М or НКР-100М.

To this end, it is necessary to deliver into and install inside the wells a column of thick-wall high-pressure water-fed pipes with inner diameter 0.0032 m. A perforated filter is installed within a filtering interval of the wells. This filter consists of water-fed pipes with perforated holes, diameter 0.0005 m; perforation frequency is 40-50 holes per each meter. The water-fed column and filter are installed inside the well by pieces, 2 m each; each following piece is connected with the previous one by steel couplings.

Annular space is sealed by cement grout with consistency: Solid : Fluid = 2 : 1. The grout is injected into the well by pumping plant HB-20/10 through a special nozzle, length 1.5-2.0 m, that is installed in the well head and fastened by a clay plug and wooden wedges. The cement grout is pumped until laitance appears on the top of the mouth of the water-fed pipe column. Structure of measuring well and well sealing are shown in Figure 3.

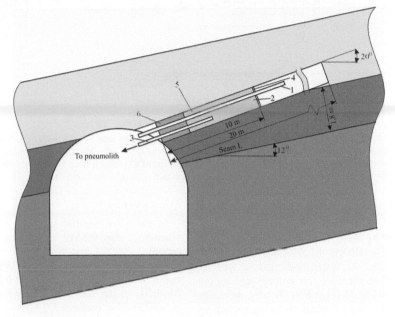

Figure 3. Structure of injection well sealing in order to study fracture thickening. 1 – injection column; 2 – cement sealing; 3 – nozzle for cement grout; 4 – limiting rubbered ring; 5 – cement plug; 6 – wooden wedges

Interval local hydraulic fracturing (LHF) should be produced only when the cement grout has been hardened and not less than 48 hours from the moment when the process of sealing has been completed.

The set of injection equipment and devices for local hydraulic fracturing of the coal-rock massif and further thickening of the fractures includes pumping plant УН-35 (1) that can, within a short period of time, pump working water pressure up to 35 MPa. This pressure is enough for hydraulic fracturing of siltstone and sandstone.

The pumping plant УН-35 with capacity up to 100 l / min. used in the experiments makes the fractures grown up to their entering the seam (Figure 3). In the mine experiments, high-pressure hoses of the longwall set of equipment can be used as a flexible pipeline. Rate of water injection is 1.0-2.0 MPa / s; the rate is regulated by throttles (4, 7) and ball cock (5). Fluid is discharged from the "injection chamber – fixed pipeline" system (Figure 4) with the help of a backwater gate (6), throttles (4, 7) and flexible high-pressure pipeline (8) in the control line.

Water pressure in the "fixed pipeline – injection chamber" system is regulated by a standard pressure manometer (10) with limit of effective range 60 MPa; the manometer is connected with the hydraulic circuit by a T-pipe (9). The fixed pipeline (11, 12, 13) is made of high-pressure pipes with outer diameter 42 mm. Length of the sector (11) is 1 m, sector (12) is 2 m, and sector (13) is 2.5 m. All sectors of the pipes are connected to each other by fittings with coupling nuts taken off from any commercial flexible hoses, and sector (13) is connected by couplings.

The WSMP plant was tested in the 11th west belt heading of seam l_1 in A.F. Zasaydko mine with the aim to choose optimal force and mode parameters for thickening the fractures created due to the hydraulic fracturing and optimal rate of the fracture filling with gas-conductive materials. To this end, a raise well, length 20 m, diameter 0.0078 m, was drilled in the picket ПК 109 at distance 0.6 m from the seal floor. Then this well was reamed to diameter 0.0112 m. In this testing of the WSMP plant, parameters of hydraulic fracturing were the following:

$P_{max} = 15$ MPa, $P_{stab} = 7$ MPa (stabilization pressure when the fluid is injected into the fractures).

Working fluid pressure was indicated by manometers 2, 10 (Figure 4).

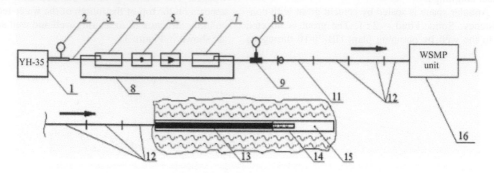

Figure 4. Set of hydraulic equipment to create fractures in the rock massif by way of local hydraulic fracturing and thickening of the fractures: 1 – high-lift pumping plant УН-35; 2 – manometer; 3 – flexible high-head pipeline; 4, 7 – throttles; 5 – ball cock; 6 – backwater gate; 8 – flexible high-pressure pipeline of the control line; 9 – T-pipe; 10 – manometer; 11, 12 – fixed pipelines; 13 – injection pipeline of the well; 14 – injection well; 15 – injection chamber; 16 – WSMP unit.

After the hydraulic fracturing had been produced, and stabilization pressure had been keeping at $P_{stab} = 0.7$ MPa for fifteen minutes water started to intensively penetrate via the fractures in the seam floor. After this, the water-sand mixture was injected both from the mine fire column and by the pumping plant УН-35 (Figure 4). Totally, sand was loaded into the WSMP plant 6 times, each time

$V = 0.0037$ m^3.

In the mine, similar to tests in the hydraulics workshop, the WSMP plant was tested in two modes:

Mode I – water was fed to the WSMP plant from the mine fire column at pressure 1.0 MPa;

Mode II – water was fed to the WSMP plant by the pump УН-35 at pressure 3.0 MPa.

Table 2. Gas-dynamics parameters of the coal-rock massif, seam l_1, A.F. Zasyadko mine, before and after hydraulic fracturing and after filling of fractures with loose gas-conductive material.

Date of measure	Place of measure and content of CH_4, %		Speed of gas release υ_1, m / s		υ_{cp}, m / s	Gas flow rate in the well	
	well head	1.5 m from the well head	υ_{min}	υ_{max}		$Q_{чвс}$, m^3 / h	Q_{cym}, m^3 / day
a) before hydraulic fracturing							
18.02.04 I-st shift	1%	6%					
b) after hydraulic fracturing							
18.02.04 I-st shift	1.5%	100%	0.1	0.15	0.125	3.7	88.8
в) after fracture thickening							
18.02.04 II-nd shift	1.8%	100%	0.1	0.2	0.15	4.5	108.0
19.02.04 I-st shift	1.8%	100%	0.1	0.25	0.175	5.2	124.8
20.02.04 I-st shift	1.5%	100%	0.1	0.3	0.2	6.0	144.0
21.02.04 I-st shift	2.3%	100%	0.1	0.3	0.2	6.0	144.0
22.02.04 I-st shift	2.5%	100%	0.1	0.25	0.175	5.2	124.8
23.02.04 I-st shift	2.3%	100%	0.1	0.25	0.175	5.2	124.8

There were 3 cycles of the chamber filling by sand per each mode. Average time duration for the sand fully discharged from the chamber was 2.0 minutes (mode I) and 1.0 minute (mode I).

Gas content in the experimental well was measured with the help of gas detector ШИ-11 and apparatus Luga (Germany) before and after hydraulic fracturing and after water-sand mixture was injected into the fractures, i.e. after the fractures were thickened.

Results of the measuring are shown in Table 2.

Data in the Table 2 show that before the hydraulic fracturing of the coal-rock massif there were practically no gas in the experimental well that was drilled into the seam l_1 argillite (floor).

After the interval local hydraulic fracturing had been produced hourly and daily methane flow in the well was $Q_{hr} = 3.7 \text{ m}^3 / \text{h}$ and $Q_{day} = 88.8 \text{ m}^3 / \text{day}$, correspondingly.

After six cycles of the water-sand mixture injection hourly and daily methane flow in the well was increased to maximal levels: $Q_{hr} = 6.0 \text{ m}^3 / \text{h}$ and $Q_{day} = 144.0 \text{ m}^3 / \text{day}$, correspondingly, i.e. showed 1.6 increase.

However, as our study progressed exceeding the six-day period no further increase of the methane flow happened. To our judgment, the reason is that the fractures, after their thickening, were connected with the longwall surface, i.e. were not isolated.

Thus, results of these experimental studies conducted in the A.F. Zasyadko mine confirm working efficiency and effectiveness of: 1) the method, patented by the authors, to create gas-conductive fractures in the coal-rock massif, and 2) the WSMP plant that helps to intensify gas drainage from the wells thanks to interval hydraulic fracturing of the rocks and filling the fractures with the loose gas-conductive material.

REFERENCES

Patent 67863 Ukraine, MPC E 21F7/00. 2006. Method to Create a Net of Gas-Conductive Fractures in the Rock Massif. Bulat, A.F., Zvyagilskyy Yu.L., Efremov, I.O. Perepelitsa, V.G., Bokiy, B.V., Shmatovskyy, L.D. Sergiichenko, G.L. Bul. 9.

Technical and Geoinformational Systems in Mining – Pivnyak, Bondarenko & Kovalevs'ka (eds)
© 2011 Taylor & Francis Group, London, ISBN 978-0-415-68877-2

Automation of drill and blast design

O. Khomenko, D. Rudakov & M. Kononenko
National Mining University, Dnipropetrovs'k, Ukraine

ABSTRACT: The available techniques for drill and blast design in underground mining were analyzed. The advantages and disadvantages of the widely used blast-hole patterns, types of cuts, and explosives are estimated. The algorithm and the program for automatic drill and blast design by optimization of spacing among blast-holes have been developed and tested. The calculations carried out for different cross-section forms have shown the advantages of the algorithm comparing to the available procedures for design.

1 INTRODUCTION

The contemporary period of mining industry development in many countries is characterized by concentrating of production facilities and improving the extraction technologies due to, first of all, utilization of new equipment. Advanced engineering solutions are efficiently implemented through using of modern mining machines. The leading manufacturers and a number of small companies are projecting new technologies that already provide essential profits to mining companies. The machines working under minimal control of personnel can be considered as the real future of the world mining industry (Khomenko, Kononenko & Mal'tsev 2010).

Regrettably, the drilling and loading equipment produced by the leading world manufacturers such as "Atlas Copco" (Sweden) and "Sandvik Tamrock" (Finland) is purchased to Ukraine, as a rule, without automatic control devices and supporting software. Just these units could connect machines working separately and create a mine e-network, which would be highly useful to increase production efficiency, reduce injuring of miners, and provide extraction under minimal personnel control. The rejection of automatic controlling units was due to their high costs affordable for wholesale buyers only. Under limited funds the owners and managers give much less priority to personnel preparation to handle the new machinery working under software support. The guaranty and post-guaranty technical services are provided rather successfully, on the contrary, the software development for routine operations is usually behind other works. This leads to disparity between technical characteristics of equipment and its actual utilization, and eventually, to ineffective usage of available capacities and facilities (Khomenko, Kononenko & Maltsev 2005).

Underground mining of black and color metal deposits in the most of countries is accompanied in 90% cases with drilling and blasting works (DBW) to make excavations, which is explained by high strength of rocks exceeding 120 MPa. The rate of making underground excavations depends on proper design of DBW. The analysis of the available algorithms used for designing DBW shows their major disadvantage such as the absence of both a universal technique and automation for main stages of projecting, which decreases the mining outcome.

Advanced mining equipment and different techniques of DBW will be ineffective if not supported by algorithms and software for automatic positioning of blast-holes over the face area. So the design of such patterns becomes the high-importance task for many mining enterprises in the world (Khomenko, Kononenko & Dolgyy 2006).

2 ANALYSIS OF USED BLAST-HOLE PATTERNS

The main stage of DBW projecting is determination of the blast-hole number N_{bh} for the excavation face and design of optimal patterns for explosive distribution over the cross-section. The parameter N_{bh} used as the key criterion of DBW efficiency depends linearly on the explosive mass needed to break up rocks on the excavation face. The fundamentals of breaking rocks by explosion were outlined in studies of M. Protodiakonov, N. Pokrovsky, O. Mindely, B. Kutuzov, M. Sadovsky, V. Rodionov and others. However, till now any generally accepted method for DBW design is not yet developed, though it could be highly useful to reduce testing explosions for each excavation type and on-site studies.

Therefore, the used now simplified technique is aimed at determination of rational specific discharge of explosive material in blast-holes distributed over the cross-section regarding to the practical experience. The explosive specific discharge controls the number of blast-holes to be installed on the excavation face. Usually the cut is placed near the excavation geometrical center; the rest of cross-section area is covered with other blast-hole groups (Shevtsov et al. 2003; Merkulov, Siltchenko, Skorikov 2002; Grebeniuk, Pyzhianov, Yefremov 1983).

The correct choice of a pattern to place drilling holes in the cross-section enables reaching the maximal coefficient of blast-hole usage, which controls the speed of making excavations. The efficient DBW design determined by cut type, pattern, blast-hole number and explosive material depends on geological and mining conditions. The excavation cuts used nowadays can be classified by their position in relation to the face surface; they are sloped cuts that have tearing impact on rocks and straight cuts that have breaking effect. The latter are applied to make excavations in monolithic bedrocks of high strength.

The cut position influences significantly on DBW effectiveness. In accordance with the instructions approved for Ukrainian mining enterprises the cut can be placed shifted in relation to the excavation geometrical center. After the blast-holes within the cut area were installed those auxiliary breaking have to be placed around. As the key parameter for placing the breaking blast-holes the thickness of a layer to be broken is considered. It is denoted by W and called also as the minimal resistance line to the just formed rock surface after explosion. The parameter W has to be made precisely after three testing explosions.

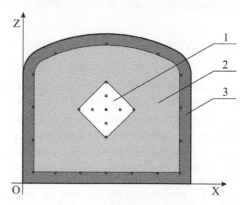

Figure 1. A scheme for zoning of blast-hole groups: 1 – excavation cut; 2 – zone of breaking blast-holes; 3 – zone of contouring blast-holes.

Regarding to the practical experience the distance a among blast-holes is calculated as $a = k_{bh} \cdot W$. The following values for the coefficient k_{bh} are recommended: $k_{bh} = 1.0...1.3$ for breaking blast-holes, $k_{bh} = 0.75$ for those contouring the bottom, $k_{bh} = 0.85$ for those contouring the roof below, $k_{bh} = 0.95$ for those contouring the sides (Figure 1). Projecting and design of the patterns for placing blast-holes proved to be the most crucial and labor-consuming operation.

According to the approaches generally accepted in the mining practice blast-holes are positioned either on a grid or along breaking contours. Despite simplicity both approaches have some disadvantages. If blast-holes are placed on a grid the cross-section top part is covered rarely; on the contrary, if the breaking contours are used the side and bottom areas are covered insufficiently. Tries to combine both methods lead to significant variations of W between adjacent blast-hole rows and complicate drilling by machines working without automatic control. As a result, the pattern disadvantages lead to explosive over-expenditure, over-use of drilling equipment, and sometimes threaten to miners' safety.

Till now the rectangular-vaulted and arched forms of the excavation cross-section are widely applied in iron-ore mining; the circular form is typical for manganese ore deposits, and that trapezoidal is characteristic for deposits where thin seems are mined. The comparative analysis of DBW designing techniques for cross-sections of different forms and size has shown that correct choice of patterns is crucial for optimal blast-hole number.

Consider as an example two commonly used forms of the excavation cross-section such as arched and rectangular-vaulted (Figure 2). The first is applied widely in Kryvyi Rig iron ore basin (Dnipropetrovsk region), the second is typical for Yuzhno-Bilozirske iron ore deposit (Zaporizhzhia region). Blast-holes are placed over the arched cross-section along breaking contours and over the rectangular-vaulted section on the grid.

The areas of these cross-sections are approximately equal; they make 11.6 and 12.2 m^2 respectively at the difference of 5%, both having 3.6 m height and 3.6 m width. The parameter W and blast-hole positions were calculated according to the strength limit of rocks ranging from 90 to 120 MPa. The blast-hole number N_{bh} for cross-sections "a" and "b" equals 43 and 51 respectively at the same $W = 0.6$ m and the distance between excavation contours and nearest blast-holes 0.2 m. The difference between N_{bh} for two sections makes 16%.

The comparative analysis shows that N_{bh} depends intricately on the cross-section form, patterns for placing blast-holes, form and position of the cut.

These parameters are not considered together in the available techniques used to design DBW; hence, the optimal pattern geometry and blast-hole spacing is determined by engineers usually on site regarding to accumulated experience. Evidently the automation of DBW design is becoming urgent due to complication of mining and geological conditions, sophistication of equipment and technologies, enhancement of the personnel qualification. Supporting software can be used much efficiently through creation of libraries containing blast-hole patterns, types and properties of explosive materials, aw well as description of technical and geological conditions on different deposits.

(a)   (b)

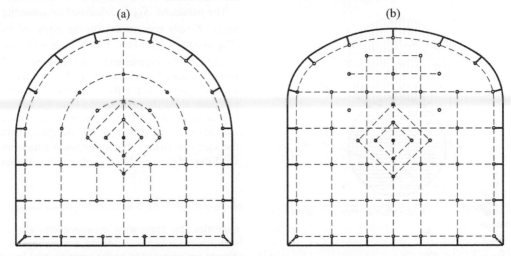

Figure 2. Patterns for blast-holes in cross-sections of arched form (a) and rectangular-vaulted form (b).

3 PROCEDURE OF DESIGNING DBW

The authors propose the following sequence to accomplish the main projecting tasks.

1. Contouring the areas for placing different blast-hole groups in the excavation cross-section.

2. Calculation of the blast-hole coordinates in the zones "1" and "3" according to the available recommendations.

3. Calculation of the blast-hole coordinates in the zone "2" following to the algorithm outlined below.

4. Automatic design of DBW using the calculated coordinates.

The spacing among contouring blast-holes are calculated according to the coefficient k_{bh}. The blast-hole positions within the cut (zone "1") are determined regarding to cut type and mining and geological conditions.

The area of the zone "2" for placing the breaking blast-holes is calculated as

$$S_2 = S_{ex} - S_1 - S_3, \qquad (1)$$

where S_{ex} – cross-section area of the excavation to be made; S_1 – cut area; S_3 – contouring area.

The zone "2" to place the breaking blast-holes is covered with the grid of steps Δx and Δz along Ox and Oz axes. It is reasonable to use the equal step grid at $\Delta x = \Delta z = d$, where d is blast-hole diameter. The 2D array of indices is filled where each cell is marked as (1) a blast-hole, (2) a rectangle within its influence area, or (3) a rectangle outside any other influence areas (Figure 3).

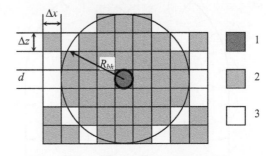

Figure 3. A scheme to determine the grid cells in the excavation cross-section: 1 – blast-holes, 2 – areas of blast-hole influence, 3 – areas outside blast-hole influence.

Around each blast-hole in the zone "2" the circular influence area is contoured, with its radius R_{bh} being equal to $a/2$ calculated at the coefficient k_{bh}.

The circular form is based on the assumption of rock homogeneity; otherwise another form of the influence area can be applied f. e. that elliptical.

The blast-hole coordinates are calculated regarding to the condition of minimal overlay among adjacent influence areas (Figure 4).

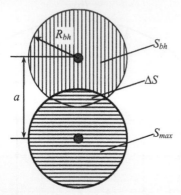

Figure 4. Overlay of the influence areas around the closest blast-holes.

$$|S_{bh} - S_{max}| < \Delta S, \qquad (2)$$

where S_{bh} – blast-hole influence area outside the just determined influence areas, S_{max} – the influence area around a single blast-hole without overlay ($S_{max} = \pi a^2 /4$), ΔS – acceptable overlay area

calculated as

$$\Delta S = \delta \cdot S_{max}, \qquad (3)$$

where δ – optimization parameter varying in different parts of the cross-section. It controls the partial overlay of circular influence areas around closest blast-holes.

The parameter S_{bh} is calculated by summing the areas of cells contoured with the circle of radius R_{bh} around a blast-hole. Firstly blast-holes are positioned in the bottom and contouring parts of the zone "2"; then the holes are placed in rest free parts of this zone consequently by looking for the best positions along the rays coming from the cut center. The order of searching for better blast-hole positions changes the final pattern. While designing patterns the user can change the optimization parameter δ for different parts of the excavation cross-section.

4 CALCULATION EXAMPLE

Consider the excavation cross-sections similar to that shown on Figure 2. Mining and geological conditions are assumed to be the same as used in calculation according to the available techniques. The positions of blast-holes calculated at $\delta = 0.1$ are shown on Figure 5.

(a)

(b)

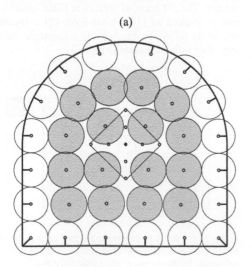
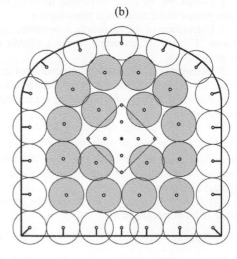

Figure 5. Positions of breaking blast-holes calculated by the developed algorithm for arched cross-section (a) and rectangular-vaulted cross-section (b). Circles are influence areas around blast holes.

Comparison of Figure 2 and 5 demonstrates more regular covering the cross-section with the blast-hole influence areas in case of their positioning by the proposed algorithm. At the same time the breaking blast-hole number makes 16 for both variants. The optimized breaking blast-hole number is at least 30% less than in the patterns designed according the to-day technique. Thus, the proposed approach makes possible to avoid over-smashing of rocks lower the cut and insufficient smashing of rocks near contours, with spending minimal explosive mass.

5 CONCLUSIONS

The algorithm to calculate the blast-hole positions for the cross-section of horizontal excavations being made with explosive materials is developed. It is based on rectangular grid discretization of the area where blast-holes have to be placed. The created software enables designing patterns for different cross-section types on the principle of optimal positioning the circular influence areas around blast-holes. The algorithm can generate both symmetrical and non-symmetrical patterns in the cross-section, which makes possible more regular breaking up of rocks. The proposed approach can take into account the stress and strains in rocks near the face of excavations by variation of the coefficient correcting spacing among blast-holes.

REFERENCES

Grebeniuk, V., Pizhianov, Ya & Yefremov, I. 1983. *Handbook on ore mining*. Moscow: Nedra.

Khomenko, O., Kononenko, M. & Mal'tsev, D. 2010. *Mining equipment for underground mining of ore deposits*. Handbook. Dnipropetrovs'k: National Mining University.

Khomenko, O., Kononenko, M. & Mal'tsev, D.. 2005. *Review of the world market of drilling and loading equipment for mining ore deposits*. Scientific Bulletin of National Mining University, 12: 5-7.

Khomenko, O., Kononenko, M., & Dolgyy, O. 2006. *Experience of using drilling, loading and auxiliary (supporting) equipment on ore mines of the world*. Scientific Bulletin of National Mining University, 1: 18-21.

Merkulov, A., Siltchenko, Yu., & Skorikov, V. 2002. *Projecting of drilling and blasting works in making excavations*. Shachtinski institute YuRGTU. Novocherkassk: YuRGTU.

Shevtsov, N., Taranov, P., Levit, V. & Gudz' A. 2003. *Destruction of rocks by explosion*. Donetsk.

Technical and Geoinformational Systems in Mining – Pivnyak, Bondarenko & Kovalevs'ka (eds)
© *2011 Taylor & Francis Group, London, ISBN 978-0-415-68877-2*

Model of waste management from hard coal mining industry in Poland

J. Grabowski, B. Bialecka
Central Mining Institute, Katowice, Poland

ABSTRACT: Mining industry of hard coal is one of the most important producers of waste in Poland. Till now, storage was basic practice of procedure with waste. The growing ecological awareness of the society, implementation of European Union's law and development of the market economy caused, in the last years, an increase in the interest in waste that could constitute energy from raw materials.

The article contains: the balance of waste from mining industry of hard coal in Poland and an analysis of possibilities of waste utilization. Moreover, essential aspects determining the construction of a complex system of production and utilization of waste were presented.

1 INTRODUCTION

Processes of exploration, mining and processing of hard coal are accompanied by production of waste. Rational and effective management of hard coal deposits incorporates also actions performed for the purpose of economic utilization of all waste, produced in the course of first working, coal mining and enrichment.

These wastes, in compliance with the Act on Waste in force, in the first place ought to be subject to reprocessing at the place of their origin.

In 2007 wastes originated in the course of hard coal production processes constituted ca. 40% of its output and it amounted to 34.4 million tons. Additionally, it is estimated that almost 550 million Mg of this sort of waste has already been deposited in the environment (Priorytetowe i innowacyjne technologie 2009).

Such high level of waste in mass of excavated material is a result of specific character of exploited deposits, technologies implemented in the processes of raw material extraction, policy of deposit management, technologies of enriching raw materials as well as increased requirements by recipients of the final product. Waste produced in the course of mining and processing of hard coal, ca. 94% constitute mine refuse, which is rock material extracted with run of mine (Lutyński & Blaschke 2009). This material is generated in the processes of mineral enrichment, thus in the course of coal preparation.

Mineral-petrographic characteristic of mine refuse is connected predominantly with the place of hard coal deposit exploitation. Moreover, there is a highly significant preponderance of carbonaceous shale, mudstone and sandstone in mine refuse.

2 TENDENCIES IN MINING WASTE UTILIZATION

Waste quality i.e. its physio-mechanical properties in great measure arise from their petrographic composition and contaminant content, mainly coal.

The fundamental norms of conduct with waste from hard coal mining include:

• reduction "at source" – at the stage of exploitation project and through optimization of adopted technologies of first working;

• recovery – in mine under the ground or on the surface;

• neutralization by means of storage.

Properties of waste, including content of coal and petrographic composition depend on location of deposit excavation site and point of waste collection in technological line, which in a decisive way influences its further adaptation and ways of application. Sequence of actions connected with its adaptation:

• coal recovery (provided it is economically substantiated);

• production of aggregate or mixtures from raw materials of top properties;

• adaptation of the remaining waste.

Former ways of managing the above mentioned waste, as it arises from sales analyses, primarily constitute engineering works connected with the leveling or remediation of areas degraded by industrial activity, mining activity included.

The analysis of the accessible data shows that around 92% of waste, produced during mining and processing of minerals is economically utilized – first and foremost in order to avoid necessity of paying charges for their storage. Out of this amount almost 70% is used for leveling of grounds, engineering works or so-called "earthen structures", while 30% is intended for other purposes (Priorytetowe i innowacyjne technologie 2009).

Currently, legal conditions of recovery of waste from hard coal mining are stipulated by the Directive of the Minister for the Environment of 21 March 2006 on recovery and neutralization of waste outside systems or devices (Rozporządzenie 2006).

This directive enumerates 6 main trends in mining waste recovery:

1. Filling terrains unfavourably transformed (such as: swallow holes, unexploited opencast excavations or exploited parts of these excavations);

2. Terrain surface hardening, of surfaces whose owner possesses title deed;

3. Utilization in underground mining techniques;

4. Utilization in arranging and protection from water and soil erosion of slopes and surfaces of closed mine waste dump's top or its part;

5. Building embankments, railway and road embankments, roads and motorway impermeable linings and settling ponds, core of hydraulic engineering structures as well as other buildings and building structures including foundations;

6. Elimination of hire hazards such as spontaneous ignition on active, closed down dumping grounds of gangue from hard coal mining.

Further, potential ways of mine waste disposal refer mainly to building engineering (hydraulic engineering, communication), revitalization of post-industrial areas, production of building materials as well as mining engineering.

The least appropriate method of waste neutralization is its storage, however, taking into account exploitation of old waste dumps, it can be assumed that part of waste dispatched to waste dumps is stored there only temporarily.

One of the future ways of mine waste management, particularly in case of worse quality waste, is production of hydraulically stabilized mixtures. Development of production technologies of hydraulic binding agents enables bonding of materials with higher content of organic compounds, which capacitates forming a structure resistant to environmental conditions. Development of stabilization technologies also gives the means to its embodying into road structures – not only in embankments and banks, but also in foundations in frost penetrable layers.

It ought to be mentioned that there is a growing number of technologies dealing with superfine waste disposal (mine sludge) and in practice, this group of waste can be fully utilized (coal recovery, coal briquette production, fuel for combined heat and power plants, production of artificial aggregate, tightening of waste dumps).

3 PROBLEMS CONNECTED WITH MINE WASTE MANAGEMENT

In spite of a high level of mine waste recovery and hard coal processing, the present, dire state of mine waste management, resulting mainly from deposition of unprocessed waste, is influenced by a number of factors which encompass:

• insufficient number of technically, ecologically and economically proven technologies of mineral processing and recovery of raw materials from waste;

• lack of sufficient, economical mechanisms to the advantage of recovery of raw materials from waste;

• capital barrier while introducing modern technological solutions in exploitation and processing of minerals.

Institutional analysis of companies dealing with recovery and processing of hard coal mine waste, but also research into mine waste market have shown a number of problems which have an effect on the current state of waste management, among others including:

1. *Low level of innovative character of adapted solutions (products, technologies, materials).*

Level of technology in a sector connected with making use of mine waste is not innovative, it is based on appropriate management of unprocessed waste. Recovery of waste originated from mining and processing of coal is recently very high and it oscillates from 95% to 97% in 2007. Such state of recovery is made up of various forms of this waste disposal, including engineering works in the main. In the last five years, amount of waste utilized in engineering works fluctuates within the limits of 30 million tons displaying, however, only a slight downward trend.

2. *There is a necessity of adjusting products offer to market possibilities and creating appropriate demand for them. Products in the sector are not diversified, and their hitherto application is relatively limited.*

Mine waste is in the first place utilized in various types of engineering works, involving remediation and leveling of degraded areas, roadwork, in cement and mining industry (hydraulic filling or removal of abandoned workings). Relatively huge amount of them, reaching periodically even up to 10%, lands in waste dumps. Aggregate obtained as a result of mine waste processing is predominantly utilized for remediation (both, technical and biological) or leveling of degraded areas, thus in a similar way to unprocessed waste.

3. *There is visible downward trend of waste disposal on the market of underground mining tech-*

niques, including backfilling, and also on the market of remediation works, due to declining number of areas requiring remediation.

After a sharp decrease in the amount of waste utilized in filling, which took place at the end of the 90s of 20th century, one can observe it has come to a standstill, and in the last 5 years, amount of waste added to packing and filling in abandoned workings at the level of 1.0-1.3 million tons annually. Comparable values (ca. 1.7 million tons / year) refer to recovery of minerals. It ought to be noticed that not as far back as in the mid of 90s, more than 10% of excavated coal was from mining by means of a system with roof filling, whereas in the year 2007 – 5.5%, and in the year 2006 – 4.9%. It is, among other things, due to mines being placed into liquidation or complete shutdown of many mines.

Unfortunately, year after year, there is a drop or actually lack of interest of mine waste disposal for this aim. Currently, in the Śląskie Province, only 5 mines extracts coal with the use of so called filling, which encompasses for ca. 2% of extraction. The remaining mines extract coal by means of breaking down.

In total, in the area of the Śląskie Province there are ca. 60 reas for remediation to be found, which make up a potential capacity of ca. 55 million m³. It is estimated that for remediation for these lands ca. 100 million Mg of aggregate is needed, which causes that this method of waste utilization is going to gradually lose its dominance.

4. *There is a considerable capital barrier while introducing modern technological solutions in mining and processing of minerals and what is more, lack of knowledge about opportunities of taking advantage of various financial instruments.*

It ought to be added that current demand conditions and customer's influence indicate an important requirement to ensure a stable – definite and reproducible quality of qualified products.

At present, low trust and vague awareness are observed in the scope of mine waste product management in investment processes and site planning.

One can notice that local authorities and enterprises which realize building investments do not consider products manufactured from mine waste as full value products.

To sum up, it should be stated that in order to maximize waste disposal in a long-term, it is obligatory to develop a comprehensive model of cooperation between all participants involved in a process of production and disposal of hard coal mine waste, and in consequence a model of management of mine waste from hard coal mining.

4 MODEL OF WASTE MANAGEMENT

Model of mine waste management to date, first of all based on mines supervised by coal partnerships (Kompania Węglowa, Jastrzębska Spółka Węglowa, Katowicki Holding Węglowy) as well as economic entities (partially also belonging to coal partnerships) ensured high level of waste disposal (in substantial part unprocessed) with the use of technologies not innovative in character, mainly engineering works, including remediation.

Forecasted changes in the methods of waste disposal to date, resulting for the most part form legislative changes connected with introducing the Act on mine waste and with gradual running out of potential associated with areas for leveling or remediation are going to cause that disposal of masses of waste by means of currently employed methods will become increasingly difficult. In consequence, masses of waste to be disposed of will appear which will require new method of management and application of new innovative technologies.

There are many possibilities of mine waste management, either by mines or by external bodies which guarantee maintaining/increasing level of mine waste disposal.

It is worth mentioning that at present there are no monopolists present on the waste disposal market, and competition is rather scarce and not capable to provide complex waste disposal.

One of possibilities enabling further maintenance of the level of mine waste disposal is coming onto the market of "operator/operators" – economic entities dealing with comprehensive disposal of waste on the basis of new innovative technologies and optimization of existing, proven on the market solutions, activity of which will support mines (Opracowanie strategii 2010).

In conditions of the Upper Silesian Industrial Region (Polish: Górnośląski Okręg Przemysłowy) actions taken by such operator/operators need to take into consideration:

- very high dependence on leading supplier;
- very high dependence of the final product on quality and quantity of supplies;
- requirement to ensure a stable – definite and reproducible quality;
- low trust and vague awareness in the scope of mine waste product management in investment processes and site planning;
- high transport costs, which limit the product's sale practically to the local market;
- necessity of implementing new innovative technologies.

Operator needs to posses wide range of products to offer and new innovative products should strengthen the existing portfolio of products.

Dependence of operator's income on sale of various products should guarantee proper diversification of income and prevent their periodical fluctuations.

In addition, operator should also posses a developed information system aiding strategic mine waste disposal and new, introduced technologies should be available in technical and economical meaning (Kozioł & Piotrowski 2009). Hitherto collected from producers and recipients, fragmentary quantity-quality characteristics of waste do not permit to take actions aiming at developing new recipes and production technologies. Knowledge about raw materials may in a considerable way contribute to a development of products which will find recipients e.g. on the market of building investments.

What should be emphasized is a lack of updated database about raw materials which considerably hinders obtaining adequate and quick information concerning quality parameters of raw materials, thus it does not provide sufficient data regarding possibilities of shaping quality stabilization of the final product.

At present, there is also a lack of detailed market analysis with respect to relevant raw materials, products and their potential recipients.

Lack of databases about raw materials and products offered by operator will make it impossible to react quickly and satisfy recipients needs and at the same time it will cause difficulties in finding new recipients.

Therefore, establishing and development of such database and detailed market analysis should be a vital element of potential operator's activity.

What should be considered is the proprietary structure of a new body, specific structure creates possibilities of obtaining raw material on preferential terms, additionally it ensures peculiar stabilization – minimization of a threat of being taken over by competition.

Development of operators on the basis of the outlined organizational assumptions of their functioning should allow him to

• rise to new regulations which capacitate sale increase;

• create optimal portfolio of recipients owing to modern products;

• rise to intense competition and meet recipients' requirements;

• introduce innovative technologies allowing an increase of sale and cost reduction.

A model of management with the concept of operator's location and connections taken into account is presented in Figure 1.

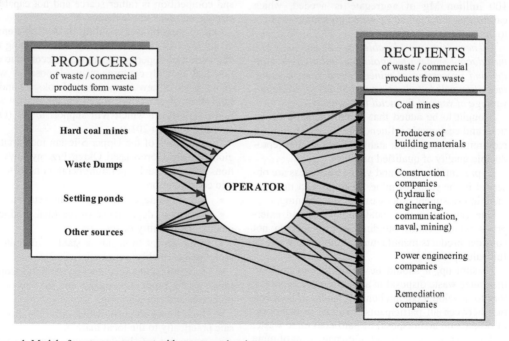

Figure 1. Model of waste management with operator taken into account. Own elaboration.

A necessary condition, giving grounds for the proposed model of mine waste disposal is also a possibility of operator's acquiring financial means for investments and research, obtained from banks, environmental protection funds or other sources.

5 CONCLUSIONS

In recent years there has been a marked change in attitude towards mine waste which begins to be considered as a raw material for utilization in various production activities, leveling and protective works, highway engineering etc. Although such possibilities of waste disposal have been known for years, currently this waste, having undergone the process of modern technologies of their conditioning, neutralization and creation of mixtures with other waste or materials are becoming to be considered as raw material on a large scale.

Throughout the world, mine waste from hard coal mines and other mineral raw materials are utilized first and foremost to:

• fill in underground and surface headings originated as a result of mining activity,

• terrain leveling, recultivation of waste dumps as well as building embankments and various types of dams,

• production of raw materials for underground building industry and other economic activities.

Ways of development of mine waste recovery in hard coal mines worldwide indicate possible increase in their utilization as filling or sealing material. Unfortunately, all significant producers of hard coal all over the world, still neutralize a considerable part of waste by means of storage on the surface which is a result of a lack of abilities or possibilities of their recovery.

An analysis of possibilities of waste recovery e.g. from preparatory works points out to many ways of its utilization. Recovery can take place directly in mines or in specialist external companies (operators).

Current tendencies towards effective use of mine waste from hard coal mining in connection with strong demand for aggregate on the side of highway engineering and building materials create great possibilities of development of companies dealing with recovery and reprocessing of this group of waste. This situation generates favourable conditions to sort out and facilitate market of producers and recipients of certified products, developed on the basis of mine waste.

One can observe positive trends supporting the position of companies functioning in waste sector, including:

• creating conditions which enable companies to specialize and take advantage of market niche situations,

• creating financial instruments within Structural Funds, which enable support of undertaking connected with improvement of innovation and competition of companies,

• development of capital market financing introduction of new innovative technologies.

Minor changes in former methods of waste management, observed in recent times, are bound to be intensified after the year 2012 by introduction of new regulations included in the Act on mine waste as well as executive regulations to this Act.

It will cause, until a perspective of the year 2020, a development of new methods of waste disposal as well as reduction of hitherto prevailing method, that is remediation and leveling of lands. It created possibilities of entering on the market entities (operators), who through implementation of new innovative technologies of waste utilization and modern techniques of managing company, fill in a hole on a future market of waste disposal.

In the article, assumption of a concept of operator has been proposed, enabling effective management of mine waste disposal of waste from hard coal mining, which will constitute a significant supplement to activities taken by producers of coal.

REFERENCES

Priorytetowe i innowacyjne technologie zagospodarowania odpadów pochodzących z hard coal minin. 2009. Etap II – *Badanie i diagnoza stanu obecnego rozwoju technologii w zakresie zagospodarowania odpadów w górnictwie.* Warszawa: IMBiGS, Politechnika Śląska, AGH.

Rozporządzenie Min. Środ z dnia 21.03.2006 w sprawie odzysku i unieszkodliwiania odpadów poza instalacjami i urządzeniami. 2006. Dz.U.z, 49: 356.

Opracowanie strategii gospodarowania odpadami powęglowymi z uwzględnieniem krajowych i europejskich wymagań prawnych na lata 2010-2020. 2010. Główny Instytut Górnictwa not published material.

Lutyński, A. & Blaschke, W. 2009. *Aktualne kierunki zagospodarowania odpadów przeróbczych węgla kamiennego,* Przegląd Górniczy.

Kozioł, W. & Piotrowski, Z. 2009. *Aktualne kierunki zagospodarowania odpadów z udostępniania węgla kamiennego.* Przegląd Górniczy.

Technical and Geoinformational Systems in Mining – Pivnyak, Bondarenko & Kovalevs'ka (eds)
© 2011 Taylor & Francis Group, London, ISBN 978-0-415-68877-2

Human resources management as factor of increase of competitiveness of enterprise

L. Iurchyshyna
National Mining University, Dnipropetrovs'k, Ukraine

ABSTRACT: Theoretical summarizing has been proposed and new solving of scientific aim given, which is consisted in explanation of conceptual states about forming organizational-economic process in management of enterprises' competitiveness as subjects' ownership on principles of their human potential, also mechanism of developing and quality progress of human resource in Ukraine as the constitutive factor of international competitiveness of countries in conditions of globalization is proved.

1 INTRODUCTION

The world economy at the modern stage is going via the processes of fundamental transformation which are manifested, on the one hand, in profounding into internationalization of production and exchange, socialization of economy, increasing countries' interdependency and their regional's integration, forming the global regulatory system, confirming universal standards of living, and on the other, – in asymmetry's intension of global development, dispute's aggravation among countries for ownership's monopoly of economical resources and competitive struggle on international markets. It is conditioned by existing inequality of fund's accumulation in different parts of the world's economy, on the one hand, and formation of qualitatively new competitive conditions of subjects' activity on international stage and changing hierarchy factors of international competitiveness of national economics, on the other. As experience of advanced develop countries shows, during the second half of 20th century traditional sources of their national power – are the dimension of the country's territory, it's natural resources, population, military potential, economical level of development and geostrategic benefits – gradually had been losing their "exclusiveness" in ensuring of high competitive status of national economics, but, at the same time, such factors as existence of powerful innovative potential of countries, mastering rate of science-technological model of economical development, and quality of human resource and conditions of their reproduction were being proposed as foregrounded.

It is no coincidence, that in countries' rating of global competitiveness, leading positions are occupied by the countries which have reached high rates of human development, firstly, based on large-scale investments of capital into human fund and intellectual "saturation" common professions. It provides countries with: 1) rising productivity of work in different areas of national economy; 2) accumulating of intellectual potential and stable temps of economical rising; 3) permanent economical structure modernization according to changing of conjecture on the world markets.

Contemporary world's integration processes objectively gaining currency of the problem of development competitive benefits of national economy. It is connected with necessity of appearance of enterprises on the world market, rising of competition and interest in raising such rates as profit, profitability, effectiveness of economical activity.

The function of management, administrative culture, management of human capital in contemporary conditions has been raised. Therefore, the problem of management is needed permanent studying and solving. It is explained not so much by degree of problem's suspense, as by changes of its contest under the influence of scientifically-technological progress, innovative policy in production, changes in productive terms and in the environment.

Important positive factor, which influences on making effective, more complete contemporary managements' decision, is enterprises' competitiveness, which is mainly defined by formulation and development of human potential. Because of it, demands of development of methodic and methodological bases of management on the background of rising of enterprises' competitiveness and formulating of human potential are being become more actual. The beginning of new century is directed at rash transformation into the human's age, culture of management and intelligence. In social production the main task is to light creative potential of the person, it means that the person becomes the main productive power. In economical activity a great attention is spared to the quality of labor force and staff's qualification, which favour the initiative of staff and integration of their joint efforts. All these things are needed changes in staff's training. At first, new, appropriate to new conditions, methodology of education, which is used for training staff, must be created.

Complex analysis on the background of conception of humans' resource management with a purpose of improving the process of management of enterprise's competitiveness using human resource is the aim of this research.

2 FORMULATING THE PROBLEM

Correction of the economical essence of competitiveness and management of human resource, creation of conceptual approaches for providing company's competitiveness on principles of using staff's potential, estimation of effectiveness of human's potential competitiveness – are directions, which needed further research in contemporary conditions of reforms' transformation.

In our opinion, rate of aspects of present scientific problem have not been solved enough yet. It means, that it is needed subsequent research for developing of international competitiveness of national economics, such indicator like human resource, in conditions of developing of world's economy, transforming into the main factor of ensuring of high competitive countries' status. At the present time there is no scientific definition of the category "human resource" and its interdependence with close categories hasn't been covered. It means that there is necessity in estimation of contemporary state of human resource in Ukraine and proving ways of management of human resource in the condition of contemporary rising of competitiveness of the country.

3 MATERIALS UNDER ANALYSIS

Problems of enterprises' competitiveness and rising of effectiveness, using of human resources, conditions and terms of international competitiveness of national economics are researched by foreign and native researchers like: A. Smith, George K. Galbraith, M. Porter, Cambell R. McConnell, Stanley L. Brue, P. Doyle, P.R. Dikson, M. Kirtsner, F. Hayek, R. Hays, S. Uilrate, D. Clark, D. Campbell, George Stonehouse, B. Houston, R.A. Fatkhut-

dinov, L.S. Shevchenko, G.L. Azoiv, Mikailyan M., B.I. Kholod, A.A. Ponomarev, A.V. Brushlinskiy and others. Classical and modern theories of competitiveness are exposed in their works. These theories characterize contemporary global competitive surrounding and its influence on methods of forms of competitiveness. Regional, innovative and social components of countries' competitiveness are also exposed there.

The main attention in these works are paying to studying general competitive situation in a branch and on the market, diagnosing company's competitiveness and its competitors, creation and realization of strategies of competitive management, and estimation of company's competitiveness.

In 1990's in staff's management of world-known companies, a change of conceptual model of management was happened. Staff was started to consider as the main source of company, which is showing its competitiveness.

Theory, which considered staff as outgoings which must be dismissed was replaced by a new theory. It was the theory of management of human resources, where staff was one of the sources of the company. A new approach is showed in Table 1.

To agree with, ensuring skilled labor, degree of motivation, distribution of responsibility and forms of work, which is defined of effectiveness of using staff, became the main factors of enterprise's competitiveness. It means that new approaches in working with staff should have systematized character in using them, wider usage of elements of planning and standardization of work's spending, using of individual forms of work, rising of means, which are needed for motivation and other work with stuff.

The analyzing of written literary information help to determine, that in Ukraine and other post soviet countries creative skills are still used insufficiently, little attention payed to social aim – achieving assigned degree for satisfactioning social needs of employees: standard labour conditions (maintenancing standards conditions of work, juridical immunity, providing of social infrastructure and others). Because of it in the countries with market economy a great attention is payed to the main function of management (Figure 1) (Kachalina 2006).

Foreign experts, paying great attention to the style of staff's management, emphasize disadvantages of the authoritarian style and the benefits of, so-called, cooperative styles. They describe two approaches to performing the functions of staff's management: the old "Locomotive for achieving the aims" and the new "Team Cooperation, as more productive" (Figure 2) (Kachalina 2006).

A leading German expert in the field of staff's management I. Henttse, formulating a common vi-

sion and approach to structuring the functional division of labor in the field of staff's management, identifies the following functional blocks:
  – identification of necessity in staff;
  – providing staff;
  – staff development;
  – usage of staff;
  – motivation of work results and behavior of staff;
  – juridical and Informational support of staff's management (Prokoshenko & Nortak 2001).

According to the Association of Consultants in Economics and Management, there are two stages of understanding of causes of low competitiveness of an enterprise.

At the first level managers explain the low competitiveness of the enterprise because of the environmental conditions (high taxes and inflation, economic instability, etc.) or defects of the internal environment of enterprises (lack of financial resources, technological backwardness, poor organization of production and management, etc.).

Table 1. Features of staff's management (Prokoshenko & Nortak 2001).

Old approach	New approach
1. Low quality – is the result of bad people's activity. Solving of this problem – is automation	1. Low quality – is the result of unsatisfactory management of people. Solving of the problem – is respect, recognition of people
2. Good "labour morality" – is a national feature.	2. Good "labour morality" – is not a national feature
3. Better quality needs expensive spendings	3. The best quality gives high profit. Our aim is good quality
4.Quality of the result is provided by following control	4. Quality of the result is planned. If deflection is founded, the process of research is changed or completely renovated
5. The control of quality – is the a task for a special branch (specialist), which only register facts	5. Providing quality is a task for every worker. General management of quality includes all necessary functions of the workers indeed
6. Profit is primary, quality is secondary. Because of profit, spending for quality must be reduced as much as possible	6. Investments for providing quality – is a guarantee of future profit. Losses because of "defects" are higher than economy for quality analyze
7. Find out a mistake and register it. Use multiple control	7. Do the best at first. Good quality that you will have – will be for free
8. Buy cheaper from the trader. Competitiveness among the traders reduces spending	8. Quality and reliability should be the main reasons for buying. "We are not so rich, to buy cheap goods"
9. Work with experts	9. Staff must have widespread knowledge
10. Failings of quality are exposed by the management	10. The management always corporate with the staff for prevention failings
11. Subordinates are responsible for quality	11. The system of management – is the cause of many failings in quality
12. Statistic is an exclusive method of specialists for providing quality. Use the main blanks for finding failings	12. Every employee acquainted with the statistic of the quality. It can help to detect and solve problems

The second level of understanding by managers of these problems explains their existence is mostly in a lack of vision due to poor knowledge of market factors, which can define competitive advantages of enterprises, inadequate training, etc.

It means that modern foreign practice of staff's management operates of the three major factors: people, financial policy, engineering and technology. The leader place is for people's labor (human factor), the effective activity of people, which is affected by the world, needs, interests, experiences, etc. Inattention to the problems of workers leads to conflicts, staff's changes, reducing productive quality and productivity, and generally reduces the efficiency of company's management and, as a result, to the loss of its competitiveness.

To the human factors we should include:
– skill level of workers;

– the level of professional training and general training of a person;
– manufacturing initiative;
– civil liability;
– social activity;
– health, physical training.

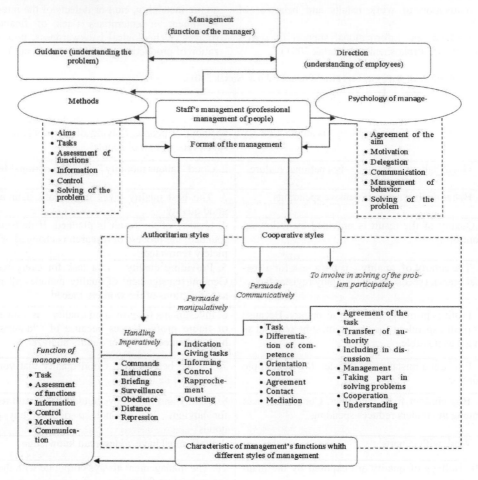

Figure 1. Management and style of management of staff.

Understanding that qualification of workers, and their desire to work well become the main productive power and locomotive start of production, resulted a reorientation of management strategies of firms on the motivation of the labor, education, development of initiative and staff's success.

Creative abilities of a person are manifested via his competence, which includes:
– skills (degree of education, the knowledge, expert skills, experience in a certain area, etc.);
– personal qualities (creativity, communication skills, reliability, etc.);

– motivation (a circle of professional and personal interests, the desire to make a career).

In the current conditions of globalization, flexible management of a firm are required. Effective management of stuff means increase of financial incentive, combination of the usual wage, with participation of workers in a company's profit, in a result 10-40% reduction of spenfing for wages and production costs. Career service of an employee depends on successfull and quality performance of their professional duties as a member of the team. At the same time the solution of promotion an employee based generally on three assessments: by the worker, by the manager he works with; the managers of the company, who present the interests of the company.

An important task of staff's management is a regular evaluation of its activity. Evaluating the activity of staff intended to:

– improving staff's performance;
– the appointment of adequate remuneration for the work of staff;
– making decision which is connected with career of an employee.

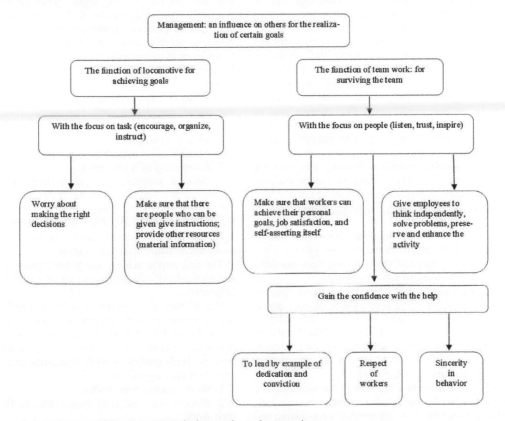

Figure 2. Functions of staff's management: the locomotive and cooperation.

4 STAGES OF HUMAN RESOURCE`S MANAGEMENT

Human resource's management has a great importance for all organizations: large and small, commercial and non-commercial, industrial and household sector. Without good experts any company cannot achieve its goals and survive in the market economy. Therefore, human resource's management is one of the most important aspects of the theory and practice in the management.

In most modern European companies, human resource's management involves the following steps:

1. Resource planning: development of plans, meeting the future needs for human resources;

2. Recruitment of staff: creation a rate of potencies candidates to fill all positions;

3. Selection: assessing of candidates for selection of job and the best of them;

4. Quantification of wage, and benefits to attract the most skilled experts;

5. Vocational guidance and adaptation: entering of hied workers into the sub-organization of employees for formatting understanding, what the organization expects of them and what kind of job will be most appreciated;

1. Education: development of training programs for young workers to master the most effective skills for job;

2. Evaluation of work activity: creating methodologies for assessing activity and presentation of them to every employee;

3. Increase, decrease, transfer, release: creation of methods for displacement workers from the one post

to another, depending on their level of professional experience, and stop retainer agreement;

4. Manager's trainings: manage the process, advancement to managements' position, creation of programs which aim to develop skills and improve labor management personnel.

A firm has a constant need to ensure high productivity of all workers. However, many organizations care about overall quality of human resources. One of the meathods for achieving it, is recruitment and selection the best and well educated staff. However, management should organize systematic training for young workers to help to find out their potential.

Training – is teaching workers of skills, which allow to increase productivity. The ultimate goal of education is to provide own organization with sufficient number of experts who can effectively solve problems.

Unfortunately, sometimes training programs are developed and applied without sufficient analysis, which occurs on performance negatively.

Study useful and necessary in following cases:

1. When a person enters the organization.

2. When the employee promoted to a new post or when he or she was entrusted to a new job.

3. When is detected that a person does not have certain skills to do his job effectively.

Management must determine how many employees are engaged in the performance of each operation. In addition, management can rate the quality of work of their employees. Planning of staff's amount is an important stage, to achieve accurate and promising targets of staff. And after it the program for achieving goals are determined.

The most important problem is the unreasonable expectations of people during appointment to a post. The principal methods are interviewing, testing and organization centers of evaluation where modeling techniques are used, which are very effective but quite expensive.

Interviews are used quite widely, but there are many problems: results can be not accurate. Quality of work life depends on the position and other factors. Good wages and additional benefits can help to involve good experts. Assessment of labor of an employee makes administration. Study shows, that the manager should create an atmosphere of mutual trust, mutual responsibility. The focus concentrates on the results of operations and mutually purposes.

For effective training people should be interested in it. To do it, favorable atmosphere and terms of organization of various courses, seminars, etc. are created. To estimate the most effective performance, managers need to collect information, how does the employee do his duties. If he is not doing well, he must have a chance to do it once more. Assessment

of performance helps to detect the best experts and get rid from worst.

Mainly, assessment of performance of employees has the following objectives: administrative, informational, motivational, which are closely connected. Information which describes the administrative decision for granting higher positions should positively motivate people. Official program of management process recruitment, promotion allows people to perceive their work in the organization as "a series of displacement" from one post to another. It must facilitate company's development and individual.

Planning of human resources - is the usage of procedures of completing staff. For convenience, we can assume that the planning process includes the following three stages:

1. Assessment of available resources.

2. Assessment of future needs.

3. Develop programs to gratification of these needs.

Management should decide about staff's timetable. If it is necessary, consult with recruitment agencies; add an announcement in the Internet, newspapers, radio, television and etc.

The first step to achieve the greatest productivity, is a professional orientation and social adaptation in collective.

One of the major developments in management of human resources is a creation of programs to improve the quality of working standards. This problem is in the center of attention of every developed country. High quality of work standards characterizes the following indicators:

1. Work should be interesting.

2. Workers must get reall wages and benefits for their work.

3. Work environment should be clean, low vibration, noise and sufficient lighting.

4. Supervision of management should be minimal, but must be made whenever it is important to do it.

5. Employees should participate in making decisions which are connected with their work.

6. Guarantees for the work and friendly relationships with colleagues should be provided.

7. They should be provided with household and medical supplies.

We think, that the concept of human resource management goes out from these terms, that moral climate at an enterprise, as satisfaction of employee, are the results of creative solutions to problems. The manager of the company must look at the whole system of management in general. With this aim, he makes solutions about main problems. He must promt employee to making right decision and show prospects, to deploy the main area, rationally distribute management's work, to coordinate their actions, be expert in organization and control of manufacturing, technology and economy, know the legal basis for management, bases of pshychology and pedagogy, to assess critically achievements and be able to persuade, be strict, value the time, build relationships with supreme organizations.

Successful solution of socio-economic problems depends on activity and educational influence on each employee. In the workplace not only team work is organized, but personality is formed, consisting labor relations and mutual cooperation, the main task is solved – to achieve goals with the help of creative attitude. The administration, which is based on order makes the meaningless.

Human Resource's Management as a special field of activity that peculiar to the economy of growing standards of production, becomes the main and necessary. It is supported by the following reasons:

1. Establishment of stock companies, where management is proceeded to a special group of people – managers, who are not owners of the property, but they can use it effectively in the interests of shareholders.

2. Complications of technological production and the rising cost of equipment are needed such business organization, to be guaranteed responsibility for the results of every.

3. The intensification of competition requires the participation of every interested person.

Scientists have developed the theory and practice of effective human resources management. Dale Carnegi gives such recommendations which generalize the experience of success in management of people. He thinks to win the sympathy of people, it is necessary:

1. To reveal sincerely interest to them.

2. To smile, to keep them positively impressed.

3. To contact them by name. For a person a sound of its name is the sweetest and the most attractive.

4. Be attentive listener and encourage others to talk about themselves, to be pleasant interlocutor.

5. Rouse person's interest in speaking on topics, which he is interested in.

6. Give people a sense of their importance and do it sincerely.

In our opinion, advices of D. Carnegi are very useful for everyone, who is engage in human resources management.

5 EFFECT OF HUMAN RESOURCES TO ENHANCE ENTERPRISE`S COMPETITIVENESS

The basis for the operation of any enterprise is its supply of resources. Human resources rather than other factors of production determine the strategic success of an enterprise, are the basis of its competitiveness.

Competitiveness – one of the most important factors that ensure sustainable development of each enterprise. In modern conditions, competitiveness is determined by many factors and terms. One of the factors which determine the achievement of competitive advantage now, is the availability of sufficient human resources and effective management.

Starting the 80th of 20th century, model of staff's management is changing; transformation process is going in the direction of staff's management to human resources management.

One of the most important characteristics of human resources that affect the competitiveness of enterprises is the educational level of staff. Improving the educational level of staff leads to an increase in labor productivity, improving product quality, increase efficiency indexes of business, stabilize its financial situation. All this leads to strengthening the competitive position of enterprises.

Improving the educational level of staff leads to increase of labor productivity, improving product quality, increase of indexes in business, stabilization of its financial situation. All these factors lead to increase of competitive position of enterprises.

Doubtless is the fact that education – is a component of human capital, which is practically inexhaustible factor of economic rising. For enterprises human capital is a mix of different qualifications, physical and professional skills of all employees.

In XXI century, at the time of appearance of unknown technologies, reducing of good's life cycle, specification of human capital is manifested in increasing demands for its quality, acceleration of studying new skills by workers, knowledge and skills. So if you consider the social-historical development of educational business in industrialized countries, you can find the fact of their entering into a phase which aims at training areas. To ensure competitiveness of enterprises necessary investments in human capital should take place, and it is a prerogative not only of large but also a need of small and medium enterprises.

We want to note, that the theory of human capital comes to the fact, that a person is in the center of the concept of investment policy, is regarded to be the highest value for the enterprise. Therefore, the management should be focused on the development of various skills of employees to maximize their usage during work, and at the same time an employee would make everything to development of organization where he works. Investments in education are necessary for both, employees and the employer, but for the employer they are more important. Profitability of investment in human resources is always optimal. As a result of professional development, workers get higher wages, but their efficiency from their activities is more useful. It means, that investments in education, training and mobility significantly raise the cost and price of employees labor and produce significant revenue, and ensure the competitiveness of workers in the labor market.

6 CONCLUSIONS

In the article, according to a study of theoretical concepts of human resource, the definition and role in managing staff and ensuring competitive advantage are presented.

Among the factors, that form the competitiveness of enterprises, one of the leading places are taken by human resources. Qualitative characteristics of human resources that affect the competitiveness of enterprises is educational level of the staff. Characteristics of the educational level of staff are due to the requirements of employers. They include personal qualities, professional skills and ethnical, social and psychological qualities which determine the general cultural level of the employee.

Knowledge and education in the market economy are changed into a commodity, which determines the standard of living standard, competitiveness and profits of the enterprise, opportunities for economic growth. The importance of human resources in mean of knowledge, skills, abilities contribute to the formation of human capital.

The analysis of the effectiveness of investments in education shows that the effectiveness of such investments is growing as terms of study growing, and return of such investment far exceeds return on investment in physical capital.

The quality of human resources depends on enterprising performance, and its competitiveness.

REFERENCES

Prokoshenko, N. & Nortak, K. *Performance management and quality*. 2001. Modular program. Moscow: Part 1, 2.
Kachalina, L.N. *Competitive management*. 2006. Moscow: Publishing House Exmo: 464.
Didkovska, L.G. & Gordienko P.L. 2007. *Management*. Alerta, CST: 516.
Porter, M. 2001. *Competition*. Trans. from english. Moscow: Williams.

The usage of rubber-air reinforced lining (RARL) during maintenance of an in-seam working in the Western Donbas mines

V. Medyanik, V. Pochepov, V. Fomichov & L. Fomichova
National Mining University, Dnipropetrovs'k, Ukraine

ABSTRACT: One of the most important constituent of coal first cost during the process of its production in the Western Donbas mines is the maintenance of an in-seam working inputs. A variety of mining and geological factors which have influence on the output stability of mines in this region cannot be applied the same type of lining with equal success in all cases. Rubber-air reinforced lining combines useful technological characteristics of different types of supports. The application of this lining can significantly reduce the impact on the stability of working. The application technology of this lining provides a high standard of production and does not require the miners' specific skills and knowledge.

1 INTRODUCTION

Coal mining method requires carrying out a large number of mine openings. In modern Ukrainian shafts the total length of supported mine openings in one mine can reach hundred or more kilometers. Relatively small part in this volume constitutes permanent workings. That is why the maintenance of temporary workings becomes a major task, the solution of which greatly effects on the first cost of the produced coal (Usachenko, Kirichenko & Shmigol' 1992).

While maintaining the in-seam working in the Western Donbas mines it is mainly used a frame and frame-anchor bolts, based on different technical solutions. However, in some cases, the application of classical lining configurations in permanent workings does not give the necessary effect, which leads to a significant increase of expenses for the drifting works, in connection with rising metal prices and maintaining of already passed courses (Mel'nikov 1988 & Farmer 1990).

The effectiveness of the lining usage based on the classical technological solutions can be divided into several groups: first one is the deviation of erected lining parameters from the parameters of lining mentioned in the technical documentation, the second – are the technical features of the separate lining elements; the third – mining and geological features of rock mass; the fourth – local discontinuity in the structure of rock mass.

Lining of the permanent workings can have a high safety margin and in normal conditions does not require dismantling. Moreover, it may at any time grow by all available means. Characteristics of the lining in temporary working, first of all, must correspond to the rule of technological balance. According to this rule accepted lining safety factor provides an available minimum of industrial expenses for carrying out and maintenance of working (Reference book... 1976).

In the mines of Western Donbas in the construction of frame and frame-anchor bolts it is extremely difficult to achieve strict compliance with design documents. Because of difficult mining and geological conditions lining erection should be done rapidly to avoid large deformations of rocks massif which surrounds working. Nevertheless the used types of linings prevent from widespread using of automation during the assembly and disassembly.

The presence of corrosive medium, local inrushes and swelling of ground often results in damage or destruction of individual lining elements, consequently lining completely or partially loses its structural strength, and therefore can no longer fulfill its functions (Safety rule...2010 & Operational regulation...2006).

From above-mentioned, we can make a conclusion about the need for development and introduction of new types of lining which will be capable to a significant improvement of the maintaining temporary workings quality. This could be an RARL. Its structural features and technology of application allow you to strike out the first group of reasons for reducing of lining effectiveness and provide an absolutely new approach for solving problems concerning struggle with the local discontinuity of rock mass (the fourth group).

2 DETAILS OF RARL SUPPORT

Physically RARL represents a rubber block reinforced with metal lattice (Figure 1) in what follows named as a module. Logically, the module consists of two parts. The first part provides the overall stiffness of support and exchange efforts between neighbor modules. The second ensures flexibility and stable contact of the lining with the surface of working. Physically the first part of the lining represents metal lattice and the rubber stiffened slab and the second part consists of a cellular structured balloon made of noncombustible rubber of different stiffness.

Figure 1. The general form of a separate module RARL.

The main load elements of the RARL module consist of metal lattice. Variable characteristics of the lattice are the diameter of metal rod and the width of a square cell (a prototype car ramp). Lattice with the help of rubber vulcanization is placed inside a square block. On the perimeter of the block is executed profile which is used as an element of the lock during the combination of individual lining elements in the common load-carrying structure (Figure 2). By varying the characteristics of this block are the side length, thickness and profile of the side sections.

To the rear of the rubber block is abutted air balloon. The balloon is implemented as a set of square cells. The lower balloon open part abuts to the stiff rubber plate. The upper part during the lining assembly have to abut to the surface of the rock mass, that's why it consists of a dense stiff rubber, which can be enhanced by a small metal cord. The middle part of the balloon has high elasticity, in the course of lining module running, experiencing the greatest hyperplastic deformations.

Directly into a stiff rubber block placed air valve, which is used to pump air into the balloon of the RARL module. This valve is equipped by a visual indicator of the three-level air pressure. Each level of the indicator determines one of the states of the air lining balloon: the low level of internal pressure, normal pressure (which corresponds to the specifications); surplus pressure. Under certain running conditions this valve provides a release of surplus air mixture from the lining balloon.

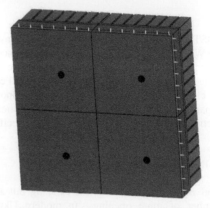

Figure 2. An example of combining 4-way ARRL modules.

Using this air valve at any given time, the worker can change the air pressure inside the balloon.

The choice of final configuration of the RARL module is based on a system of mathematical expressions, the solution of which allows to get a set of allowed values for all the changes of the lining characteristics. For specific mining and geological conditions erected fragment of lining may look like in Figure 2. In this case, the inside section of working has a form of rectangle. Surrounding rocks of the working heavily flooded and have developed jointing. The disposition of working in the rock mass is such that lining should take considerable static loads.

The structure of the lining module balloon consists of 49 cells the size of which corresponds to the average step of jointing rock mass which are abutted to the contour of working. The height of cells equals to five sides of the square that forms the cross-section of the cell. This choice was determined by the two features: a large watering rock mass and the value of the static loads which influence on the lining. In such a way the following tasks are solving: maintaining of the lining contact continuity and rocks; providing of water drain without release of water to working; smoothing of the local excessive pressure on the entire cross section of working.

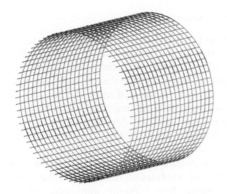

Figure 3. Interior power cage of the lining assembled from RARL modules in the circular cross-section of working.

The net section formation of the working by making the RARL modules is carried out at the expense of the power framework elements. The form of a power cage for working out a circular cross section is shown on the Figure 3. As it can be seen after combining several RARL modules generates close to ideal carrier gauze form. In this case, due to the redistribution of efforts in air balloons of the lining this gauze is almost uniformly loaded. This increases its total constructive stability. Hard rubber blocks increase its resistance to local cross motions of the lining force cage elements.

In such a way, individual elements of the RARL module provide specific characteristics of the lining: the force cage – is a resistance to rock pressure, the maintenance of the working section; balloons – are the transmission and equalization of stresses, the overall lining compliance. In addition to the basic characteristics inherent only in the classical types of temporary linings, this one can be used to solve such problems as struggle with water inflow, etc.

3 APPLICATION TECHNOLOGIES OF RARL LININGS

Rubber-air reinforced lining can be used to maintain different arbitrary cross-section workings which are operated in complex geological conditions under which the use of the classical frame and frame-roof bolting is not practical. Due to its structural features RARL can be mounted both hand and mechanized methods.

During the erection of lining on the basis of RARL modules a high level of contact between separate elements of lining and between the elements of lining and rock massif is always achieved. This is reached by the help of lining mounting technology. For example, consider the sequence of lin-

ing erection on the basis of six RARL modules in the working of cross-circular section (Figure 4).

In the first stage RARL modules are mounted in the following way: the first module is placed on the working ground in such a way that the corbelled ends of the metal lattice and side profile rubber plate must fit into the corresponding cutouts of already mounted RARL modules. In this case, the balloon cells of the installed module should bear against the balloons of already erected lining. Then, in a similar way, but with a glance of the already installed module, the left and right bottom lining modules are attached.

After that, on the one of the open sides of the RARL modules is installed striker plate, which subsequently will be used for closure of the force cage in cross section of working. Then the upper RARL modules are set and striker plate is fixed by special bayonet lock. The first phase of lining assembly is over.

After assembling of the force cage balloons of installed RARL modules are inflated with air mixture. Pumping of balloons on the perimeter has the several stages. At each stage, in balloons of opposite RARL modules successively a certain pressure value which is specified by the technical conditions is pumped. This continues until the desired requirement pressure is receipted in all balloons of installed modules. Then you can proceed to install the next group of RARL modules

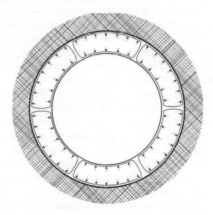

Figure 4. Mechanical model of interaction of RARL modules and rock mass adjacent to the working contour.

During drivage under the real conditions it is difficult to sustain given by designers cross-section. Building of the working lining with application of the RARL modules is not a problem no more. As shown in Figure 5 at the expense to the cells balloons RARL modules elasticity it becomes possible to provide high-quality contact between the force lining cage and working contour which has no regular form. In this case, the structural characteristics of RARL modules may be the identical throughout the length and perimeter of the working.

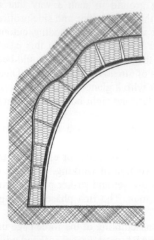

Figure 5. An example of interaction modules RARS with contour working under real conditions.

The possibility of mine water export which comes into the working from an arch realized for improvement of maintenance conditions of RARL modules is presented in Figure 5. As follows from the figure between the balloons of RARL modules intentionally is made an air gap, for moving by the flow of water which percolates from the arch and sides of the working. Continuous and dense locks of the stiff part of the RARL module do not allow this water to fall further in the working. A snug engagement inside the cells of balloons in the surface contour of working reduces the volume of water inflow.

In some cases, loss of stability of working contour, when carrying out emergency works, or for amplification strength characteristics if support on the individuals emergency sections modules RARS can be filled out, under great pressure, quick-setting, expanding at solidification solution. In this case, such modules cannot be dismantled, but they can ensure stability of the working is comparable with indicators of support in the permanent workings.

4 ADVANTAGES AND DISADVANTAGES RARS

To estimate the need for application specific type of lining is possible only after a comparative analysis. This analysis is based on a comparison of advantages and disadvantages of selected types of lining with a glance of the characteristics of environment in which these supports should operate.

4.1 Advantages

High speed of construction and it has low requirements for the adhearance to exact shape of the working contour it is the main advantage of this type of lining. In fact, during the lining installation miners can build it directly behind the plane of working without giving time for mountain rocks to move into working. And due to the compliance of balloons of RARL modules observance high precision of contour working is not required.

The culture of production during the application of RARL modules increases due to the system of its assembly. It is virtually impossible to build this support over 1-2 meters without precise tight fit of separate modules. The use of the bayonet lock allows you to combine RARL modules quickly and in compliance with high quality contact.

Due to design features of lining the distribution of the external load is not only the cross section of working, but also along it. It can reduce in a number of violations of the rock mass integrity on the working contour the concentration of efforts on a single element by support the distribution of internal stresses around RARL force cage.

So long as the competent operation of RARL modules not get damaged affecting their technological characteristics, they are characterized by a high degree of reuse. Easy assembly and disassembly allows you to save the full functionality of the RARL modules after several cycles of usage. That is, one module can be installed multiple times in different temporary workings the geological conditions of which are identical or similar. But even after the destruction of the RARL module its individual components can be used to produce new lining elements.

Since the operating conditions of lining ensure its ongoing continuity and the lining elements don't have a high fragility, it provides an effective protection against small fall or collapse. That pretty much can reduce injuries caused by these nonperiodical events, the prediction of which is a difficult analytical task.

Finally, another advantage is the possibility of RARL lining with its modules, is holding and disposal of mine waters. In cases of high water content due to technical features of the RARL balloons modules becomes possible to manage the process of filtering and storage of mine water, while maintaining the basic technological characteristics of the lining RARL all along supported the working.

4.2 Disadvantages

The main drawback of RARL is the relatively high cost of modules production. It is connected with the usage of rubber, the chemical composition of which should provide resistance lining material to high-temperature fire. In addition, individual elements of the lining module have different characteristics of strength and elasticity, which in turn also increases the cost of a separate module.

When you manually install the RARL module weight is a limiting factor in speed of erection lining working. When installing the unit in manual mode, it may be damaged, which can lead to the need for its extraordinary change. Besides, in some cases for certain sizes of cross sections workings, with manual installation of RARL modules, it becomes necessary to use additional mounting structures, which must be located directly in the area of lining collecting, as well as individual modules for the tabular in-seam working.

Perform installation of RARL modules requires high-pressure compressor near the zone of works by lining installing. For supplying the best operating conditions of lining the balloon pressure of RARL modules should at most correspond to the values specified in the specifications for the maintenance of a particular working. It requires certain skills and knowledge.

During the modules operation there is a need for constant monitoring and carrying out control measurements of air pressure in balloons of individual modules. Changes in the state of the surrounding rock mass of working can cause significant deformation and redistribution of stresses along the working contour. The pressure in RARL balloons individual modules can be changed either in the direction of increasing or decreasing the side. Such changes lead to a reduction in the quality of contact between contour working and RARL force cage. Therefore, in such cases it is necessary to recover the given technical conditions the pressure in the balloons of lining.

5 CONCLUSIONS

Complicated mining and geological conditions specific to the mines of Western Donbas make a number of problems in stabilization of temporary workings. Usually these problems are solved by a quantitative increase of the standard lining elements to a one running meter. The part of an incipient problems are connected with low-speed and quality of erected lining, and some simply cannot be solved by using of classical lining schemes of an in-seam workings. The elimination of most of these problems requires a considerable amount of engineering time and money.

Rubber-air reinforced lining allows air to avoid most of the problems which appear during the maintenance of in-seam workings. Its design philosophy makes it possible to strengthen the production culture, reduce the maintenance cost of one running meter of temporary working and in some cases can increase the period of working operation. Wide application of this lining for maintenance of the temporary workings in the mines of Western Donbas will provide an opportunity to overcome main drawback – high cost.

REFERENCES

Usachenko, B.M., Kirichenko, V.Y. & Shmigol', A.V. 1992. *Development openings protection of deep mines in the Western Donbass*: Obzor/ CNIEIugol'. Moscow.
Mel'nikov, N.I. 1988. *Drifting and timbering of mine openings*. Moscow: Nedra: 343.
Farmer, Y. 1990. *Collieries' openings*: Transl. From engl. /Transl. E.A. Mel'nikov. Moscow: Nedra: 269.
Reference book about timbering of mine openings. 1976. Edited by. M.N. Geleskula. Izd.2, revised and supplemented edition. Moscow: Nedra: 508.
Safety rules for collieries. 2010. NPAOP 10.0-1.01-10. Kyiv: DP "RG protection of labor": 430.
Operational regulations of collieries. 2006. Standard of Ukrainian Minvugleprom. Kyiv-Donetsk: DonNDVI: 360.

The system approach to increase an effective protection of mine workings

O. Vladyko
National Mining University, Dnipropetrovs'k, Ukraine

ABSTRACT: Hydro-geological conditions influencing stability of underground mine workings are observed. The mine working analysis as complete system is made. The cognitive map is developed and imitating simulation of geotechnological parameters of a mine working taking into account external factors is performed.

1 INTRODUCTION

Mining of deposits by underground way in comparison with other ways have a row of the features essentially complicating technology of mining. It is connected with complex mining and geological conditions of burial, tectonic disturbance of rocks and complex hydro-geological conditions. Additional problems are created by the vast space of underground mine workings and impossibility of complete unwatering of a rock mass. It demands the system approach to the organization of optimum performance of a branching underground mine workings complex. The system analysis at mine workings development was used by G.V. Babijuk, V.V. Pershin, E.I. Rogov. System research of interacting of underground mine workings with a support carried out V.I. Bondarenko, I.A. Kovalevsky, A.N. Shashenko and etc. They researched the interaction between rock mass and lining, joints' attitude and inflows into mine workings and pointed the random character of interaction between rock mass and mine working (Shashenko 1988). Proceeding from the casual character of interacting the system "mine working-rock mass" exact forecasting of affecting a rock pressure and water inflows in developments is strongly complicated. As a rule, interactings in system "mine working-rock mass" have nonlinear regularity and these dependences for each mine working are individual. The insufficient account of these factors at mining operations often leads to destruction of mine workings, inrushes of rocks and water inrushes.

2 SYSTEM ANALYSIS AT DEVELPMENT OF MINE WORKINGS

The offered approach is based on complete vision of installations, the phenomena or processes, and also revealing of their structure and functions. The system structure is fathomed as resistant to orderliness in space and in a time of elements and connections between them defining functional layout of system and its' affecting with environment. To structural elements we refer the space location and mine working caliber, disposing of the equipment and an order of mining operations performance.

The application of system approach opens the greatest prospects by consideration the problems connected with the mine workings protection. Its basis is determination the system regularities which define methods and stages of analysis and design protection methods for mine workings.

In our case system (mine working) N consists from edge massif n_1, a rock outcrop n_2 and lining n_3. In system and between subsystems the important role is carried out by connections which connect separate elements among themselves. Connection we name important for ultimate goals of consideration an exchange between elements, for example, substance, energy, the information. As the elementary act of connection for a mine working is the affecting of element n_1 onto element n_2 through x_{12}, and an element n_2 onto element n_1 – through x_{21}.

Observing a mine working as the system can be recorded (Gubanov 1988):

$$\Sigma : \{(N),(x),F\}, \qquad (1)$$

where Σ – system; $\{N\}$ – assemblage of elements in it; $[x]$ – assemblage of connections; F – function new property of system.

The purpose of mine workings development and functioning is transportation and mineral exploration, ventilation etc. Efficiency of mine working functioning is defined by expenses for maintenance one in working order for all period of life cycle. External environment is the surroundings of system (mine working).

The purpose of a construction and functioning of developments is transportation and mineral investigation, ventilation etc. Efficiency of functioning of development is defined by expenses for maintenance in working order for all period of life cycle of development. Environment acts as an environment of system (development). Thus external factors act as the active reasons which render on it influences. Environment renders influences by fields of pressure and a filtration which depend from and factors of concentration of pressure and water inflows.

External and internal environments of system is in interdependence and mutual conditionality (Figure 1). If through external environment the field of pressure, and in internal – protection (development fastening) and partial deformation of development influences. From outer side the filtrational field, and with internal protection against waters (nabyz-concrete) and a partial drainage is enclosed. Change of the internal environment of development lead to environment changes (fields of pressure and a filtration).

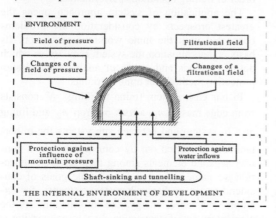

Figure. 1. Model of system "development – environment".

Subsystem (development) cooperating with environment is in dynamic balance. The filtrational field and a field of pressure constantly cooperate with support developments and protection against water inflows. Infringements of dynamic balance leads to catastrophic consequences (to inrushes, the big water inflows in a short space of time etc.).

The construction of mountain development and its maintenance in an efficient condition throughout all life cycle depends from shaft-sinking and tunneling works (STW). The basic criteria of efficiency STW is speed of carrying out of development and labor input of carried out works, and also expenses for constructions of development and its mainte-

nance in an efficient condition. Efficiency STW depends on a kind of the applied equipment and the organization of works.

3 REALIZATION OF THE SYSTEM APPROACH WITH APPLICATION OF IMITATING MODELLING

The block diagram of offered imitating model is presented on Figure 2. In model the statistical method of Monte-Carlo with which help is used is reproduced industrial process described on likelihood laws.

In structure of imitating model it is conditionally allocated five blocks:

1) geological and hydro-geological conditions of drivage; 2) technological operations; 3) the equipment, 4) idle times; 5) decision-making.

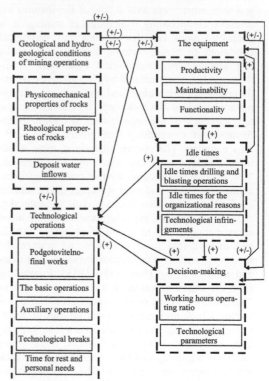

Figure 2. The structural diagram of relationships of cause and effect of model.

As a calculation example carrying out of preparatory developments on the South Belozersky deposit of Joint-Stock Company "Zaporozhye iron-ore industrial complex" has been considered. Influence of idle times at carrying out of developments on operating ratio of working hours and technological parameters was investigated.

For the block of geological and hydro-geological conditions were used: slates of average fissuring with a fortress 7-9 and quartzite's of average stability and average fissuring a fortress 14-15. For the block of technological operations the given organizations of works were used at carrying out of developments. For the equipment block installation boring mine "Axera", load-haul-dumper PDM PNE-2500 (1700) was considered. As in model investigated influence of idle times on development carrying out the data on the enterprise and the accepted

conditional data was used. For reception of steady result of statistical model 1000 change of parameters has been spent.

4 RESULTS OF MODELLING

As a result of the spent modeling it has been established at idle times for the various reasons within 15-19% average coefficient working hours uses has made at a dispersion Figure 3 and 4.

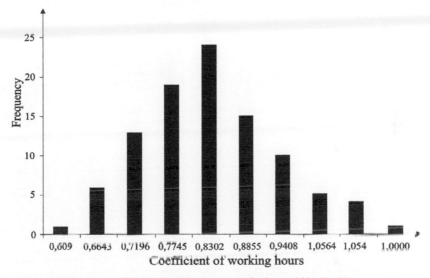

Figure 3. Frequency of distribution coefficient of working hours: at idle times within 15-19%.

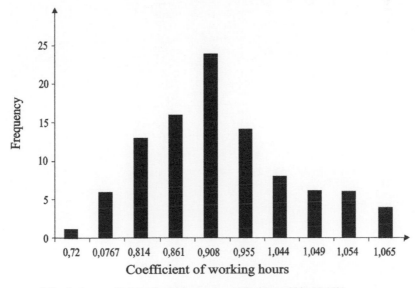

Figure 4. Frequency of distribution coefficient of working hours: at idle times within 10-15%.

Reduction of idle times to limits of 10-15% increases coefficient uses to at a dispersion Figure 4. Speed of carrying out of development has increased on 10%. Reliability level – 95%.

CONCLUSIONS

1. The model connecting hydro-geological conditions and technological parameters of carrying out of mountain developments is developed.

2. On an example of test calculations for conditions South Belozersky Joint-Stock Company deposit "Zaporozhye iron-ore industrial complex" dependences of technological parameters on external factors are established considering a casual component.

REFERENCES

Shashenko, A.N. 1988. *Stability of underground workings in the nonuniform-native array:* Thes. ... doctor. tekhn. science. Dnipropetrovs'k: 495.

Gubanov, V.A., Zakharov, V.V. & Kovalenko, A.N. 1988. *Introduction to systems analysis.* Because of the Leningrad University: 232.

Concept of valorisation in geotechnological, geological and social processes of mining enterprises development

N. Sobko & A. Melnikov

National Mining University, Dnipropetrovs'k, Ukraine

ABSTRACT: The main task of the mechanism of interaction between state and business is formulated. The concept of valorization in three directions is examined. The main directions of the mechanism of mud valorisation process and the mechanism of social valorisation of the mining companies is proposed.

1 INTRODUCTION

In today's challenging conditions of mud development of domestic mining industry because of the need of technological upgrading of enterprises an important place should be given to the creation of effective mechanism of innovative and investment development processes of mud mining enterprises. In order to achieve the best efficiency of this mechanism, the mechanism of interaction between state mining companies, which provides for the co-ordination of state interests and the business of technology and technological development of enterprises, research and evaluation mud factors and risks when making investment decisions in order to attract investments in the mining industry and receive as a result of additional economic benefits of the state and business must be laid fundamentally (Pilov 2009).

2 RESEARCH ANALYSIS

Today, focusing on global and national trends in the mining industry under the provisions of the Concept of development of the coal industry by 2030, technically and technologically mud processes of mining companies must be a fundamental reform of the mining industry in Ukraine, which will increase production, improve product quality, reduce costs on production and negative environmental impact of mining activities of enterprises, enhance social security of workers and enterprises in the industry to increase competitiveness of domestic mining companies in the world market.

The main directions of improving the mechanism of interaction between state and business should be:

1) the improving of the regulatory framework;
2) the development strategy for the development of mining industry;
3) the state support of mining companies;
4) the improving the mechanisms for attracting investment in mining;
5) the search for new sources of investment in the development of enterprises;
6) the developing of effective mechanism mud evaluation of mineral deposits;
7) the formation mechanism of investment decisions and more.

All these areas should be combined by the single concept – the concept of valorisation – the concept of interaction between state and business (mining companies) for more economic benefits and minimize risk.

In scientific papers in the analysis of developments alluvial deposits of natural and industrial valorisation mud means the measures of complex geological, technological, technical, social and environmental factors to establish valorisation index, which allows all options to obtain an objective valuation of alluvial deposits in the current period (Bolohovitinova 2003). However, considering the current trends of Ukraine's economy, industry and mining in particular, trends in mud processes of mining companies listed concept should be expanded and improved in order to develop and use effective strategies like the concept of valorisation of mining enterprises.

Valorisation – an economic category, which means increasing value of resources of society by optimizing the benefits of state and business when they are used in a rational distribution and promotion of social production. The object of valorisation is the processes and mechanisms of increasing of resources value of society. The subject of valorisation is conceptual and methodological approaches to optimizing the benefits of state and business community with resources in the economy and encourage risk of social production. The aim of valorisation lies in optimizing benefits and state entities on the basis of conceptual approaches to establish a rational use of resources of society and stimulating impact of investment in the development of social production.

Because the mining enterprise can be regarded as a complete range of Household, within which the interaction between human development, engineering and technology, geological processes and the nature take place, the concept of valorisation should be considered in three areas (mud, social and environmental), which leads to the concept of valorisation division into three separate parts:

1) mud valorisation (as the concept of adding value to geological, technical and technological resources of mining companies through the optimal interaction between state and business);

2) social valorisation (as the concept of adding value to the labor potential of mining companies, security and the guarantee of high social standards for employees of mining companies through the optimal interaction between state and business);

3) environmental valorisation (as the concept of adding value to natural resources, mining companies, the stabilization of the ecological situation in the country through the optimal interaction between state and business).

3 GEOTECHNOLOGICAL VALORISATION

Geotechnological valorisation processes of mining companies - increasing value of resources and processes of mud mining enterprises by optimizing the interests and economic benefits of the state and business during the investigation and evaluation of geotechnological factors and risk factors in decision making in the field of subsoil management and stimulating effect on investment in mining enterprises.(Bolybah 2006).

The concept of geotechnological valorisation aimed at reconciling the interests of state and business process and technical development of mining companies and provides geotechnological evaluation factors and risk factors in making economic decisions aimed at attracting investment to mining companies and receive as a result of additional economic benefits of the state and business.

Economic Content of geotechnological valorisation of mining companies:

1. Getting more economic benefits of the state and business by optimizing common interests.

2. Reducing the risk of investing in the develop-

ment of mud mining companies through the allocation of risk between the government and mining companies.

3. Involving additional investment funds to upgrade and modernize the production base of mining companies, development of innovative geotechnological processes of mining enterprises.

4. Improving the investment in the development processes of mud mining enterprises.

5. Development of the economy in general, and mining industries in particular.

The tax mechanism that provides the implementation processes of valorisation mud mining enterprises of the state provides mechanisms by means of tax measures aimed at improving conditions of geotechnological mining companies provided directing funds mobilized by mining companies to develop innovative businesses in order to obtain the joint economic benefits of mining companies and state and minimize the risk of mud modernization of enterprises. Based on the analysis of tax incentive tools geotechnological of mining companies and the conditions that are created in the industry it is encouraged to implement the following tax incentives geotechnological of mining enterprises: the providing of investment tax credit, providing tax rebate from income tax.

So the key areas of mechanism of geotechnological valorisation processes of the mining companies are:

1. State financial support.

2. Organizational and administrative interaction between state and mining enterprises.

3. Innovation and investment mechanism of geotechnological valorisation processes of mining enterprises.

4. The mechanism of effective evaluation of mineral deposits.

5. Organization management of mining companies.

The current mechanism for state financial assistance for development and mining enterprises needs improvement because of the inefficiency of application and the need to adapt it to the WTO (World Trade Organization) provisions.

4 GEOECOLOGICAL VALORISATION

Ukraine needs a new environmental policy, a significant role in which must be given to the greening of social production based on modern energy-saving and environmentally friendly technologies. For their implementation requires the development of new scientific and methodological approaches to ecological and economic evaluation of technology solutions, implementation of which will contribute to the greening of the manufacturing process in general. This can be achieved by the interaction of factors values in resource allocation of economic risk, to stimulate the production and sales entities (Mining Law of Ukraine 1999).

Geoecological situation in Ukraine as a result of human impact on the environment, especially in areas with excessive concentration of industry (Dnieper, Donbas). Therefore, the need for effective legal security is a critical issue in the regulation of geo-ecological processes in social and industrial complexes.

In recent years a significant growth of interest in environmental issues in connection with the concept of sustainability and better management in harmony with nature is noted. The growing interest connected with it comes from new legislation that comes from national and international sources, such as the European Commission, an organization that seeks to influence the relationship between development projects and the environment.

Impact of mining and metallurgical production on the environment to detect the majority of technological solutions ensure compliance of environmental requirements at the level of standards that were 10-15 years ago. However, the ecological situation in Ukraine, and the transition to international quality standards ISO 14000 require continuous review of environmental norms and standards. Improving the environmental and economic evaluations allows for selection of technology solutions to their planning stages. The basis of this selection based on the principle of ecological safety of production as of today and the future.

Lawmaking of the legislative body of government (parliament – Verkhovna Rada of Ukraine) and executive bodies should be based on results of analysis of the impact of any production environment, so that there was a systematic review of environmental norms and standards in various industries, especially in mining steel industry. In turn, the subjects of mining relations, legal and natural persons of Ukraine, foreign legal and natural persons, stateless persons engaged in geological exploration of mineral deposits, design, construction (reconstruction), operation, liquidation of accidents and the liquidation or conservation enterprises mining and mineral processing, and transmitting the mining, must comply the mining legislation (Dvoryashuna 2005).

Due to the necessity of a new quality of social development, the further development of processes of globalization in recent decades appear to be stable tendency to increase the role of environmental factors in ensuring competitiveness and sustainable economic growth. Environmental factors take value system, they increasingly define the strategic future of national economies, their subjects. It causes companies to the need for more fully integrating their activities in a modern perspective and the role of environmental factors. Emerging environmental threats to competitiveness, but at the same time there are new conditions, the chances for improving the competitive strategies of enterprises.

The company must complete enough to realize the possible directions, strength and nature of threats to its market position. The entities of threat agents that initiate and make changes in the competitive environment may include: consumers, businesses on the market, competitors ("actual competitors"); companies that can enter or are already on the market ("potential competitors"), producers of substitute products ("related competitors"); suppliers; state.

Geoecological valorisation concept is reduced to increase the value of raw resource of nature by optimizing the interaction of natural processes and ecological and economic factors in stimulating the development social and industrial complexes. Natural processes is the reaction of the environment on society, aimed at using the elements of nature in shaping their welfare. Ecological and economic factors are the force of nature and economic relations in society, seeking to maintain a balance in the development of human civilization. Incentive is the creating of favorable conditions for the functioning of business entities (enterprises).

Geoecological valorisation provides a scientific approach that aims to:
- developing a methodology for determining the degree of environmental risk to the environment caused by industrial activity of environmentally dangerous object;
- introduction of modern scientific achievements, and energy saving technologies, low-waste, waste-free and environmentally friendly manufacturing processes;
- improving environmental and economic mechanisms to encourage environmentally sound technologies and environmental systems, the widespread use of environmental auditing and certification of production;
- establishment of effective environmental management to control technological activity, natural resource management and allocation of productive forces;
- introduction of eco-efficient methods of production, corporate social responsibility, cleaner production to reduce emissions and discharges, minimize waste and comprehensive use of raw material resources, including recycled;
- improving waste accounting and statistical re-

ports on the creation, storage and disposal of waste;

• developing of the national system for monitoring the environment.

Researches, designs, combining the interests of state and business in a market economy are extremely important. Therefore, optimization of geoecological protection of state interests and maximizing profit business, through research in the field of mining and metallurgical industry, are the part of geoecological valorisation.

Natural processes are the reaction to the environment of society, aimed at using the elements of this environment for their welfare. Ecological and economic factors are the forces of nature and economic relations in society (economics), which strive to create balance in the development of human civilization (Tshustov 2000). The development strategy of a perspective plan for enhancing the economic entity in the market, find new markets for the production of additional or new products and solving of problems in terms that take account of changing its internal or external environment. Incentive is a creation of favorable conditions for the functioning of an economic entity.

Existing problems geoecological valorisation related to the paucity of financial resources that could be obtained for the financing of environmental activities (Kobyshko 2007). Implementation of geoecological valorisation shown in Figure 1.

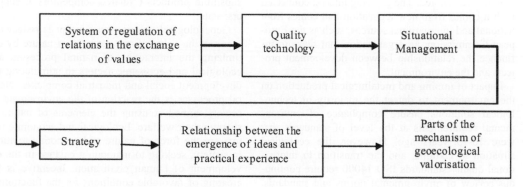

Figure 1. Implementation of geoecological valorisation.

State regulation of the geo-ecological processes in the industrial complex is extremely multifaceted, but not sufficiently corrected according to contemporary needs of environmental protection, is the need to establish reasonable rules and regulations for the rent of natural resources will appear objective necessity. The actions of the state mining complex in the last ten years that could not be regarded as such that the national interests of Ukraine. First of all concerned adopted the laws on privatization lobbying for a song all the mining base of Ukraine and some smelters. Ukraine lost hundreds of billions of dollars for that. Anti-state acts were carried out under the socialist slogans of support for domestic (national) manufacturer, saving jobs, labor, creating conditions for attracting investments and other myths. The whole this bluff now is clear even for inexperienced. The owners of most attractive companies "half strategy" received in private hands for a pittance under certain investment commitments not fulfilled, and even eliminated any obligation. The state has not received money from privatization attractive enterprises of mining and metallurgical complex (MMC) that could save the industry when the global economic crisis.

5 SOCIAL VALORISATION

Sustainable economic development is impossible without the formation of socially prosperous society. The combination of problems of market relations with increased attention to social issues creates the necessary preconditions for economic recovery, sustainable grasslands formed the man throughout the employment and after its completion (Resolution of CMU 2004).

Social valorisation means better quality of life by optimizing staff relations, benefits the state and businesses to establish the relationship of social processes and changes in environmental factors. The system of social valorisation in the broadest sense is a system of legal, socio-economic and political guarantees, which represent conditions for livelihood: able-bodied citizens – at the expense of labor input, economic independence and entrepreneurship, socially vulnerable segments – at state expense, but not below the statutory minimum subsistence level. In the narrow sense of social valorisation is a system of legislative, social, economic and moral-psychological guarantees, means and measures by which to create equal conditions of society that prevent adverse impacts on human environment (in particular, employee labor organizations) provide decent and socially acceptable quality of life.

The concept of the mechanism of social valorisation process provides better quality of life of employees in material production based on the rationalization of state and business relationships and establishing a relationship of social processes and changes in factors internal and external environment. This concept is aimed at addressing the creation of the state and the business of such material, institutional, technological and safe environment for participation of workers in manufacturing, which provide physical and moral decline and technological deterioration of the human body, extension of working age, quality of life and providing career opportunities.

Activities of states to ensure social needs and development goals of the interests of different social groups in society is made in the form of government regulation. The term "regulation" has several semantic values. In broad terms it is a state intervention in the economy in general and in the narrow it is administrative - legal regulation of economic and social relations (Project of law 2009).

So, specifically state regulation can be defined as state influence on the reproduction processes in the economy appropriate means to target businesses and individuals to achieve goals and policy priorities of social development. State regulation – it's almost all public functions related to economic and social-economic activities and conditions designed to ensure a functioning market economic system.

There are basic functions of social valorisation:

• economic – encouraging various– kinds of social and economic activities effectively;

• social – to create conditions– of social stability through social and economic guarantees to the population;

• backup, strategy – a task which – "create economic and social conditions for the preservation and development of "human potential", to provide for all social groups the necessary level of life".

So social valorisation of course improve the quality of life of workers by optimizing the relationship of state and business benefits of establishing a relationship based on social processes and changes in environmental factors. The system of social valorisation in the broadest sense is a system of legal, socio-economic and political guarantees, which represent conditions for livelihood: able-bodied citizens – at the expense of labor input, economic independence and entrepreneurship, socially vulnerable segments – at state expense, but not below the statutory minimum subsistence level. In the narrow sense of social valorisation is a system of legislative, social, economic and moral-psychological guarantees, means and measures by which to create equal conditions of society that prevent adverse impacts on human environment (in particular, employee labor organizations) provide decent and socially acceptable quality of life.

Currently, it is important to create favorable conditions to hardworking man, realizing their talents, achieved high employment figures, which must meet high income – the same person will have the opportunity to improve their lives and provide decent living for his family. The main purpose of business, besides profits should be the creation of such material, institutional, environmental and psychological conditions of human participation in production, which would provide for reduction of physical and moral-psychological deterioration of the human body, extension of working age, quality of life and opportunities for enhancing and career growth. However, business is developing harmoniously only legitimate space where the state encourages entrepreneurial initiative.

The state must create and regulate social conditions of society to the welfare of society, eliminate the negative impact the functioning of market processes, social justice and socio-political stability.

So the social valorisation of problems can be attributed to stimulate economic growth and the subordination of the interests of consumption, increased employment and business motivation proper maintenance of living standards and social protection, preservation of cultural and natural heritage.

Areas of social valorisation:

1) social security, which is the state;

2) social security, which is the employer.

In its full manifestation of social valorisation should pursue the following objectives (Philipova 2007):

• providing members of society living wage and providing material aid to those due to objective reasons it needed protection against the factors that affect the standard of living;

• creation of conditions that enable citizens to earn their money freely for the life of any law is contradictory ways;

• creation of conditions that ensure a certain minimum satisfaction (the extent of social opportunities and taking into account national and historical specificity) needs of citizens in education, health care, etc.;

• roviding favorable conditions for hired workers, protecting them from negative impacts of industrial production and create conditions that would exclude injury and occupational diseases;

- environmental safety of society;
- creation of conditions that exclude social and ethnic armed conflicts;
- ensuring freedom of spiritual life, protection from ideological pressure;
- create a favorable psychological climate in society as a whole, and in structures, protection from psychological pressures;
- ensuring the maximum possible stability of society.

Creating safe working conditions and ensuring human life – a complex problem and should always decide on a national and local levels. Therefore the question of social security for workers should be implemented through the following sources: State budget and funds. It is also provides social valorisation.

Any social security is based on social guarantees, which also represent a certain system. Guarantees should:

- regulate the labor market by the confusion arising disputes between its supply and demand;
- promote full employment of all able-bodied members of society (no one should be against his will remain out of work);
- pay all categories of the working population, with emphasis at this need.
- The safeguards system should take into account the population structure of a particular region, on the employment potential can be divided into the following groups:
- suitable for work in modern conditions;
- those who can only be used after appropriate training;
- those who can work only when creating the appropriate conditions;
- those who do not work;
- those not seeking work.

The most important principles of social valorisation are:

- social responsibility of state and society by loving care to each individual, the realization of their rights to work, choice of profession, job, protecting the health and life, disability compensation;
- social justice in employment – equal pay for equal work, the right to occupational safety and health, preservation of health and ability of citizens to social benefits in the event of illness, the high level of compensation for lost capacity, providing medical, social and vocational rehabilitation affected the production;
- comprehensive and mandatory protection of workers from social and professional risks, ensuring the right to social security as the main socio-

economic development;

- the minimum level of social and occupational risks, availability and transparency of relevant information to ensure safety and availability of social understanding in society about the establishment of professional and social risks, social security, namely their minimization and compensation;
- subjects of many social protection, which subjects should act state (represented by ministries and agencies), employers, professional associations and organizations (companies with insurance), regional authorities;
- state guarantees associated with the social protection of the simultaneous independence and self-government non-defense systems and programs;
- all major subjects of interest protection (state, entrepreneurs, social insurance companies and a wide range of professional organizations of workers) in the formation and improvement of relevant systems;
- solidarity of all subjects of social protection based on "social contracts" relating to the distribution of the financial burden of compensation and minimize the social and professional risks;
- economic and social freedom of workers in labor – the choice of profession and acceptable levels of professional and social risks, the availability of vocational training, employment, freedom of association, is the right to join trade unions, associations and other similar organizations to protect their rights;
- responsibility of employees for maintaining their health and disability;
- multilevel methods differently directed and social protection – from state guarantees for all workers to the narrowly focused measures for specific categories and their professional groups to implement the differentiated approach to various categories to be protected;
- differently directed and social protection.

Social valorisation is a system of activities of the both the state and entrepreneurs. For the first availability of social security is a criterion for effective functioning. Because from that, how secure a society depends welfare state, the degree of conflict and vulnerability of the environment. For the protection necessary social entrepreneur as one way of motivating the work, as it will satisfy not only physiological but also social needs.

From the perspective of a systems approach to manufacturing enterprise is a complex open active socio – economic system which takes into account the economic component of its economic resources (financial, employment, innovation, information, technical and technological), and social – the internal relationships between team members and external communications relate between the company and potential or actual customers, competitors, partners, government agencies, local governments, other contact groups. This system characterized by common goals, an organizational structure and define rules of interaction elements. Due to its openness and social-enterprise economic system characterized by a continuous exchange between its elements and the environment matter (goods, services, materials),

energy (resources, skills, experience) and information (Filipova 2007).

So social valorisation can be considered, which is represented in Figure 3.

Thus, the main directions of the mechanism of valorisation of technological, environmental, social processes of mining enterprises and improvement of existing mechanisms of cooperation between the State and mining companies.

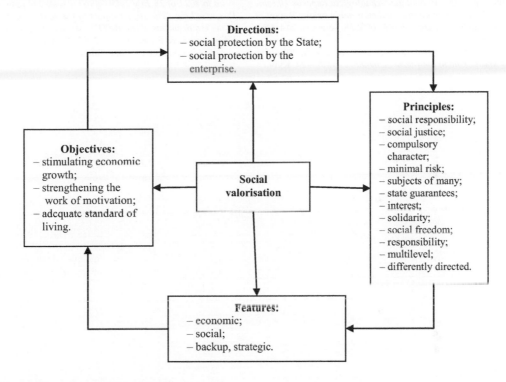

Figure. 2. The system of social valorisation

6 CONCLUSIONS

For the first time concepts: valorisation mud, as the concept of adding value to geological, technical and technological resources of mining companies through the optimal interaction between state and business geoecological valorisation as raw resource value increase at the expense of state and business benefits of establishing a relationship based on natural processes and ecological and economic factors stimulating the development of mining industry, social valorisation as improving the quality of life of workers by optimizing the relationship of state and business benefits of establishing a relationship based on social processes and changes in environmental factors.

Basic directions of the mechanism of geotech-

nological valorisation processes of the mining companies are state financial support of mining companies, organizational-administrative cooperation and state mining enterprise, innovation and investment mechanism geotechnological valorisation processes of mining companies, an effective mechanism for evaluation of mineral resources, organization management of mining companies.

The first time the realization geoecological valorisation through components of the mechanism geoecological situations such as: the regulation of relations in the exchange of values, quality technology, situational management, strategy, idea and experience.

The mechanism of social valorisation, which includes working together government, business and workers to achieve mutual benefits. Sustainable economic development is impossible without the formation of socially prosperous society. Solving problems with the functioning of the social sphere, is the prerogative of the state, but this issue and business should be socially responsible.

REFERENCES

Pilov, P. 2009. *Alluvial geotechnological deposits Ukraine valorisation of natural and man-made origin*: Report of researcher-research #0108U000558 Sciences: 214. Dnipropetrovs'k.

Bolohovitinova, O., Kvasnyuk, B. & Kireev, S. 2003. *The state's role in long-term economic growth*: 424. Kyiv.

Mining Law of Ukraine on. 1999. #1127-XIV. //www.rada.gov.ua.

Dvoryashyna, N., 2005 *Current issues of economic assessment of forest resources*. Scientific Bulletin.

Tshustov, S., Nikiforov, A. & Kutsenko, T. 2000. And others State Regulation of Economy: 316. Kyiv.

Kobushko, I. 2007 *Financial and economic mechanism of ecological industry*: Science: 08.00.06.

Resolution of "The approving the State Development and Reform of mining and smelting complex for the period up to 2011 of 28 July 2004" #967 ///www.rada.gov.ua.

Progect of Law "The state support of the coal industry with the requirements of the WTO"/// www.rada.gov.ua

Filippova, S. & Simakova, N. 2007. *Sources of financing of fixed capital of Ukraine*. Odesa.

Research of the electrets' effect on the fibers of the polypropylene filtering materials

S. Cheberyachko & D. Radchuk
National Mining University, Dnipropetrovs'k, Ukraine

ABSTRACT: The question of workers respiratory organs defense from aerosols is considered. It is certain that for effective defense it is necessary to use porous filtering materials that conform to the standards requirements and doesn't have an influence on the capacity of workers. It becomes possible at presence of electric forces on the surface of filtering material. Investigated the charging process of filtering material and described methods of researches realization. Basic dependences of filtering material charge are certain according to the parameters of environment and charging process.

1 INTRODUCTION

The problem of respiratory protection of workers from the solid dispersion pollutants, including industrial aerosols, is very important. Harmful impurities not only reduce working capacity, but also, after prolonged inhalation, lead to disease of pneumoconiosis and dust bronchitis. Solution of this problem is in the creation of respiratory protective devices (first of all respirators) with filtering layers that separate the particles of pollution from the air phase. Relevance is to develop new filter materials and the searching technological scheme for manufacturing of high efficiency filter, which would meet the ever increasing working, economic and environmental requirements.

The traditional air cleaning technique is used in different porous materials. The simplest, reliable and cost-effective way to clean air and gases from aerosols is the highly fibrous filters. Since the mid of XX century air filters based on synthetic fibers are widely used which are made of cellulose and its esters, asbestos, fiberglass, lavsan.

In 50-ies of XX century in the Soviet Union, ultra-thin fibrous materials was created made of polymer solutions – Petryanov's filtering cloth (FPP), and they still occupy an important place in the technique of filtering (Petryanov 1984). Their main distinguishing feature is the high efficiency of the delay micronsized particles, with a low hydrodynamic resistance. Note that the fibers FPP present an electrostatic charge, which allows additional filtering mechanisms related to the work force of electrostatic attraction (Filatov 1997). It is possible to create a lightweight and easy to use respirators, which are very well-proven and some brands are used today.

The main disadvantage of FPP cloth is that fiber production is quite expensive and environmentally unsafe (according to fire safety corresponds to the group "B", by the level of explosion because it uses solvents are among the most dangerous category of "B1" and sanitary-hygienic characteristics to the class 3 environmental hazards). Therefore, the filter Petryanov's replacement to cheaper, technologically, but no less effective filtering materials is the actual problem.

2 MATERIALS AND METHODS

The presence of electrical forces on the surface of the filtering material can reduce the pressure drop on respirators while providing a sufficiently high level of cleaning polluted air and long term use. This is because the filtering material can be produced with lower packing density of fibers. The principle of such filters is to capture charged particles Coulomb forces, and neutral particles that get listed in the electrets charge.

Electrets filters are obtained by charging the fibers in the corona discharge. Polarization happens on the special aggregate in Figure 1, which consists of a corona electrodes system 3, which voltage is applied to the transformer unit 1 and control block 2. Filtering material 5, located on the bench 6 which is connected to ground through adjustable support 4, to ensure stable power supply.

Distance a is selected as little as possible to reduce the size of the air layer, and it is 0.5-2 mm, while the b value is trying to make as much as possible in order to create greater tension on the electrodes, and it usually is 30-70 mm.

The process of charging electrets is widely presented in the literature. Thus, the authors in (Rychkov 2000) studied the time of the charge drain from the filtering material surface to determine life time of respiratory protection devices (RPD). In publications (Jiang 1996 & Kravtsov 2003) was an attempt to solve the problem of increasing the life time of electrets by putting in the structure of a filter various composite additives. At the same time research was conducted to find a regulation of the filters electrets properties and choice of the optimal electrification method. This study presents data to assess the impact of time finding the filtering material in the various sizes of corona discharge (Halyhanov 2003).

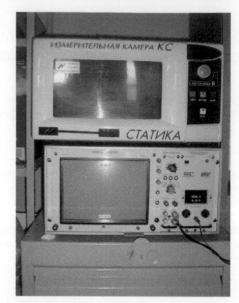

Figure 2. General view of apparatus "Static".

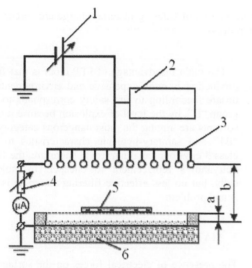

Figure 1. Aggregate arrangement of charging filtering material.

The aim of this work is to determine the optimal parameters of the external environment and the voltage that applied to the corona electrode to obtain a stable surface charge density on the filter.

The study was performed on samples of two-layer filtering material "Eleflen" that is made of polypropylene melt: fibers with a 1-3 microns diameter, 4 microns fibers thickness and the 0.035 packing density. Experimental samples were the squares with sides of 150 mm and have a marking depending on the type of the test.

Samples in turn placed on the bench 6. Tests were conducted at different temperatures from 10 to 40 °C and humidity from 50 to 100% at voltage of 10 kV to 50 kV on the corona electrode. Quantification of the surface charge density was carried out by the method described in GOST 25209-82 "Plastics and polymer envelope. Methods for determining the surface electrets charge".

Its essence lies in measuring the voltage that is fed to the measuring electrode to compensate the electrets field that triggered charges. Device for measuring the surface electrets' potential is shown in Figure 2. It consists of two units of measurement chamber, which is a measure of the effective stress on the electrets surface to 1000 V and display unit. The principle of operation is based on the use as a zero-indicator of electron-beam vobulyator, displaying compensation voltage on the digital voltmeter. Sample was placed in the chamber and put down on it measuring electrode. The magnitude and polarity of compensative voltage was determined by setting a zero rate on the digital voltmeter. Register compensative voltage volume surface electrets potential (GOST 25209-82 1982). Surface charge density of samples is determined by the formula

$$e_{ef} = \varepsilon\varepsilon_2 U_e / S , \text{Kl} / \text{m}^2,$$

where ε – dielectric air penetration, ε_2 – dielectric penetration of electrets material; U_e – surface electrets potential, V; S – area of the measuring electrode, m^2.

3 RESULTS AND DISCUSSION

Our results showed that with increasing of corona electrodes voltage increases and the potential for surface charges. At least 50 kV it's a linear dependence in Figure 3.

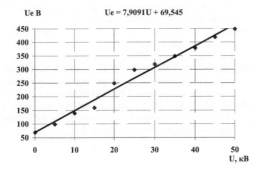

Figure 3. Dependence on the surface potential of electrets' corona electrode voltage.

It was also found that the longer an electrets' is under corona discharge the more potential in it. Clearly, the optimal velocity of filtering material corona charge installation provides maximum power from the finished production on the one hand and a sufficient amount of charge on the other.

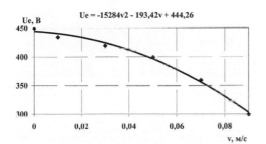

Figure 4. Dependence of potential on the filtering material surface on the aggregate moving speed (corona electrode voltage 35 kV).

As seen from Figure 4 it is necessary to set speed in the range 0.02-0.04 m / s to provide high surface charge density of fibers.

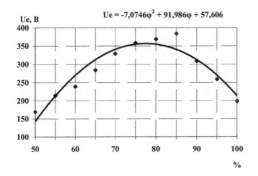

Figure 5. Dependence of filtering material surface potential of humidity at a 35 kV voltage of corona electrode.

Interesting fact is regarding the data about evaluation of the impact of humidity on the charging process of the filtering layer. Humidity in the charging electrets room imitated by using a humidifier brands Boneco. As in Figure 5 there is a certain level of humidity in which there is greatest potential value at the material surface, at the same voltage of 35 kV on corona electrode.

The similar results were obtained at assessment of the influence of environmental temperature on the value of surface potential on filtering material in Figure 6.

Figure 6. Dependence of filtering material surface potential at air temperature at a 35 kV voltage of corona electrode.

Therefore one should regulate environmental parameters at electrets charging and maintain their keeping in the optimum mode.

4 CONCLUSIONS

One of the effective ways from dispersed harmful effects of pollutants on the respiratory tract is using respirators, but there is a need to develop and implement new filtering materials. In data filtering materials – to improve the security performance using a static charge. Used of the filtering material charging method is an effective and depends on the established potential value on the corona electrode surface. Pattern was established between environmental parameters and value of the surface potential on polypropylene filtering material. It was determined that the optimal rate for material charging is 0.02-0.04 m/s. It was found that the maximum surface potential at polypropylene material corona charging is observed at temperature of 20-25 °C and humidity 75-80%. Therefore, maintaining the established environmental parameters and charging process will improve the protective properties of the polypropylene filtering materials.

REFERENCES

Petryanov, I. & Kascheev, B. et al. 1984. *Lepestok. Low-weight respirators.* Moscow: Nauka: 168.

Filatov, N. 1997. *Electroformation of fibers materials (EFM-process).* Moscow: Oil and Gas: 297.

Rychkov A., Boytsov V. 2000. *Electret's effect in the structures of polymer-metal:* Monograph. St Petersburg: Publishing. RHPU them. A. I. Herzen: 250.

Jiang, J., Xia, Z., Zhang, H. et al. 1996. *Charge storage and transport in high density polyethylene and low density polyethylene.* Shanghai, China: Proceedings of 9 International Symposium on Electrets: 128.

Kravtsov, A., Holdade, V. & Zotov, S. 2003. *Polymer electrets filtering material for protect respiratory organs.* Gomel: YMMS NASB: 204.

Halyhanov, M. 2003. *Effect of fillers on the polarizability of the polar polymers in the corona discharge.* Kazan: Journal for technological Kazan University, 2: 374-378.

GOST 25209-82. 1982. *Plastics and polymer envelope. Methods for determining the surface electrets charge.* Moscow: Edition of the Standards: 6.

Technical and Geoinformational Systems in Mining – Pivnyak, Bondarenko & Kovalevs'ka (eds)
© 2011 Taylor & Francis Group, London, ISBN 978-0-415-68877-2

Application of mathematical simulation method for solving the task focused on efficiency increase of hydraulic influence process on coal seam

S. Grebyonkin, V. Pavlysh & O. Grebyonkina
Donetsk National Technical University, Donetsk, Ukraine

ABSTRACT: Based on the mathematical simulation results of technological schemes of hydraulic influence on coal seam, the method of process efficiency increase as means of its conditions management is substantiated.

1 ACTUALITY OF THE THEME

Preliminary treatment of coal seams by liquids under various modes is an important way of solving the problem of dangerous phenomena fight in mines (DNAOP 1.1.30-1.XX-04 2004). One of the basic advantages of this method is prior and irreversible change of coal massif state allowing to prevent occurrence of dangerous phenomena during conduction of mining operations (Bulat 2003).

The results of scientific-research and design-engineering developments have provided the basis for corresponding sections of normative documents regulating necessity and order of liquid preliminary injection to fight dust formation and gas-dynamic phenomena in coal mines (DNAOP 1.1.30-1.XX-04 2004 & Bulat 2003).

Analysis of liquid injection existing methods into the coal seam to control its state and fight with basic dangers shows that the necessary condition of a high efficiency of influence is uniformity of hydraulic treatment of a coal seam. Nonuniformity of moisture distribution within the massif is one of the basic causes of moistening insufficient efficiency that is caused by predominant movement of a liquid within large cracks and leading to insufficient moisture in coal blocks, whereas at other areas of a seam the moisture content growth can exceed required values.

Basic reason of treatment insufficient uniformity is filtration anisotropy of a coal seam. Moisture quality growth is an important way to increase hydraulic influence efficiency.

In connection with that, the given work is current.

One of the important ways to improve methods of hydraulic influence on the coal seam is the research of liquids movement processes within the massif and determination rational technological schemes, modes and injection parameters based on it.

Work experience in given trend shows that together with experimental methods, the significant results in this field can be achieved with help of mathematical modeling of the processes occurring during injection of materials possessing fluidity (fluids, gases, aerosols, suspensions) into a seam coal. Application of this method allows to establish characteristics of fluid and air filtration within the massif that is not available for studying under natural conditions, to investigate influence of coal seam properties and parameters of interaction in a wide range of their change on distribution of injected into a seam matters avoiding significant labor and material costs needed for mine experiments conduction. In connection with that, the expediency of further expanding of mathematical modeling application processes of hydraulic influence on coal seam and linkage of received results with practice demands is obvious (Pavlysh 2005).

Let us consider process of a seam hydro-treatment.

Coal massif as an object of hydraulic influence is characterized by the ability to receive and filtrate water and gas, and also accumulate their definite amount. These abilities are characterized by such natural indices as permeability and porosity.

Coal seam is presented as a fracture-porous structure located on a big depth under the load of above-situated layers of rocks. Equation, providing the basis for mathematical model of liquid injection process in the XOY plane has the following form:

$$\frac{\partial p}{\partial t} = \chi_x \frac{\partial}{\partial x}\left[k_x(p)\frac{\partial p}{\partial x}\right] + \chi_y \frac{\partial}{\partial y}\left[k_y(p)\frac{\partial p}{\partial y}\right]. \quad (1)$$

In order to form mathematical model, the equation is supplemented by boundary conditions which are defined by the process technology. Set boundary task is calculated by a computer using finite difference method. As was mentioned before, the basic reason of insufficient influence is filtration anisotropy of a seam that is proved by way of the modeling process when setting variable values of permeability coefficient in discrete points of the seam treatment zone. As a result of computer modeling, overcoming method of negative influence of anisotropy on the seam treatment quality has been substantiated.

This method consists of application of a cascade injection system.

Essence of a cascade technology consists of uniform treatment provision of a seam working area at the expense of liquid counter flows during simultaneous work of adjacent injection boreholes. Development of cascade method efficiency evaluation criteria compared to implemented injection through single boreholes is based on the provision of uniform saturation of the treatment area by the working liquid.

Treatment nonuniformity is quite fully defined by the area of untreated sections and by degree of spread of values of moisture growth in the project zone of influence. In connection with that, the following indices are chosen for quality evaluation:

– index of relative value of untreated areas

$$v = \frac{S_U}{S_{PR}} \cdot 100 , \% \qquad (2)$$

where S_U – untreated sections; S_{PR} – the influence project zone.

– variation value of moisture relative growth defined in each point of discrete area:

$$V_R = \frac{\sqrt{\dfrac{1}{N-1} \sum_{i=1}^{N} \left(\Delta W_i - \Delta \bar{W} \right)^2}}{\Delta \bar{W}} \cdot 100 , \% \qquad (3)$$

where ΔW_i и $\Delta \bar{W}$ – values of moisture growth correspondingly in the i-th point and average within the treated zone

$$\Delta W_i = \frac{\Delta W_{cal.i}}{\Delta W_{max}} , \qquad (4)$$

where $\Delta W_{cal.i}$ – moisture growth in the i-th point; ΔW_{max} – maximal moisture increase.

To evaluate efficiency of cascade influence, two indices will be used:

– value of relative decrease of an untreated area

$$\Theta_S = \frac{S_{u.a} - S_{u.c}}{S_{u.a}} \cdot 100 , \% \qquad (5)$$

where $S_{u.a}$ and $S_{u.c}$ – areas of untreated sections during injection through single borehole and cascade;

– value of variation decrease of moisture growth:

$$\Theta_v = \frac{V_R^0}{V_R^k} , \qquad (6)$$

where V_R^0, V_R^k – variation values correspondingly for single borehole and cascade.

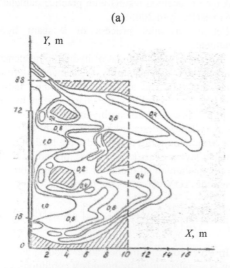

(a)

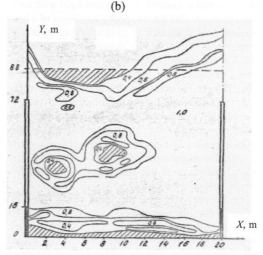

(b)

Figure 1. Moisture growth distribution during injection through long boreholes under constant pressure single borehole (a); cascade of boreholes (b).

Technology development of cascade injection has to be directed to decrease of the influence degree of factors reducing treatment uniformity, considering requirements of minimal complexity, labor-intensiveness and use of serially produced equipment.

Figure 1 presents the results of moisture growth simulation during moistening by traditional scheme through single borehole and through cascade of two boreholes, and Table 1 shows the research results of cascade technology efficiency.

Table 1. Cascade technology efficiency indices.

№ of the variant	Treatment uniformity				Efficiency compared to a single borehole			
	$q_1 \neq q_2$		$q_1 = q_2$		$q_1 \neq q_2$		$q_1 = q_2$	
	v , %	V_R^0 , %	v , %	V_R^k , %	$Э_S$, %	$Э_V$	$Э_S$, %	$Э_V$
Long boreholes								
1	4.2	27.3	3.0	24.3	68.4	1.8	77.5	2.0
2	4.8	31.2	2.2	25.7	71.0	1.9	86.7	2.4
3	5.1	30.5	2.3	24.8	71.3	2.0	87.0	2.5
4	3.4	25.1	2.6	22.2	77.1	2.4	82.4	2.7
5	4.3	30.2	2.9	25.3	76.2	2.1	84.0	2.5
6	6.3	30.6	4.6	28.1	67.0	1.8	75.8	2.0
7	4.5	31.4	4.1	26.2	74.1	1.9	76.4	2.3
8	4.4	33.0	2.3	28.0	77.4	2.0	88.4	2.3
9	8.5	42.6	4.2	35.8	62.3	1.6	81.4	2.0
10	8.1	40.1	3.8	31.3	65.7	1.6	83.9	2.0
Average indices	5.4	32.2	3.2	27.1	70.5	1.9	82.6	2.2
Short boreholes								
1	9.5	46.2	7.0	40.3	60.7	1.7	71.0	2.0
2	6.7	40.3	5.3	36.2	69.2	1.6	75.6	1.9
3	8.9	44.0	6.1	35.5	67.7	1.8	77.8	2.3
4	10.7	45.8	7.5	41.6	57.5	1.6	70.2	1.8
5	12.3	50.1	8.4	47.0	61.2	1.6	75.5	1.7
Average indices	9.6	45.3	6.9	40.1	63.8	1.7	74.1	1.9

Computer researches of the method efficiency for various conditions has shown that during liquid injection during cascade boreholes, the 50-80% reduction of the untreated area occurs and moisture growth variation decreases 1.5-2 times compared to a single borehole. Experimental conditions were conducted under industrial conditions at "Khrustal'naya" mine of "Donbassanthracite" enterprise (preliminary moistening in filtration mode through advance long boreholes parallel to the bottomhole plane), at "Mine named after Skochyns'kyy" (hydro-mechanical influence in mode of hydro-loosening through short boreholes perpendicular to the bottomhole plane with efficiency control in terms of gas emission), at mine named after V.M. Bazhanov – "Makeevugol" (hydraulic influence through boreholes in niche), at mine named after M.I. Kalinin – "Artemugol" (hydraulic influence in mode of filtration during development working drivage).

The experiments results to some extent harmonize with the results of mathematical modeling (divergence of theoretical and experimental data does not exceed 15%).

2 CONCLUSIONS

Using mathematical simulation for liquid filtration process in anisotropic coal seam during injection through a single borehole and cascade of boreholes, the fundamental possibility of overcoming the filtration anisotropy of a seam is proved at the expense of counter flows of liquids interaction at cascade injection.

REFERENCES

DNAOP 1.1.30-1.XX-04. 2004. *Safe conduction of mining operations at seams prone to gas-dynamic phenomena (1st edition)*. Kiev: Mintopenergo of Ukraine: 268.

Bulat, A.F., Sophiskiy, K.K., Silin, D.P., Muchnik, E.I., Baradulin, E.G., Zhytlenok, D.M., Zhmykhov, V.N., Vorob'yov, E.A., Kalphakchiyan, A.P. 2003. *Hydrodynamic influence on gas-saturated coal seams*. Dnipropetrovs'k: 220.

Pavlysh, V.N. 2005. *Theory development and technology improvement of influence processes on the coal seams*. Monograph. Donetsk: 347.

A new comprehensive method for designing roof bolt support system for underground coal mines

Avi Dutt
B. Tech, Part IVth, Department of Mining Engineering, ITBHU, India

ABSTRACT: Since its development in the 1920s, bolting has become the most dominant support method in underground excavations. However, because of the geological environment, the design process for roof bolt systems is an art rather than a science. An attempt has been made to develop a comprehensive approach for designing rock bolt supports by optimizing bolt length, bolt density, and bolt pretension during installation.

Roof bolts are the first line of defense protecting mineworkers from the hazards of ground falls. Because roof bolts utilize the inherent strength of the rock mass, they have many advantages when compared with earlier standing support systems. Due to of their central importance, roof bolts have received more research attention than any other ground control topic.

Roof bolt design consists in specifying the proper bolt type, capacity, length, and pattern for a particular roof rock, stress level, and application. The interactions between these variables are very complex, and our understanding of their mechanics remains imperfect. Numerous roof bolt design methods have been proposed over the years, but none has gained widespread acceptance by the coal mining industry.

The following text presents a conglomerate of the outcomes of numerous experiments and the resulting theories of roof bolting when applied to the Indian coal measure strata and geomining conditions, for devising an effective roof bolting support system

As in the design of any other support system, the design of a roof bolt support system depends on: the nature of rock mass and that of discontinuities; magnitude and distribution of induced stresses; support requirements considering the acceptable deformation and lifetime of openings; and shape and size of openings. For a complete and appropriate roof bolting support system design, the following parameters need to be properly determined (Luo et al., 1998):

Bolt Type.
Bolt Length.
Pattern and Spacing.
Bolt Diameter and Anchor capacity.
Magnitude of Pre-Tension required (in case).
The above stated factors are to be considered one by one as below:

Bolt Type: The coal measure strata are composed of stratified layers of sedimentary origin and are chara-cterized by numerous geological discontinuities viz. faults, joints, slips, slickenside, kettle-bottoms, cleats etc. Of the various types of rock bolts developed so far, the Full Column Grouted rock Bolts have been found to provide an effective BEAM BUILDING as well as an efficient KEYING of the geological discontinuities by providing significant frictional forces along fractures, cracks, and weak as well as bedding planes and thereby preventing sliding and/or separation along the interface.

Bolt Length:

$$L_b = (I_s / 13)(log 10H)((100 - CMRR)/100)^{1.5}$$

(Mark 2001),

where I_s – actual intersection span (average of the sum-of-the diagonals, ft; H – depth below surface; $CMRR$ – coal mine roof rating.

Bolt Diameter (Coats & Cochrane 1971):

$$R_{max} = \alpha A; \quad P = R_{max}/(SF * n);$$

$$P = (\pi * \alpha * d^2)/(SF); \quad C = \pi * G * d^2/4,$$

where $R_{max} = max$. Bearing capacity of bolt.

$P = max$. Allowable tension/axial load in bolt.

SF – safety factor (generally taken as 2 to 4); C – yield capacity of bolt; G – grade of steel; d – bolt diameter; A – cross sectional area of bolt. Also $C \infty d^2$; So as the bolt diameter increases, the bearing capacity increases and also; Bolt stiffness ∞d^2.

Annulus Design: The annulus between the bolt & rock should be of appropriate size and type, so as to ensure an effective load transfer between rock-grout-rock interfaces. It is recommended that the hole wall should have helicoidal grooves oriented at an appropriate angle to the hole axis. This technique increases bolt resistance and its bearing capacity and enhances load transfer between rock and bolt. As is clear from the following figure, grooved hole transfers shear as well as normal forces and prevents shear failure at grout-rock interface as against the smooth walled hole, which transfers load through shear friction only.

T = Bolt Tension Force
S = Grout Shear Force
N = Grout Normal Force
N' = Rock Tension Force in Bolt's Vicinity
S' = Rock Shear Force in Bolt's Vicinity

From the above figure:

For non-grooved hole: $T = S + N\cos\alpha - S'$ or $S' = S + N\cos\alpha - T$.

For grooved hole:

$T = S + N\cos\alpha - S' - N'\sin\alpha$ or
$S' = S + N\cos\alpha - T - N'\sin\alpha$,

where $(90° - \alpha)$ – orientation of groove w.r.t hole axis.

Thus, the shear friction force at grout-rock interface decreases and prevents shear failure of bolt at this interface.

Case Study: 9th Seam, Digwadih Colliery.
General Data about 9th Seam.

Helical Drag Bit Technology has been successfully tested for drilling such holes.

Bolt Spacing and Pattern:

The Bolt Spacing and pattern for **Roof support** can be determined on the basis of following relation:

$PRSUP = L_b * N_b * C / 14.5(S_b * W_e)$ (Mark 2001),

where $PRSUP$ – support factor; L_b – bolt length, m; N_b – number of bolts per set; S_b – spacing between rows, m; W_e – gallery span, m; C – capacity of bolt, kN.

$PRSUP = 17.8 - 0.23 * RMR$ (high and moderate depth) (Mark 2001); $PRSUP = 15.5 - 0.23 * RMR$ (shallow deposit).

The reinforcement factor of the bolted roof is given by:

$R.F. = 1/(1 - D)$,

where D – strain reduction produced in roof beam due to friction effect of bolting.

$$D = 0.265\{[bL]^{-0.5}\}\{[NP(\{h/t\} - 1)/w]^{0.33}\};$$

$$P = w * L * X * h / [(X/b) + 1][(L/a) + 1],$$

where P – Bolt Tension Force, lb; $N = N_0$ of bolts per set; B – spacing between sets, ft; t – average layer thickness for roof rock, ft; h – Bolt length, ft; L – gallery span, ft; w – unit weight of roof rock, lb/ft3; X – length of immediate roof; a – spacing between the bolts in a row.

Table 1. The general information about geology, development working dimensions and surrounding rock properties and strata conditions are as given.

DIGWADIH COLLIERY, 9TH SEAM	
General	
Parameter	Value
Seam Thickness	2.94 m
Dip	1 in 7 due $S76°50'W$
Method of Development	B and P, along the floor
Gallery Width	4.8 m
Gallery Height	2.9 m
Pillar Size	50x50 m, c to c
Depth of Cover	420 m
Parting between XIth and IXth seams	60 m
Average Vertical stress at the Working Horizon	8.240 MPa
Average Horizontal stress at the Working Horizon	3.125 MPa

RMR DETERMINATION (CMRI)	
Major Coal Measure Rock Type in Immediate Roof	Carbonaceous Shale
Discontinuities types in Roof	Joints , Slips, Faults
Number of Main/Prominent Joint Sets	2
Orientation (strike) of prominent Joint Sets	$N\,10°$ and $N\,100°$
Orientation of Prominent Slips	$N\,10°$ and $N\,240°$
Type of Slips	Slickensided
Average Layer Thickness in Coal	7 cms.
Water Condition in Roof	Moist

RATINGS OF DIFFERENT PARAMETERS IN RMR* DETERMIATION		
Parameter	Description	Rating
Layer Thickness	7cms.	13
Structural Features	Joints, Slips / Slickensides (Indices – 13)	07
Weather ability	98.81%	15
Compressive Strength	482 kg / cm^2 (47.236 MPa)	09
Ground Water	Moist	09
RMR		53
Adjusted Value of RMR		43
Adjustment factor		0.9*0.9#
Roof is classified as IIIrd A Fair		

10% reduction for solid blasting and 10% reduction for depth of working.
RMR determination is based on Geomining Conditions.

Vertical stress – $P_v = w * h$ P_V;

Horizontal stress – $P_H = (v/1-v)P_V$,

where w – unit weight of rock; h – depth from surface; v – Poisson's Ratio.

Rock Load Determination:
Rock Load in Galleries –
$B * D * [1.7 - 0.037 * RMR +$
$0.0002 * RMR^2 = 4.80$ t / m^2;

Rock Load at Junctions –

$5 * B^{0.3} * D$ [1-RMR/100]$^2 = 5.43$ t / m^2,

where D – Density of Roof Rock = 2.09 t / m^2 = 2.09 g / cc = 20.5 kN / m^3; B – Width of Gallery / roadway span = 4.8 m.

Rock Properties* for 9th Seam:
Table 2. Rock mass properties and post-peak characteristics.

Rock mass properties	
Intact rock strength s*ci*	47.23 MPa
Hoek-Brown constant *mi*	10
Geological Strength Index GSI	40-50
Friction angle *f¢*	28.5°
Cohesive strength *c¢*	2.025 MPa

Rock mass compressive strength *scm*	7.35 MPa
Rock mass tensile strength *stm*	-0.08 MPa
Deformation modulus *Em*	52000 MPa
Poisson's ratio *n*	0.275
Dilation angle *a*	-
Post-peak characteristics	
Friction angle *ff¢*	-
Cohesive strength *cf¢*	-
Deformation modulus *Efm*	-
Intact Elastic Modulus (shale)	14 GPa

*these values are not precise(except the intact rock strength) and have been roughly approximated for present analysis of rock material. The correct values require large scale laboratory analyses over a long period of time with the help of sophisticated techniques.

The immediate roof of the 9th seam is composed of thin shale layers. These are full of geological discontinuities and the most prominent and unstable weakness zone extends up to a height of 4 ft out of which the lower thickness of about 45 centimeters is even more unstable and is a source of majority of roof falls and overhangs.

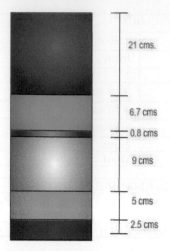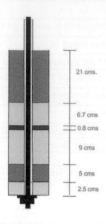

21 cms.

6.7 cms

0.8 cms

9 cms

5 cms

2.5 cms

Figure 1. A section of the immediate roof, of thickness 45 centimeters.

In majority of the cases, separated rocks blocks are found hanging from the outbye end of the roof bolt. Where roof bolt was not present these blocks readily fell down & hence served as main sources of fatality. In other cases the block separated from either the lower or upper roof rock at the 0.8 centimeter thin slice interface and fell down.

The block sliding and slipping is further assisted by lowering of sliding friction at beds' interfaces & discontinuity faces, due to the presence of granular sand that has probably been formed due to interbed sliding during deflection and deformation phase as well as due to erosion. Presence of groundwater is another contributing factor.

At some places where the block separated at a height more than 45 cms. (approx.) the confining deviatoric stresses at the block faces coupled with gravity effects led to the bending of roof bolt, which led to non-uniform loading of bearing plate resulting in plate buckling.

Besides the roof comprises of numerous joints and slips .The intersection of these discontinuity planes results in the formation of rock wedges, which easily separate from the roof rock if adequate support is not provided in time.

Roof Bolt Support Design ForIX[th] Seam.
Bolt Type:

The strata of 9th seam is layered (layers of carbonaceous shale: Compressive strength = 47.236 MPa, $RMR = 43$, IIIrdA Fair Category) with average layer thickness of 7.5 cms and is full of joints. For such sedimentary (stratified) rocks FULL COLUMN GROUTED rock bolts have successfully been used for effective **BEAM BUILDING AND**

KEYING. So, all the bolts are of full column grouted type.

Bolt Length: he bolt lengths of 1.8 m and 1.95 m (excluding threaded lengths) have been suggested/used/is being used.

Bolt Diameter: For the present steel grade & bolt length the commonly used diameter is 20 mm and the drill hole diameter is 30 mm.

Annulus Design: It is recommended that the hole wall should have helicoidal grooves oriented at less than 40 degrees to the hole axis. This technique increases bolt resistance by a factor 1.3-1.62; its bearing capacity by about 3 times and enhances load transfer between rock and bolt.

Bolt Pattern and Spacing:

The Bolt Spacing and pattern for *Roof support* has been suggested on the basis of following relation:

$PRSUP = L_b*N_b*C/14.5(S_b*W_e)$.

$PRSUP = 17.8 - 0.23*RMR$.

$R.F. = 1/(1-D)$.

$D = 0.265\left\{[bL]^{-0.5}\right\}\left\{\left[NP(\{h/t\}-1)/w\right]^{0.33}\right\}$.

$P = w*L*X*h/[(X/b)+1][(L/a)+1]$.

Taking different values for the quantities in the above expressions, the results are shown in Table 3. Values of constants: $L = 4.8$ m = 16 ft; $t = 7.5$ cm = 3 in.; $X = 9$ m = 30 ft; $w = 2.09$ t / m^3 = 130.475 lb / ft^3.

Table showing Bolt length (only grouted length) required for above Roof Bolt Patterns (for 50 m gallery length).

Table 3. Showing comparison between different possible bolting patterns.

s.no	h	N	a	b	P	D	RF
1[@]	6	5	3	3	4660	0.594	2.46
2*	6	6	3	3.2	4946	0.625	2.66
3	6	6	3	3	4665.24	0.631	2.71
4	6	4	3	3	4665.24	0.552	2.23
5	6	5	2.67	2.67	3790	0.589	2.44
6	6	5	2.67	3	4220.2	0.575	2.35
7	6.5	4	3	2.33	3911	0.609	2.56
8	6.5	5	3	2.83	4799	0.636	2.74
9	6.5	5	3	3	5053.7	0.628	2.69

[@] present pattern * suggested pattern.

Table 4. Bolt length.

s.n	No.of Rows	Bolts/set	Total No of bolts	Total Length of bolt (m)	
1[@]	56	5	56x5	56x5x1.8	= 504
2*	53	6	[#]53x(1.8*4+1.2*2)	53x(1.8*4+1.2*2)	= 508.8
3	56	6	56x6	56x6x1.8	= 604.8
4	56	4	56x4	56x4x1.8	= 403.2
5	63	5	63x5	63x5x1.8	= 567
6	56	5	56x5	56x5x1.8	= 504
7	72	4	72x4	72x4x1.95	= 561.6
8	59	5	59x5	59x5x1.95	= 575.25
9	56	5	56x5	56x5x1.95	= 546

for this pattern: in a row, 4 bolts are 1.8 m long and rest 2 are 1.2 m long.

Roof bolt pattern 2 is the best of all the patterns that offer greater reinforcement factor than present pattern, for the following reasons:

1) Nominal Increase in Total Bolt Length from present pattern, against patterns 3, 7, 8, 9.

2) Reasonably high reinforcement factor.

3) Greater Bolt Tension, which ensures better binding of layers.

4) Greater "P/RF" ratio than patterns 3, 7, 8.

5) Deflection upto a distance of 27 cms. from the wall is zero. Present pattern leaves a distance of 60 cms. from the wall while for pattern 2, it is only 15 cms. Thus side overhang is within zero deflection region.

Explanation for Shorter (1.2 m roof Bolts) at the ends of each row in pattern 2:

Shown below is the bolt pattern for roof:

The Figure (at end) shows the mean stress profile of the immediate roof of the opening. As can be seen that a mean stress of magnitude about 5.85 MPa exists near the edge of the opening. Since this value is much smaller than the compressive strength of the rock, the rock in this zone can't fail in compression. Furthermore, as the grey line depicts, the deflection up to a distance of 26 cms from the wall is zero, so beam building in this zone is not the major concern. The role of rock bolt here, is to ensure an effective keying of discontinuities in this zone (span = 60 cms; 15+45). As the most disturbed zone/weakness zone extends up to a height of about 1.2 m above roof, so rock bolt of 1.2 m is sufficient to create a compression zone to prevent tensile failure (due to tension in this zone) as well as failures along discontinuity surfaces. Rest of the pattern for roof support is justified on the basis of Reinforcement Factor and the total length of roof bolt used as given above.

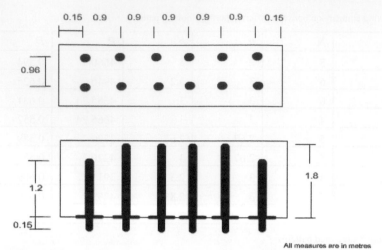

All measures are in metres

Figure 2. Bolt pattern for roof.

The Bolt Spacing and pattern for *Roadway Junction* has been suggested on the basis of Safety factor of the suggested pattern:

The applied load for a roof bolt pattern is given as: $S = n * b_c / w * sp$.

The factor of safety is then given as:

$FOS = S/Rock\ Load\ at\ Junction$.

where n – no of bolts in a row; w – width of gallery / junction; b_c – full column grouted roof bolt capacity; sp – spacing between two rows.

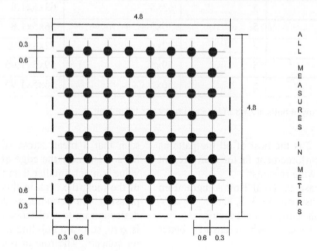

Figure 3. Bolt pattern.

The Greatest Advantage of the suggested pattern would be that, for each bolt the cylindrical shear zone around each bolt is wholly supported on the bearingplate (bearing plate is 15 cm x 15 cm). This ensures a better load transfer mechanism for each bolt, provided the bolt is installed vertically, as can be inferred from the table below.

The Bolt Spacing and pattern for *Side wall support* has been suggested on the basis of the figure given below:

As is clear from the figure A, the tensile stress zone extends up to a distance of 0.5 m from the wall side. Rock bolts of length 1.2 m with an adequate pretension are sufficient to produce a continuous compression zone along the wall to counter the tensile stresses that can cause side spalling. In case of numerous cracks, adequate area support such as wire mesh is suggested in conjunction with roof bolting to provide a continuous bearing surface to prevent spalling and further propagation of cracks and fractures. At corners of junctions, the wooden/steel props clamped to side edge for full height of the edge, are recommended to provide adequate resistance as the edges are the locations of high stress concentration.

Table 5. Comparison between the present and suggested patterns for junction support.

Parameter	Value	
	Present Pattern	Suggested Pattern
W	4.8 m	4.8 m
N	7	8
b_c	5 tonnes	5 tonnes
sp	0.8 m	0.6 m
S	9.11	13.89
FOS	1.68	2.56
Shear radius of rock column around each bolt = (sp/4)	0.2 m = 20 cms	0.15 m = 15 cms
Radius of Compression zone created around each bolt=(sp/2)	0.4 m = 40 cms	0.3 m = 30 cms

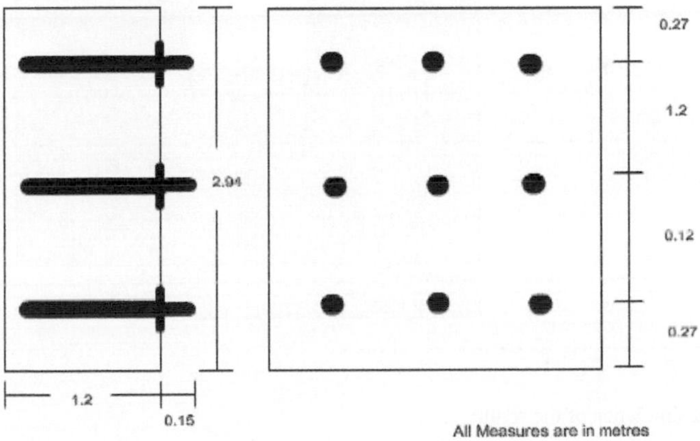

Figure 4. Pattern for side wall support.

Cost analysis of two SSRs: These have been calculated for total bolt length required for 100 m gallery length + 1 Junction. The gallery width and height are 4.8 m and 2.94 m respectively with junction of 4.8 x 4.8 m.

Length of roof bolts presently used for side walls = 1.8 m; Length of roof bolts suggested for side walls = 1.2 m {Grouted length only}.

Table 6. Comparison of the approximate total bolt length required.

Location	Present SSR	Suggested SSR
For Gallery Roof	504*2 m	508.8*2 m
For Gallery Walls	42x3x2x2x1.8 = 907.2 m	42x3x2x2x1.2 = 604.8 m
For Junction	45x1.8 = 81 m	64x1.8 = 115.2 m
Total	1996.2 m	1737.6 m

Therefore, total saving in length = (1996.2 − 1737.6) = 258.6 m

Thus, approx.259 m of bolt length is saved for every 100 m gallery length and 1 junction.

Taking cost of bolt steel per meter = Rs. 41.026 / m
The annual saving can thus be calculated as:
Taking:
Pull per Blast = 1.5 m
Blasts per shift = 2
Number of shifts per day = 3

Number of working faces = 5
Number of working days in a Year = 300
Factor for reduced production/ advance rate = 0.8
Therefore, annual advance = 1.5*2*3*5*300*0.8
= 10,800 m
No. of (100 m galleries + 1 junction) pairs = 108 (approx.)
Total Annual saving = 108*259*41.026
= Rs. 11, 47,579.00
Thus, annual saving is approx. Rs. 11, 48,000.00

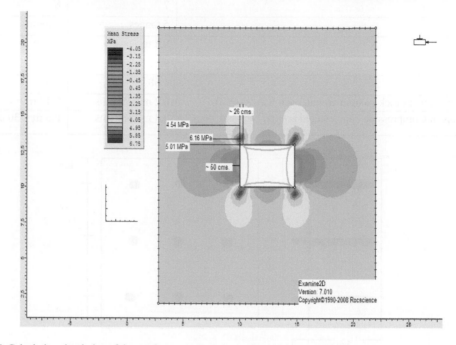

Figure 5. Calculation simulation of the results.

REFERENCES

Ismet Canbulat. 2008. *Evaluation And Design of Optimum Support Systems In South African Collieries Using The Probabilistic Design Approach*. Chapter 1-2. University of Pretoria.

Louis A. Panek. 1955. *Analysis of Roof Bolting Systems Based on Model Studies*. Transactions AIME. Mining Engineering.

Jun Lu Luo. 1999. *A New Rock Bolt Design Criterion And Knowledge-Based Expert System For Stratified Roof*. Faculty of Mining Engineering. Virginia Polytechnic Institute and State University. Blacksburg. Virginia.

Christopher Mark, Gregory M. Molinda, Dennis R. Dolinar. 2001. *Analysis of Roof Bolt Systems*. NIOSH. Pittsburgh Research Laboratory Pittsburgh. Pennsylvania USA.

ErnestoVillaescusa, Christopher R. Windsor, A. G. Thompson. Rock Support and Reinforcement Practice in Mining: 408.

C. Chunlin Li. *Stabilization of highly stressed weak and soft rocks – observations, principles and prac*tice. 11th Congress of the International Society for Rock Mechanics – Ribeiro e Sousa, Olalla & Grossmann (eds)© 2007 Taylor & Francis Group, Lisbon, ISBN 978-0-415-45084-3.

Support Design. Chapter 18, Sec. 18.1.3.2, page 1687-88, SME Mining Engineering Handbook.

Survey Section Records .2009. Digwadih Colliery. Jamadoba Group. Jharia Division. Tata Steel.

EXPLOSION-PROOF ACCUMULATOR LOCOMOTIVES AB8T

Mining Machines holding leads the Ukrainian market for producing and supplying mining equipment. Mining Machines includes:

Druzhkovka Heavy Engineering Plant

Gorlovka Mashinostroitel

Donetskgormash

Donetsk Energozavod

Kamensk Heavy Engineering Plant

Sverdlovsk Heavy Engineering Plant

Krivoy Rog Mining Equipment Repair Plant

Mining Machines Engineering and Technical centre

Mining Machines Quality System

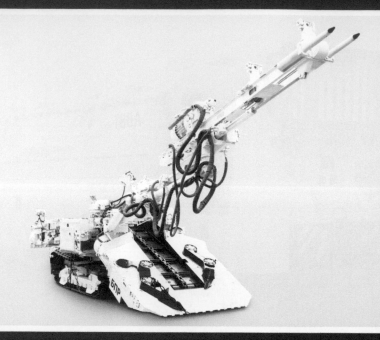

DRILLING LOADING MACHINE BPR

Mining Machines holding leads the Ukrainian market for producing and supplying mining equipment. Mining Machines includes:

Druzhkovka Heavy Engineering Plant

Gorlovka Mashinostroitel

Donetskgormash

Donetsk Energozavod

Kamensk Heavy Engineering Plant

Sverdlovsk Heavy Engineering Plant

Krivoy Rog Mining Equipment Repair Plant

Mining Machines Engineering and Technical centre

Mining Machines Quality System

Contacts
97, ul. Artyoma, Donetsk, 83001, Ukraine
Tel.: 8 (062) 381-53-45
Fax: 8 (062) 381-53-53
E-mail: sec2@mmc.kiev.ua
Web-site: www.mmc.kiev.ua

нпк ГОРНЫЕ МАШИНЫ

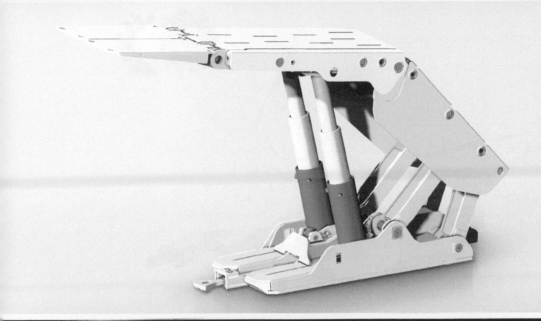

POWERED ROOF SUPPORT 09DT

Mining Machines holding leads the Ukrainian market for producing and supplying mining equipment. Mining Machines includes:

Druzhkovka Heavy Engineering Plant

Gorlovka Mashinostroitel

Donetskgormash

Donetsk Energozavod

Kamensk Heavy Engineering Plant

Sverdlovsk Heavy Engineering Plant

Krivoy Rog Mining Equipment Repair Plant

Mining Machines Engineering and Technical centre

Mining Machines Quality System

Contacts
97, ul. Artyoma, Donetsk, 83001, Ukraine
Tel.: 8 (062) 381-53-45
Fax: 8 (062) 381-53-53
E-mail: sec2@mmc.kiev.ua
Web-site: www.mmc.kiev.ua

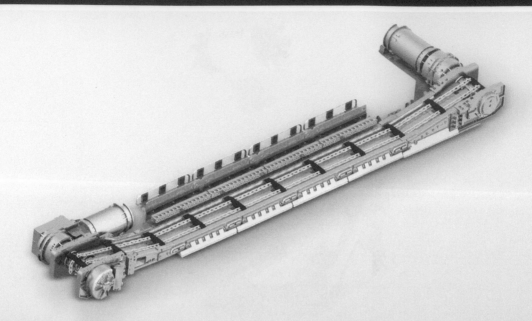

ARMORED FACE CONVEYOR SP26U

Mining Machines holding leads the Ukrainian market for producing and supplying mining equipment. Mining Machines includes:

Druzhkovka Heavy Engineering Plant

Gorlovka Mashinostroitel

Donetskgormash

Donetsk Energozavod

Kamensk Heavy Engineering Plant

Sverdlovsk Heavy Engineering Plant

Krivoy Rog Mining Equipment Repair Plant

Mining Machines Engineering and Technical centre

Mining Machines Quality System

Contacts
97, ul. Artyoma, Donetsk, 83001, Ukraine
Tel.: 8 (062) 381-53-45
Fax: 8 (062) 381-53-53
E-mail: sec2@mmc.kiev.ua
Web-site: www.mmc.kiev.ua

LONG WALL SHEARER UKD 400

Mining Machines holding leads the Ukrainian market for producing and supplying mining equipment. Mining Machines includes:

Druzhkovka Heavy Engineering Plant
Gorlovka Mashinostroitel
Donetskgormash
Donetsk Energozavod
Kamensk Heavy Engineering Plant
Sverdlovsk Heavy Engineering Plant
Krivoy Rog Mining Equipment Repair Plant
Mining Machines Engineering and Technical centre
Mining Machines Quality System

Contacts
97, ul. Artyoma, Donetsk, 83001, Ukraine
Tel.: 8 (062) 381-53-45
Fax: 8 (062) 381-53-53
E-mail: sec2@mmc.kiev.ua
Web-site: www.mmc.kiev.ua

State Higher Educational Establishment «National Mining University»

DEPARTMENT OF UNDERGROUND MINING

Underground Mining Department was established in 1900 in order to prepare specialists in underground mining.

There are 46 people working at the department nowadays, among them there are 9 professors, 18 professor associates and 9 assistants.

The department conducts preparation of the future specialists based on such areas and specialities:

Area: 0503 Mining of minerals

Speciality: 7.05030101, 8.05030101

Mining of the deposits and extraction of minerals

Specializations:

- Underground mining of stratified deposits;
- Underground mining of ore deposits;
- Projection of mines and underground structures;
- Underground mining of minerals with profound learning of information technologies;
- Underground mining of minerals with profound learning of profession-oriented English language;
- Underground mining of minerals with profound learning of management in production field.

Address:

49005, Ukraine, Dnipropetrovs'k
19, Karl-Marks ave.
SHEE «National Mining University»
Underground Mining Department

Tel.: +38 (0562) 47-23-26, 47-14-72
E-mail: v_domna@yahoo.com
olga.malova@yahoo.com

GP «SVERDLOVANTHRACITE»

«Dolzhanskaya-Kapitalnaya» mine
«Krasnyy Partizan» mine
«Centrsoyuz» mine
«Kharkovskaya» mine
«Sverdlovskaya» PP
«Centrsoyuz» MPP
«Krasnopartizanskaya» MPP

Technological raw material for:

Ferrous and non-ferrous metallurgy,
chemical and electro-technical industries

Municipal-home necessities

Microphone powder

Coal-graphite blocks

Thermo-graphite

Fuel for TPP

«For best trade mark»
Spain, Madrid, 1996

«Golden Mercury»
England, Oxford, 2002

«Highest quality»
Ukraine, Kiev, 2004

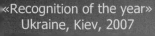

«Recognition of the year»
Ukraine, Kiev, 2007

«The best enterprises
of Ukraine», Kiev, 2009

«Leader of the branch»
Kiev, 2010

«The best enterprises
of Ukraine», Kiev, 2010

Requisites:

CORRESPONDENT BANK: SWIFT: DEUTDEFF
                    DEUTSCHE BANK
                    FRANKFURT \ DEUTDEFF, EUR
BENEFICIARIES BANK: K/acc 10094986271000
                    PJSC UKRSOTSBANK
                    STREET KOVPAKA 29, KIEV
                    UKRAINE SWIFT: UKRSUAUX
BENEFICIARY: №26009000082587
Science research Institute of mining problems AES of Ukraine
        49050, DNEPROPETROVSK , GAGARIN av, 105/44
        UKRSOTSBANK
        DNEPROPETROVSK, MECHNIKOVA, 11
        CODE BANK 300023 OKPO 00039019

We invite you to take part

in the VI International

Scientific-Practical Conference

«School Of Underground Mining - 2012»

which will be hold in September 23-29, 2012

Address of organizing committee:

49005, Ukraine, Dnipropetrovs'k
19, Karl-Marks ave.
SHEE «National Mining University»
Underground Mining Department
Tel./fax +38 (056) 374-21-84
+38 (0562) 47-14-72
+38 (0562) 46-90-47
E-mail: olga.malova@yahoo.com
vvlapko@mail.ru
http://www.msu.org.ua